T0256156

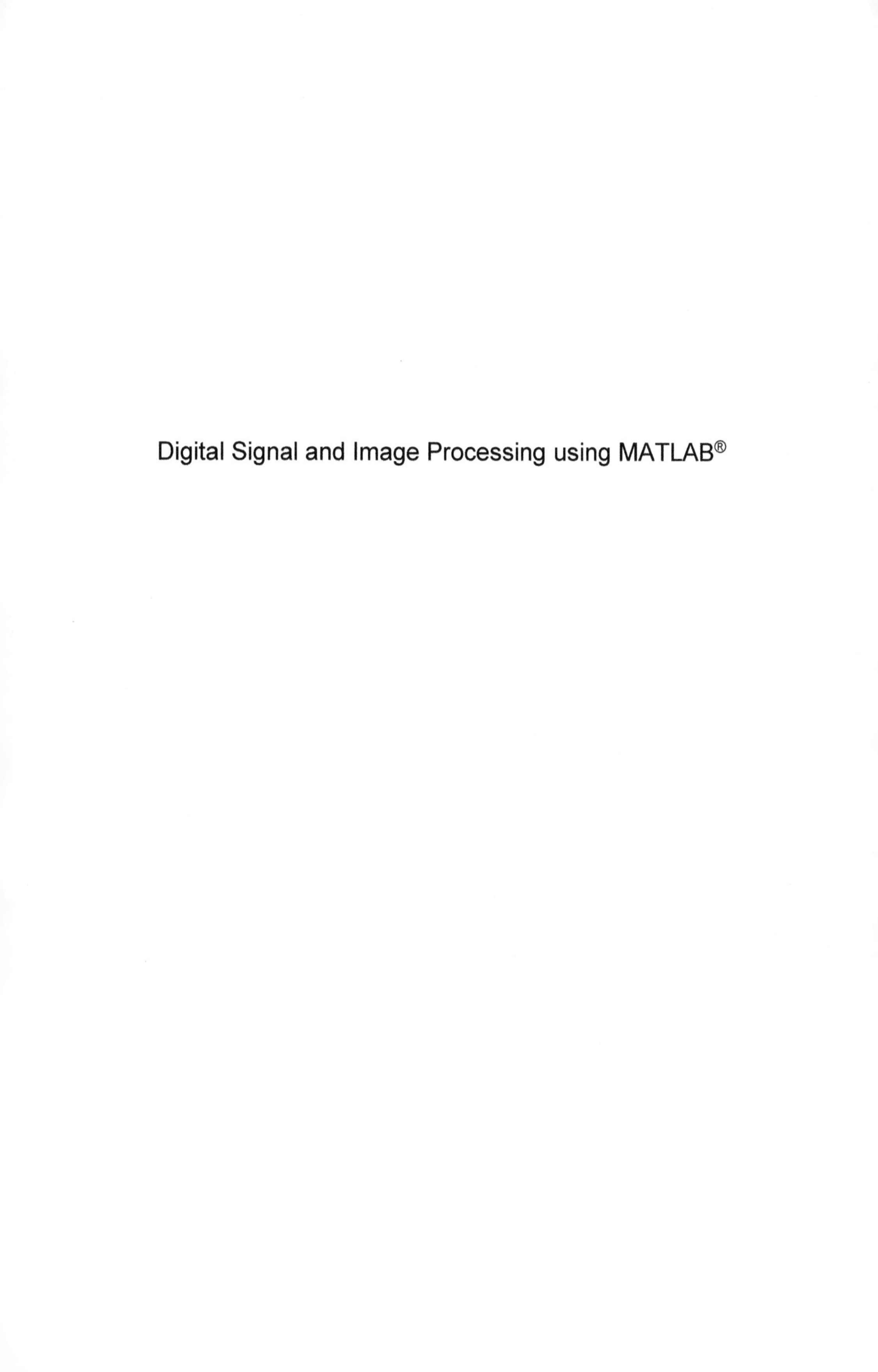

Digital Signal and Image Processing using MATLAB®

Revised and Updated 2nd Edition

Digital Signal and Image Processing using MATLAB®

Volume 3

Advances and Applications: The Stochastic Case

Gérard Blanchet
Maurice Charbit

WILEY

First published 2015 in Great Britain and the United States by ISTE Ltd and John Wiley & Sons, Inc.

ISTE Ltd
27-37 St George's Road
London SW19 4EU
UK

www.iste.co.uk

John Wiley & Sons, Inc.
111 River Street
Hoboken, NJ 07030
USA

www.wiley.com

Library of Congress Control Number: 2015948073

British Library Cataloguing-in-Publication Data
A CIP record for this book is available from the British Library
ISBN 978-1-84821-795-9

Contents

Foreword

This book is the third volume in a series on digital signal processing and associated techniques. Following on from "Fundamentals" and "Advances and Applications, The Deterministic Case", it addresses the stochastic case. We shall presume throughout that readers have a good working knowledge of MATLAB® and of basic elements of digital signal processing.

Whilst our main focus is on applications, we shall also give consideration to key principles. A certain number of the elements discussed belong more to the domain of statistics than to signal processing; this responds to current trends in signal processing, which make extensive use of this type of technique.

Over 60 solved exercises allow the reader to apply the concepts and results presented in the following chapters. There will also be examples to alleviate any demonstrations that would otherwise be quite dense. These can be found in more specialist books referenced in the bibliography. 92 programs and 49 functions will be used to support these examples and corrected exercises.

Mathematical Concepts

The first chapter begins with a brief review of probability theory, focusing on the notions of conditional probability, projection theorem and random variable transformation. A number of statistical elements will also be presented, including the law of large numbers (LLN), the limit-central theorem, or the delta-method.

Statistical inferences

The second chapter is devoted to statistical inference. Statistical inference consists of deducing interesting characteristics from a series of observations with a certain degree of reliability. A variety of techniques may be used. In this chapter, we shall discuss three broad families of techniques: hypothesis testing, parameter estimation, and the determination of confidence intervals. Key notions include Cramer-Rao bound, likelihood ratio tests, maximum likelihood approach and least square approach for linear models.

Monte-Carlo simulation

Monte-Carlo methods involve a set of algorithms which aim to calculate values using a pseudo-random generator. The quantities to calculate are typically integrals, and in practice, often represent the mathematical expectation of a function of interest. In cases using large dimensions, these methods can significantly reduce the calculation time required by deterministic methods. Monte-Carlo methods involve drawing a series of samples, distributed following a target distribution. The main generation methods, including importance sampling, the acceptance-rejection method, the Gibbs sampler, etc., will be presented. Another objective is to minimize the mean square error between the calculated and true values, and variance reduction methods will be studied using simulations.

Second order stationary process

The fourth chapter covers second order random stationary processes in the broadest sense: *Wide Sense Stationary (WSS)*. The chapter is split into three parts, beginning with empirical second order estimators, leading to the correlogram. Then follow general and detailled results on the linear prediction which is fundamental role in the WSS time series. The third part is devoted to the non-parametric spectral estimation approaches (smooth periodograms, average periodograms, etc.). A detailed discussion on the bias-variance compromise is given.

Inferences on HMM

States are directly visible in simple Markov models, and the modeling process depends exclusively on transition probabilities. In hidden-state Markov models (HMM), however, states can only be seen via observed signals which are statistically linked to these states. HMMs are particularly useful as control models, using latent variables of mixtures connected to each observation.

A wide variety of problems may be encountered in relation to inference, for example seeking the sequence most likely to have produced a given series of observations; determining the a posteriori distribution of hidden states; estimating the parameters of a model; etc. Key algorithms include the Baum-Welch algorithm and the Viterbi algorithm, to cite the two best-known examples. HMM have applications in a wide range of domains, such as speech recognition (analysis and synthesis), automatic translation, handwriting analysis, activity identification, DNA analysis, etc.

Selected Topics

The final chapter presents applications which use many of the principles and techniques described in the preceding chapters, without falling into any of the categories defined in these chapters. The first section is devoted to high resolution techniques (MUSIC and ESPRIT algorithms), whilst the second covers classic communication problems (coding, modulation, eye diagrams, etc.). The third section presents the Viterbi algorithm, and the fourth is given over to scalar and vectorial quantification.

Annexes

A certain number of functions are given in simplified form in the appendix. This includes a version of the `boxplot` function, alongside functions associated with the most common distributions (Student, χ^2 and Fischer).

Remarques

The notation used in this book is intended to conform to current usage; in cases where this is not the case, every care has been taken to remove any ambiguity as to the precise meaning of terms. On a number of occasions, we refer to nitnslseries instead of nitnsltime series or nitnslsequences to avoid confusion.

Notations and Abbreviations

$\emptyset$		empty set
$\sum_{k,n}$	$=$	$\sum_k \sum_n$
$\text{rect}_T(t)$	$=$	$\begin{cases} 1 & \text{when} \quad \lvert t\rvert < T/2 \\ 0 & \text{otherwise} \end{cases}$
$\text{sinc}(x)$	$=$	$\dfrac{\sin(\pi x)}{\pi x}$
$\mathbb{1}(x \in A)$	$=$	$\begin{cases} 1 & \text{when } x \in A \\ 0 & \text{otherwise} \end{cases}$ (indicator function of A)
$(a, b]$	$=$	$\{x : a < x \leq b\}$
$\delta(t)$		$\begin{cases} \text{Dirac distribution when } t \in \mathbb{R} \\ \text{Kronecker symbol when } t \in \mathbb{Z} \end{cases}$
$\text{Re}(z)$		real part of z
$\text{Im}(z)$		imaginary part of z
i or j	$=$	$\sqrt{-1}$
$x(t) \rightleftharpoons X(f)$		Fourier transform
$(x \star y)(t)$		continuous time convolution
	$=$	$\int_{\mathbb{R}} x(u)y(t-u)du$
$(x \star y)(t)$		discrete time convolution
	$=$	$\sum_{u \in \mathbb{Z}} x(u)y(t-u)$

$\boldsymbol{x}$ or $\underline{x}$	vector $\boldsymbol{x}$
$\boldsymbol{I}_N$	$(N \times N)$-dimension identity matrix
$\boldsymbol{A}^*$	complex conjugate of $\boldsymbol{A}$
$\boldsymbol{A}^T$	transpose of $\boldsymbol{A}$
$\boldsymbol{A}^H$	transpose-conjugate of $\boldsymbol{A}$
$\boldsymbol{A}^{-1}$	inverse matrix of $\boldsymbol{A}$
$\boldsymbol{A}^{\#}$	pseudo-inverse matrix of $\boldsymbol{A}$

$\mathbb{P}$	probability measure
$\mathbb{P}_\theta$	probability measure indexed by θ
$\mathbb{E}\{X\}$	expectation of X
$\mathbb{E}_\theta\{X\}$	expectation of X under the distribution $\mathbb{P}_\theta$
$X_c = X - \mathbb{E}\{X\}$	zero-mean random variable
$\operatorname{var}(X) = \mathbb{E}\{\lvert X_c\rvert^2\}$	variance of X
$\operatorname{cov}(X,Y) = \mathbb{E}\{X_c Y_c^*\}$	covariance of (X,Y)
$\operatorname{cov}(X) = \operatorname{cov}(X,X) = \operatorname{var}(X)$	variance of X
$\mathbb{E}\{X \mid Y\}$	conditional expectation of X given Y
$a \xrightarrow{\mathcal{L}} b$ or $a \xrightarrow{\mathcal{L}} b$	a converges in law to b or a converges in distribution to b
$a \xrightarrow{P} b$	converges in probability to b
$a \xrightarrow{a.s.} b$	a converges almost surely to b

ADC	Analog to Digital Converter
ADPCM	Adaptive Differential PCM
AMI	Alternate Mark Inversion
AR	Autoregressive
ARMA	AR and MA
BER	Bit Error Rate
bps	bits per second
cdf	cumulative distribution function
CF	Clipping Factor
CZT	Causal z-Transform
DAC	Digital to Analog Converter

DCT	Discrete Cosine Transform
d.o.f.	degree of freedom
DFT	Discrete Fourier Transform
DTFT	Discrete Time Fourier Transform
EM	Expectation Maximization
ESPRIT	Estimation of Signal Parameter via Rotational Invariance Techniques
FIR	Finite Impulse Response
FFT	Fast Fourier Transform
FT	Continuous Time Fourier Transform
GLRT	Generalized Likelihood Ratio Test
GEM	Generalized Expectation Maximization
GMM	Gaussian Mixture Model
HDB	High Density Bipolar
HMM	Hidden Markov Model
IDFT	Inverse Discrete Fourier Transform
i.i.d./iid	independent and identically distributed
IIR	Infinite Impulse Response
ISI	InterSymbol Interference
KKT	Karush-Kuhn-Tucker
LDA	Linear Discriminant Analysis
LBG	Linde, Buzzo, Gray (algorithm)
LMS	Least Mean Squares
MA	Moving Average
MSE	Mean Square Error
MUSIC	MUltiple SIgnal Charaterization
PAM	Pulse Amplitude Modulation
PCA	Principal Component Analysis
PSK	Phase Shift Keying
QAM	Quadrature Amplitude Modulation
rls	recursive least squares
rms	root mean square
ROC	Receiver Operational Characteristic

SNR	Signal to Noise Ratio
r.v./rv	random variable
STFT	Short Term Fourier Transform
TF	Transfer Function
WSS	Wide (Weak) Sense Stationary (Second Order) Process

Chapter 1

Mathematical Concepts

1.1 Basic concepts on probability

Without describing in detail the formalism used by Probability Theory, we will simply remind the reader of some useful concepts. However we advise the reader to consult some of the many books with authority on the subject [1].

Definition 1.1 (Discrete random variable) *A random variable X is said to be discrete if the set of its possible values is, at the most, countable. If $\{a_0, \ldots, a_n, \ldots\}$, where $n \in \mathbb{N}$, is the set of its values, the probability distribution of X is characterized by the sequence:*

$$p_X(n) = \Pr(X = a_n) \tag{1.1}$$

representing the probability that X is equal to the element a_n. These values are such that $0 \leq p_X(n) \leq 1$ and $\sum_{n \geq 0} p_X(n) = 1$.

This leads us to the probability for the random variable X to belong to the interval $]a, b]$. It is given by:

$$\Pr(X \in]a, b]) = \sum_{n \geq 0} p_X(n) \mathbb{1}(a_n \in]a, b])$$

The function defined for $x \in \mathbb{R}$ by:

$$\begin{aligned} F_X(x) &= \Pr(X \leq x) = \sum_{\{n : a_n \leq x\}} p_X(n) \\ &= \sum_{n \geq 0} p_X(n) \mathbb{1}(a_n \in] - \infty, x]) \end{aligned} \tag{1.2}$$

is called the *cumulative distribution function* (*cdf*) of the random variable X. It is a monotonic increasing function, and verifies $F_X(-\infty) = 0$ and $F_X(+\infty) = 1$.

Its graph resembles that of a staircase function, the jumps of which are located at x-coordinates a_n and have an amplitude of $p_X(n)$.

Definition 1.2 (Two discrete random variables) *Let X and Y be two discrete random variables, with possible values $\{a_0, \ldots, a_n, \ldots\}$ and $\{b_0, \ldots, b_k, \ldots\}$ respectively. The joint probability distribution is characterized by the sequence of positive values:*

$$p_{XY}(n,k) = \Pr(X = a_n, Y = b_k) \tag{1.3}$$

with $0 \leq p_{XY}(n,k) \leq 1$ and $\sum_{n\geq 0} \sum_{k\geq 0} p_{XY}(n,k) = 1$.

$\Pr(X = a_n, Y = b_k)$ represents the probability to *simultaneously* have $X = a_n$ and $Y = b_k$. This definition can easily be extended to the case of a finite number of random variables.

Property 1.1 (Marginal probability distribution) *Let X and Y be two discrete random variables, with possible values $\{a_0, \ldots, a_n, \ldots\}$ and $\{b_0, \ldots, b_k, \ldots\}$ respectively, and with their joint probability distribution characterized by $p_{XY}(n,k)$. We have:*

$$\begin{aligned} p_X(n) &= \Pr(X = a_n) = \sum_{k=0}^{+\infty} p_{XY}(n,k) \\ p_Y(k) &= \Pr(Y = b_k) = \sum_{n=0}^{+\infty} p_{XY}(n,k) \end{aligned} \tag{1.4}$$

$p_X(n)$ and $p_Y(k)$ denote the marginal probability distribution of X and Y respectively.

Definition 1.3 (Continuous random variable) *A random variable is said to be continuous*[1] *if its values belong to $\mathbb{R}$ and if, for any real numbers a and b, the probability that X belongs to the interval $]a,b]$ is given by:*

$$\Pr(X \in]a,b]) = \int_a^b p_X(x)dx = \int_{-\infty}^{\infty} p_X(x)\mathbb{1}(x \in]a,b])dx \tag{1.5}$$

where $p_X(x)$ is a function that must be positive or equal to zero such that $\int_{-\infty}^{+\infty} p_X(x)dx = 1$. $p_X(x)$ is called the probability density function (pdf) of X.

[1]The exact expression says that the probability distribution of X is *absolutely continuous with respect to the Lebesgue measure.*

The function defined for any $x \in \mathbb{R}$ by:

$$F_X(x) = \Pr(X \leq x) = \int_{-\infty}^{x} p_X(u)du \tag{1.6}$$

is called the *cumulative distribution function* (*cdf*) of the random variable X. It is a monotonic increasing function and it verifies $F_X(-\infty) = 0$ and $F_X(+\infty) = 1$. Notice that $p_X(x)$ also represents the derivative of $F_X(x)$ with respect to x.

Definition 1.4 (Two continuous random variables) *Let X and Y be two random variables with possible values in $\mathbb{R} \times \mathbb{R}$. They are said to be continuous if, for any domain Δ of $\mathbb{R}^2$, the probability that the pair (X, Y) belongs to Δ is given by:*

$$\Pr((X, Y) \in \Delta) = \iint_{\Delta} p_{XY}(x, y)dxdy \tag{1.7}$$

where the function $p_{XY}(x, y) \geq 0$, and is such that:

$$\iint_{\mathbb{R}^2} p_{XY}(x, y)dxdy = 1$$

$p_{XY}(x, y)$ is called the joint probability density function of the pair (X, Y).

Property 1.2 (Marginal probability distributions) *Let X and Y be two continuous random variables with a joint probability distribution characterized by $p_{XY}(x, y)$. The probability distributions of X and Y have the following marginal probability density functions:*

$$\begin{aligned} p_X(x) &= \int_{-\infty}^{+\infty} p_{XY}(x, y)dy \\ p_Y(y) &= \int_{-\infty}^{+\infty} p_{XY}(x, y)dx \end{aligned} \tag{1.8}$$

An example involving two real random variables (X, Y) is the case of a complex random variable $Z = X + jY$.

It is also possible to have a mixed situation, where one of the two variables is discrete and the other is continuous. This leads to the following:

Definition 1.5 (Mixed random variables) *Let X be a discrete random variable with possible values $\{a_0, \ldots, a_n, \ldots\}$ and Y a continuous random variable* *with possible values in* $\mathbb{R}$. *For any value* a_n, *and for any real numbers a and b, the probability:*

$$\Pr(X = a_n, Y \in]a, b]) = \int_a^b p_{XY}(n, y)dy \tag{1.9}$$

where the function $p_{XY}(n, y)$, *with* $n \in \{0, \ldots, k, \ldots\}$ *and* $y \in \mathbb{R}$, *is* ≥ 0 *and verifies* $\sum_{n\geq 0} \int_{\mathbb{R}} p_{XY}(n, y)dy = 1$.

Definition 1.6 (Two independent random variables) *Two random variables X and Y are said to be independent if and only if their joint probability distribution is the product of the marginal probability distributions. This can be expressed:*

- *for two discrete random variables:*

$$p_{XY}(n, k) = p_X(n)p_Y(k)$$

- *for two continuous random variables:*

$$p_{XY}(x, y) = p_X(x)p_Y(y)$$

- *for two mixed random variables:*

$$p_{XY}(n, y) = p_X(n)p_Y(y)$$

where the marginal probability distributions are obtained with formulae (1.4) *and* (1.8).

It is worth noting that, knowing $p_{XY}(x, y)$, we can tell whether or not X and Y are independent. To do this, we need to calculate the marginal probability distributions and to check that $p_{XY}(x, y) = p_X(x)p_Y(y)$. If that is the case, then X and Y are independent.

The following definition is more general.

Definition 1.7 (Independent random variables) *The random variables* $(X_1, \ldots, X_n)$ *are jointly independent if and only if their joint probability distribution is the product of their marginal probability distributions. This can be expressed:*

$$p_{X_1 X_2 \ldots X_n}(x_1, x_2, \ldots, x_n) = p_{X_1}(x_1)p_{X_2}(x_2)\ldots p_{X_n}(x_n) \tag{1.10}$$

where the marginal probability distributions are obtained as integrals with respect to $(n-1)$ *variables, calculated from* $p_{X_1 X_2 \ldots X_n}(x_1, x_2, \ldots, x_n)$.

For example, the marginal probability distribution of X_1 has the expression:

$$p_{X_1}(x_1) = \underbrace{\int \cdots \int}_{\mathbb{R}^{n-1}} p_{X_1 X_2 \ldots X_n}(x_1, x_2, \ldots, x_n) dx_2 \ldots dx_n$$

In practice, the following result is a simple method for determining whether or not random variables are independent: if $p_{X_1 X_2 \ldots X_n}(x_1, x_2, \ldots, x_n)$ is a product of n positive functions of the type $f_1(x_1) f_2(x_2) \ldots f_n(x_n)$, then the variables are independent.

It should be noted that if n random variables are independent of one another, it does not necessarily mean that they are jointly independent.

Definition 1.8 (Mathematical expectation) *Let X be a random variable and $f(x)$ a function. The mathematical expectation of $f(X)$ – respectively $f(X, Y)$ – is the value, denoted by $\mathbb{E}\{f(X)\}$ – respectively $\mathbb{E}\{f(X, Y)\}$ – defined:*

– *for a discrete random variable, by:*

$$\mathbb{E}\{f(X)\} = \sum_{n \geq 0} f(a_n) p_X(n)$$

– *for a continuous random variable, by:*

$$\mathbb{E}\{f(X)\} = \int_{\mathbb{R}} f(x) p_X(x) dx$$

– *for two discrete random variables, by:*

$$\mathbb{E}\{f(X, Y)\} = \sum_{n \geq 0} \sum_{k \geq 0} f(a_n, b_k) p_{XY}(n, k)$$

– *for two continuous random variables, by:*

$$\mathbb{E}\{f(X, Y)\} = \int_{\mathbb{R}} \int_{\mathbb{R}} f(x, y) p_{XY}(x, y) dx dy$$

provided that all expressions exist.

Property 1.3 *If $\{X_1, X_2, \ldots, X_n\}$ are jointly independent, then for any integrable functions $f_1, f_2, \ldots, f_n$:*

$$\mathbb{E}\left\{\prod_{k=1}^{n} f_k(X_k)\right\} = \prod_{k=1}^{n} \mathbb{E}\{f_k(X_k)\} \quad (1.11)$$

Definition 1.9 (Characteristic function) *The characteristic function of the probability distribution of the random variables $X_1, \ldots, X_n$ is the function of $(u_1, \ldots, u_n) \in \mathbb{R}^n$ defined by:*

$$\phi_{X_1 \ldots X_n}(u_1, \ldots, u_n) = \mathbb{E}\left\{e^{ju_1 X_1 + \cdots + ju_n X_n}\right\} = \mathbb{E}\left\{\prod_{k=1}^{n} e^{ju_k X_k}\right\} \tag{1.12}$$

Because $\left|e^{juX}\right| = 1$, the characteristic function exists and is continuous even if the moments $\mathbb{E}\left\{X^k\right\}$ do not exist. The Cauchy probability distribution, for example, the probability density function of which is $p_X(x) = 1/\pi(1 + x^2)$, has no moment and has the characteristic function $e^{-|u|}$. Let us notice $|\phi_{X_1 \ldots X_n}(u_1, \ldots, u_n)| \leq \phi_X(0, \ldots, 0) = 1$.

Theorem 1.1 (Fundamental) $(X_1, \ldots, X_n)$ *are independent if and only if for any point* $(u_1, u_2, \ldots, u_n)$ *of* $\mathbb{R}^n$*:*

$$\phi_{X_1 \ldots X_n}(u_1, \ldots, u_n) = \prod_{k=1}^{n} \phi_{X_k}(u_k)$$

Notice that the characteristic function $\phi_{X_k}(u_k)$ of the marginal probability distribution of X_k can be directly calculated using (1.12). We have $\phi_{X_k}(u_k) = \mathbb{E}\left\{e^{ju_k X_k}\right\} = \phi_{X_1 \ldots X_n}(0, \ldots, 0, u_k, 0, \ldots, 0)$.

Definition 1.10 (Mean, variance) *The mean of the random variable X is defined as the first order moment, that is to say $\mathbb{E}\{X\}$. If the mean is equal to zero, the random variable is said to be centered. The variance of the random variable X is the quantity defined by:*

$$\operatorname{var}(X) = \mathbb{E}\left\{(X - \mathbb{E}\{X\})^2\right\} = \mathbb{E}\left\{X^2\right\} - (\mathbb{E}\{X\})^2 \tag{1.13}$$

The variance is always positive, and its square root is called the standard deviation.

As an exercise, we are going to show that, for any constants a and b:

$$\begin{aligned} \mathbb{E}\{aX + b\} &= a\mathbb{E}\{X\} + b \quad (1.14) \\ \operatorname{var}(aX + b) &= a^2 \operatorname{var}(X) \quad (1.15) \end{aligned}$$

Expression (1.14) is a direct consequence of the integral's linearity. We assume that $Y = aX + b$, then $\operatorname{var}(Y) = \mathbb{E}\left\{(Y - \mathbb{E}\{Y\})^2\right\}$. By replacing $\mathbb{E}\{Y\} = a\mathbb{E}\{X\} + b$, we get $\operatorname{var}(Y) = \mathbb{E}\left\{a^2(X - \mathbb{E}\{X\})^2\right\} = a^2 \operatorname{var}(X)$.

A generalization of these two results to random vectors (their components are random variables) will be given by property (1.6).

Definition 1.11 (Covariance, correlation) *Let (X, Y) be two random variables*[2]*. The covariance of X and Y is the quantity defined by:*

$$\begin{aligned} \text{cov}(X, Y) &= \mathbb{E}\{(X - \mathbb{E}\{X\})(Y^* - \mathbb{E}\{Y^*\})\} \\ &= \mathbb{E}\{XY^*\} - \mathbb{E}\{X\}\mathbb{E}\{Y^*\} \end{aligned} \tag{1.16}$$

In what follows, the variance of the random variable X will be noted $\text{var}(X)$*.* $\text{cov}(X)$ *or* $\text{cov}(X, X)$ *have exactly the same meaning.*

X and Y are said to be uncorrelated if $\text{cov}(X, Y) = 0$ *that is to say if* $\mathbb{E}\{XY^*\} = \mathbb{E}\{X\}\mathbb{E}\{Y^*\}$*. The correlation coefficient is the quantity defined by:*

$$\rho(X, Y) = \frac{\text{cov}(X, Y)}{\sqrt{\text{var}(X)}\sqrt{\text{var}(Y)}} \tag{1.17}$$

Applying the Schwartz inequality gives us $|\rho(X, Y)| \leq 1$.

Definition 1.12 (Mean vector and covariance matrix) *Let $\{X_1, \ldots, X_n\}$ be n random variables with the respective means $\mathbb{E}\{X_i\}$. The mean vector is the n dimension vector with the means $\mathbb{E}\{X_i\}$ as its components. The $n \times n$ covariance matrix $\boldsymbol{C}$ is the matrix with the generating element* $C_{ij} = \text{cov}(X_i, X_j)$ *for $1 \leq i \leq n$ and $1 \leq j \leq n$.*

Matrix notation: if we write

$$\boldsymbol{X} = \begin{bmatrix} X_1 & \ldots & X_n \end{bmatrix}^T$$

to refer to the random vector with the random variable X_k as its k-th component, the mean-vector can be expressed:

$$\mathbb{E}\{\boldsymbol{X}\} = \begin{bmatrix} \mathbb{E}\{X_1\} & \ldots & \mathbb{E}\{X_n\} \end{bmatrix}^T$$

the covariance matrix:

$$\begin{aligned} \boldsymbol{C} &= \mathbb{E}\{(\boldsymbol{X} - \mathbb{E}\{\boldsymbol{X}\})(\boldsymbol{X} - \mathbb{E}\{\boldsymbol{X}\})^H\} \\ &= \mathbb{E}\left\{\boldsymbol{X}\boldsymbol{X}^H\right\} - \mathbb{E}\{\boldsymbol{X}\}\mathbb{E}\{\boldsymbol{X}\}^H \end{aligned} \tag{1.18}$$

and the *correlation matrix*

$$\boldsymbol{R} = \boldsymbol{D}\boldsymbol{C}\boldsymbol{D} \tag{1.19}$$

[2]Except in some particular cases, the random variables considered from now on will be real. However, the definitions involving the mean and the covariance can be generalized with no exceptions to complex variables by conjugating the second variable. This is indicated by a star ($*$) in the case of scalars and by the exponent H in the case of vectors.

with

$$\boldsymbol{D} = \begin{bmatrix} C_{11}^{-1/2} & 0 & \dots & 0 \\ \vdots & \ddots & \ddots & \vdots \\ \vdots & \ddots & \ddots & 0 \\ 0 & \dots & 0 & C_{nn}^{-1/2} \end{bmatrix} \tag{1.20}$$

$\boldsymbol{R}$ is obtained by dividing each element C_{ij} of $\boldsymbol{C}$ by $\sqrt{C_{ii}C_{jj}}$, provided that $C_{ii} \neq 0$. Therefore $R_{ii} = 1$ and $|R_{ij}| \leq 1$.

Notice that the diagonal elements of a covariance matrix represent the respective variances of the n random variables. They are therefore positive. *If the n random variables are uncorrelated, their covariance matrix is diagonal and their correlation matrix is the identity matrix.*

Property 1.4 (Positivity of the covariance matrix) *Any covariance matrix is positive, meaning that for any vector $\boldsymbol{a} \in \mathbb{C}^n$, we have $\boldsymbol{a}^H \boldsymbol{C} \boldsymbol{a} \geq 0$.*

Property 1.5 (Bilinearity of the covariance) *Let $X_1, \dots, X_m, Y_1, \dots, Y_n$ be random variables, and $v_1, \dots, v_m, w_1, \dots, w_n$ be constants. Hence:*

$$\mathrm{cov}\left(\sum_{i=1}^{m} v_i^* X_i, \sum_{j=1}^{n} w_j^* Y_j\right) = \sum_{i=1}^{m}\sum_{j=1}^{n} v_i^* w_j \mathrm{cov}\left(X_i, Y_j\right) \tag{1.21}$$

Let $\boldsymbol{V}$ and $\boldsymbol{W}$ be the vectors of components v_i and w_j respectively, and $A = \boldsymbol{V}^H\boldsymbol{X}$ and $B = \boldsymbol{W}^H\boldsymbol{Y}$. By definition, $\mathrm{cov}\left(A, B\right) = \{(A - \mathbb{E}\{A\})(B - \mathbb{E}\{B\})^*\}$. Replacing A and B by their respective expressions and using $\mathbb{E}\{A\} = \boldsymbol{V}^H\mathbb{E}\{\boldsymbol{X}\}$ and $\mathbb{E}\{B\} = \boldsymbol{W}^H\mathbb{E}\{\boldsymbol{Y}\}$, we obtain, successively:

$$\begin{aligned} \mathrm{cov}\left(A, B\right) &= \mathbb{E}\left\{\boldsymbol{V}^H(\boldsymbol{X} - \mathbb{E}\{\boldsymbol{X}\})(\boldsymbol{Y} - \mathbb{E}\{\boldsymbol{Y}\})^H\boldsymbol{W}\right\} \\ &= \sum_{i=1}^{m}\sum_{j=1}^{n} v_i^* w_j \mathrm{cov}\left(X_i, Y_j\right) \end{aligned}$$

thus demonstrating expression (1.21). Using matrix notation, this is written:

$$\mathrm{cov}\left(\boldsymbol{V}^H\boldsymbol{X}, \boldsymbol{W}^H\boldsymbol{Y}\right) = \boldsymbol{V}^H\boldsymbol{C}\boldsymbol{W} \tag{1.22}$$

where $\boldsymbol{C}$ designates the covariance matrix of $\boldsymbol{X}$ and $\boldsymbol{Y}$.

Property 1.6 (Linear transformation of a random vector) *Let* $\{X_1, \dots, X_n\}$ *be* n *random variables with* $\mathbb{E}\{\boldsymbol{X}\}$ *as their mean vector and* $\boldsymbol{C}_X$ *as their covariance matrix, and let* $\{Y_1, \dots, Y_q\}$ *be* q *random variables obtained by the linear transformation:*

$$\begin{bmatrix} Y_1 \\ \vdots \\ Y_q \end{bmatrix} = \boldsymbol{A} \begin{bmatrix} X_1 \\ \vdots \\ X_n \end{bmatrix} + \boldsymbol{b}$$

where $\boldsymbol{A}$ *is a matrix and* $\boldsymbol{b}$ *is a non-random vector with the adequate sizes. We then have:*

$$\begin{aligned} \mathbb{E}\{\boldsymbol{Y}\} &= \boldsymbol{A}\mathbb{E}\{\boldsymbol{X}\} + \boldsymbol{b} \\ \boldsymbol{C}_Y &= \boldsymbol{A}\boldsymbol{C}_X\boldsymbol{A}^H \end{aligned}$$

Definition 1.13 (White sequence) *Let* $\{X_1, \dots, X_n\}$ *be a set of* n *random variables. They are said to form a white sequence if* $\mathrm{var}(X_i) = \sigma^2$ *and if* $\mathrm{cov}(X_i, X_j) = 0$ *for* $i \neq j$*. Hence their covariance matrix can be expressed:*

$$\boldsymbol{C} = \sigma^2 \boldsymbol{I}_n$$

where $\boldsymbol{I}_n$ *is the* $n \times n$ *identity matrix.*

Property 1.7 (Independence $\Rightarrow$ non-correlation) *The random variables* $\{X_1, \dots, X_n\}$ *are independent, then uncorrelated, and hence their covariance matrix is diagonal. Usually the converse statement is false.*

1.2 Conditional expectation

Definition 1.14 (Conditional expectation) *We consider a random variable* X *and a random vector* $\boldsymbol{Y}$ *taking values respectively in* $\mathcal{X} \subset \mathbb{R}$ *and* $\mathcal{Y} \subset \mathbb{R}^q$ *with joint probability density* $p_{X\boldsymbol{Y}}(x, \boldsymbol{y})$*. The conditional expectation of* X *given* $\boldsymbol{Y}$*, is a (measurable) real valued function* $g(\boldsymbol{Y})$ *such that for any other real valued function* $h(\boldsymbol{Y})$ *we have:*

$$\mathbb{E}\{|X - g(\boldsymbol{Y})|^2\} \leq \mathbb{E}\{|X - h(\boldsymbol{Y})|^2\} \tag{1.23}$$

$g(\boldsymbol{Y})$ *is commonly denoted by* $\mathbb{E}\{X|\boldsymbol{Y}\}$*.*

Property 1.8 (Conditional probability distribution) *We consider a random variable* X *and a random vector* $\boldsymbol{Y}$ *taking values respectively in* $\mathcal{X} \subset \mathbb{R}$ *and* $\mathcal{Y} \subset \mathbb{R}^q$ *with joint probability density* $p_{X\boldsymbol{Y}}(x, \boldsymbol{y})$*. Then* $\mathbb{E}\{X|\boldsymbol{Y}\} = g(\boldsymbol{Y})$ *where:*

$$g(\boldsymbol{y}) = \int_{\mathcal{X}} x\, p_{X|\boldsymbol{Y}}(x, \boldsymbol{y}) dx$$

with

$$p_{X|\mathbf{Y}}(x,\boldsymbol{y}) = \frac{p_{X\mathbf{Y}}(x,\boldsymbol{y})}{p_{\mathbf{Y}}(\boldsymbol{y})} \quad \text{and} \quad p_{\mathbf{Y}}(\boldsymbol{y}) = \int_{\mathcal{X}} p_{X\mathbf{Y}}(x,\boldsymbol{y})dx \tag{1.24}$$

$p_{X|\mathbf{Y}}(x,\boldsymbol{y})$ *is known as the conditional probability distribution of* X *given* $\boldsymbol{Y}$.

Property 1.9 *The conditional expectation verifies the following properties:*

1. *linearity:* $\mathbb{E}\{a_1X_1 + a_2X_2|\mathbf{Y}\} = a_1\mathbb{E}\{X_1|\mathbf{Y}\} + a_2\mathbb{E}\{X_2|\mathbf{Y}\}$;
2. *orthogonality:* $\mathbb{E}\{(X - \mathbb{E}\{X|\mathbf{Y}\})h(\mathbf{Y})\} = 0$ *for any function* $h : \mathcal{Y} \mapsto \mathbb{R}$;
3. $\mathbb{E}\{h(\mathbf{Y})f(X)|\mathbf{Y}\} = h(\mathbf{Y})\mathbb{E}\{f(X)|\mathbf{Y}\}$, *for all functions* $f : \mathcal{X} \mapsto \mathbb{R}$ *and* $h : \mathcal{Y} \mapsto \mathbb{R}$;
4. $\mathbb{E}\{\mathbb{E}\{f(X,\boldsymbol{Y})|\boldsymbol{Y}\}\} = \mathbb{E}\{f(X,\boldsymbol{Y})\}$ *for any function* $f : \mathcal{X} \times \mathcal{Y} \mapsto \mathbb{R}$; *specifically*

   $$\mathbb{E}\{\mathbb{E}\{X|\boldsymbol{Y}\}\} = \mathbb{E}\{X\}$$

5. *refinement by conditioning: it can be shown (see page 13) that*

   $$\text{cov}\left(\mathbb{E}\{X|\boldsymbol{Y}\}\right) \leq \text{cov}\left(X\right) \tag{1.25}$$

   The variance is therefore reduced by conditioning;
6. *if* X *and* $\boldsymbol{Y}$ *are independent, then* $\mathbb{E}\{f(X)|\boldsymbol{Y}\} = \mathbb{E}\{f(X)\}$. *Specifically,* $\mathbb{E}\{X|\boldsymbol{Y}\} = \mathbb{E}\{X\}$. *The reverse is not true;*
7. $\mathbb{E}\{X|\boldsymbol{Y}\} = X$, *if and only if* X *is a function of* $\boldsymbol{Y}$.

1.3 Projection theorem

Definition 1.15 (Scalar product) *Let* $\mathcal{H}$ *be a vector space constructed over* $\mathbb{C}$. *The scalar product is an application*

$$X, Y \in \mathcal{H} \times \mathcal{H} \mapsto (X, Y) \in \mathbb{C}$$

which verifies the following properties:

- $(X,Y) = (Y,X)^*$;
- $(\alpha X + \beta Y, Z) = \alpha(X,Z) + \beta(Y,Z)$;
- $(X,X) \geq 0$. *The equality occurs if, and only if,* $X = 0$.

A vector space constructed over $\mathbb{C}$ has a Hilbert space structure if it possesses a scalar product and if it is complete[3]. The norm of X is defined by $\|X\| = \sqrt{(X,X)}$ and the distance between two elements by $d(X_1,X_2) = \|X_1-X_2\|$. Two elements X_1 and X_2 are said to be orthogonal, noted $X_1 \perp X_2$, if and only if $(X_1,X_2) = 0$. The demonstration of the following properties is trivial:

- Schwarz inequality:

$$|(X_1,X_2)| \leq \|X_1\| \, \|X_2\| \tag{1.26}$$

  the equality occurs if and only if λ exists such that $X_1 = \lambda X_2$;

- triangular inequality:

$$| \, \|X_1\| - \|X_2\| \, | \leq \|X_1 - X_2\| \leq \|X_1\| + \|X_2\| \tag{1.27}$$

- parallelogram identity:

$$\|X_1 + X_2\|^2 + \|X_1 - X_2\|^2 = 2\|X_1\|^2 + 2\|X_2\|^2 \tag{1.28}$$

In a Hilbert space, the projection theorem enables us to associate any given element from the space with its best quadratic approximation contained in a closed vector sub-space:

Theorem 1.2 (Projection theorem) *Let $\mathcal{H}$ be a Hilbert space defined over $\mathbb{C}$ and $\mathcal{C}$ a closed vector sub-space of $\mathcal{H}$. Each vector of $\mathcal{H}$ may then be associated with a unique element X_0 of $\mathcal{C}$ such that $\forall Y \in \mathcal{C}$ we have $d(X,X_0) \leq d(X,Y)$. Vector X_0 verifies, for any $Y \in \mathcal{C}$, the relationship $(X - X_0) \perp Y$.*

The relationship $(X - X_0) \perp Y$ constitutes the *orthogonality principle*.

A geometric representation of the orthogonality principle is shown in Figure 1.1. The element of $\mathcal{C}$ closest in distance to X is given by the *orthogonal projection* of X onto $\mathcal{C}$. In practice, this is the relationship which allows us to find the solution X_0.

This result is used alongside the expression of the norm of $X - X_0$, which is written:

$$\begin{aligned} \|X - X_0\|^2 &= (X, X - X_0) - (X_0, X - X_0) \\ &= \|X\|^2 - (X, X_0) \end{aligned} \tag{1.29}$$

The term $(X_0, X - X_0)$ is null due to the orthogonality principle.

[3] A definition of the term "complete" in this context may be found in mathematical textbooks. In the context of our presentation, this property plays a concealed role, for example in the existence of the orthogonal projection in theorem 1.2.

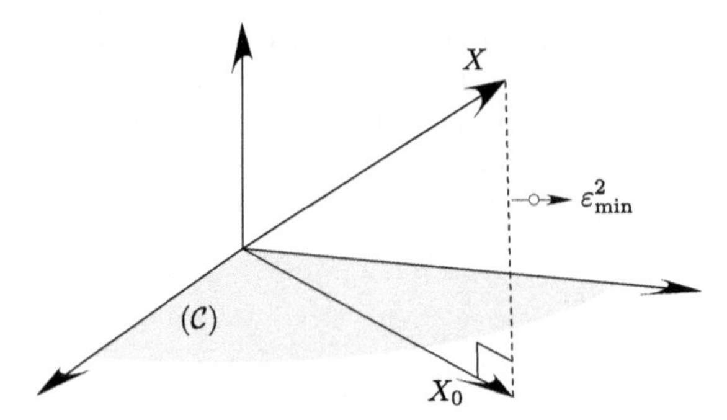

Figure 1.1 – *Orthogonality principle: the point X_0 which is the closest to X in $\mathcal{C}$ is such that $X - X_0$ is orthogonal to $\mathcal{C}$*

In what follows, the vector X_0 will be noted $(X|\mathcal{C})$, or $(X|Y_{1:n})$ when the sub-space onto which projection occurs is spanned by the linear combinations of vectors $Y_1, \ldots, Y_n$.

The simplest application of theorem 1.2 provides that for any $X \in \mathcal{C}$ and any $\varepsilon \in \mathcal{C}$:

$$(X|\varepsilon) = \frac{(X, \varepsilon)}{(\varepsilon, \varepsilon)} \varepsilon \tag{1.30}$$

The projection theorem leads us to define an application associating element X with element $(X|\mathcal{C})$. This application is known as the *orthogonal projection* of X onto $\mathcal{C}$. The orthogonal projection verifies the following properties:

1. linearity: $(\lambda X_1 + \mu X_2|\mathcal{C}) = \lambda(X_1|\mathcal{C}) + \mu(X_2|\mathcal{C})$;
2. contraction: $\|(X|\mathcal{C})\| \leq \|X\|$;
3. if $\mathcal{C}' \subset \mathcal{C}$, then $((X|\mathcal{C})|\mathcal{C}') = (X|\mathcal{C}')$;
4. if $\mathcal{C}_1 \perp \mathcal{C}_2$, then $(X|\mathcal{C}_1 \oplus \mathcal{C}_2) = (X|\mathcal{C}_1) + (X|\mathcal{C}_2)$.

The following result is fundamental:

$$(X|Y_{1:n+1}) = (X|Y_{1:n}) + (X|\varepsilon) = (X|Y_{1:n}) + \frac{(X, \varepsilon)}{(\varepsilon, \varepsilon)} \varepsilon \tag{1.31}$$

where $\varepsilon = Y_{n+1} - (Y_{n+1}|Y_{1:n})$. Because the sub-space spanned by $Y_{1:n+1}$ coincides with the sub-space spanned by $(Y_{1:,n}, \varepsilon)$ and because ε is orthogonal to the sub-space generated by $(Y_{1:n})$, then property (4) applies. To complete the proof we use (1.30).

Formula (1.31) is the basic formula used in the determination of many recursive algorithms, such as Kalman filter or Levinson recursion.

Theorem 1.3 (Square-integrable r.v.) *Let $\mathcal{L}^2_P$ be the vector space of square-integrable random variables, defined over the probability space $(\Omega, \mathcal{A}, P)$. Using the scalar product $(X, Y) = \mathbb{E}\{XY^*\}$, $\mathcal{L}^2_P$ has a Hilbert space structure.*

Conditional expectation

The conditional expectation $\mathbb{E}\{X|Y\}$ may be seen as the orthogonal projection of X onto sub-space $\mathcal{C}$ of all measurable functions of Y. Similarly, $\mathbb{E}\{X\}$ may be seen as the orthogonal projection of X onto the sub-space $\mathcal{D}$ of the constant random variables. These vectors are shown in Figure 1.2. Because $\mathcal{D} \subset \mathcal{C}$, using Pythagoras's theorem, we deduce that:

$$\operatorname{var}(X) = \|X - \mathbb{E}\{X\}\|^2 = \|X - \mathbb{E}\{X|Y\}\|^2 + \underbrace{\|\mathbb{E}\{X|Y\} - \mathbb{E}\{X\}\|^2}_{=\operatorname{var}(\mathbb{E}\{X|Y\})}$$

demonstrating $\operatorname{var}(\mathbb{E}\{X|Y\}) \leq \operatorname{var}(X)$. This can be extended to random vectors, giving the inequality (1.25) i.e. $\operatorname{cov}(\mathbb{E}\{X|\boldsymbol{Y}\}) \leq \operatorname{cov}(X)$.

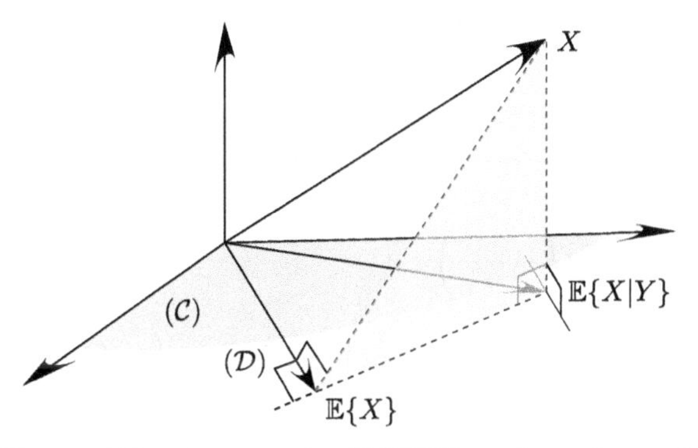

Figure 1.2 – *The conditional expectation $\mathbb{E}\{X|Y\}$ is the orthogonal projection of X onto the set $\mathcal{C}$ of measurable functions of Y. The expectation $\mathbb{E}\{X\}$ is the orthogonal projection of X onto the set $\mathcal{D}$ of constant functions. Clearly, $\mathcal{D} \subset \mathcal{C}$*

1.4 Gaussianity

Real Gaussian random variable

Definition 1.16 *A random variable X is said to be Gaussian, or normal, if all its values belong to $\mathbb{R}$ and if its characteristic function (see definition (1.9)) has the expression:*

$$\phi_X(u) = \exp\left(jmu - \frac{1}{2}\sigma^2 u^2\right) \tag{1.32}$$

where m is a real parameter and σ is a positive parameter. We check that its mean is equal to m and its variance to σ^2.

If $\sigma \neq 0$, it can be shown that the probability distribution has a probability density function with the expression:

$$p_X(x) = \frac{1}{\sigma\sqrt{2\pi}} \exp\left(-\frac{(x-m)^2}{2\sigma^2}\right) \tag{1.33}$$

Complex Gaussian random variable

In some problems, and particularly in the field of communications, the complex notation $X = U + jV$ is used, where U and V refer to two *real*, *Gaussian*, *centered*, *independent* random variables with the same variance $\sigma^2/2$. Because of independence (definition (1.7)), the joint probability distribution of the pair (U, V) has the following probability density:

$$\begin{aligned} p_{UV}(u,v) &= p_U(u)p_V(v) = \frac{1}{\sigma\sqrt{\pi}} \exp\left(-\frac{u^2}{\sigma^2}\right) \times \frac{1}{\sigma\sqrt{\pi}} \exp\left(-\frac{v^2}{\sigma^2}\right) \\ &= \frac{1}{\pi\sigma^2} \exp\left(-\frac{u^2+v^2}{\sigma^2}\right) \end{aligned}$$

If we notice that $|x|^2 = u^2 + v^2$, and if we introduce the notation $p_X(x) = p_{UV}(u,v)$, we can also write:

$$p_X(x) = \frac{1}{\pi\sigma^2} \exp\left(-\frac{|x|^2}{\sigma^2}\right) \tag{1.34}$$

Expression (1.34) is called the probability density of a *complex Gaussian* random variable. The word *circular* is sometimes added as a reminder that the isodensity contours are the circles $u^2 + v^2 =$ constant.

Note that:

$$\begin{aligned} \mathbb{E}\left\{X^2\right\} &= \mathbb{E}\left\{(U+jV)(U+jV)\right\} = 0 \\ &\text{and} \\ \mathbb{E}\left\{|X|^2\right\} &= \mathbb{E}\left\{XX^*\right\} = \mathbb{E}\left\{(U+jV)(U-jV)\right\} \\ &= \mathbb{E}\left\{U^2\right\} + \mathbb{E}\left\{V^2\right\} = \sigma^2 \end{aligned}$$

Gaussian random vectors

Definition 1.17 (Gaussian vector) *$X_1, \ldots, X_n$ are said to be n jointly Gaussian variables, or that the length n vector $\begin{bmatrix} X_1 & \ldots & X_n \end{bmatrix}^T$ is Gaussian, if any linear combination of its components, that is to say $Y = \boldsymbol{a}^H \boldsymbol{X}$ for any $\boldsymbol{a} = \begin{bmatrix} a_1 & \ldots & a_n \end{bmatrix}^T \in \mathbb{C}^n$, is a Gaussian random variable.*

This definition is applicable for vectors with real or complex components.

Theorem 1.4 (Distribution of a real Gaussian vector) *It can be shown that the probability distribution of a n Gaussian vector, with a length n mean vector $\boldsymbol{m}$ and an $(n \times n)$ covariance matrix $\boldsymbol{C}$, has the characteristic function:*

$$\phi_{\boldsymbol{X}}(u_1, \ldots, u_n) = \exp\left(j\boldsymbol{m}^T\boldsymbol{u} - \frac{1}{2}\boldsymbol{u}^T\boldsymbol{C}\boldsymbol{u}\right) \quad (1.35)$$

where $\boldsymbol{u} = (u_1, \ldots, u_n)^T \in \mathbb{R}^n$. Let $\boldsymbol{x} = (x_1, \ldots, x_n)^T$. If $\det\{\boldsymbol{C}\} \neq 0$, the probability distribution's density has the expression:

$$p_{\boldsymbol{X}}(x_1, \ldots, x_n) = \frac{1}{(2\pi)^{n/2}\sqrt{\det\{\boldsymbol{C}\}}} \exp\left(-\frac{1}{2}(\boldsymbol{x} - \boldsymbol{m})^T\boldsymbol{C}^{-1}(\boldsymbol{x} - \boldsymbol{m})\right) \quad (1.36)$$

Theorem 1.5 (Distribution of a complex Gaussian vector) *We consider a length n complex Gaussian vector, with a length n mean vector $\boldsymbol{m}$ and an $(n \times n)$ covariance matrix $\boldsymbol{C}$. If $\det\{\boldsymbol{C}\} \neq 0$, the probability distribution's density has the expression:*

$$p_{\boldsymbol{X}}(x_1, \ldots, x_n) = \frac{1}{\pi^n \det\{\boldsymbol{C}\}} \exp\left(-(\boldsymbol{x} - \boldsymbol{m})^H\boldsymbol{C}^{-1}(\boldsymbol{x} - \boldsymbol{m})\right) \quad (1.37)$$

We have

$$\mathbb{E}\left\{(\boldsymbol{x} - \boldsymbol{m})(\boldsymbol{x} - \boldsymbol{m})^H\right\} = \boldsymbol{C} \quad (1.38)$$

$$\mathbb{E}\left\{(\boldsymbol{x} - \boldsymbol{m})(\boldsymbol{x} - \boldsymbol{m})^T\right\} = \boldsymbol{0}_n \quad (1.39)$$

where $\boldsymbol{0}_n$ is the $(n \times n)$ null-matrix.

Below, the real and complex Gaussian distributions will be noted $\mathcal{N}(\boldsymbol{m}, \boldsymbol{C})$ and $\mathcal{N}_c(\boldsymbol{m}, \boldsymbol{C})$ respectively.

Theorem 1.6 (Gaussian case: non-correlation $\Rightarrow$ independence) *If n jointly Gaussian variables are uncorrelated, $\boldsymbol{C}$ is diagonal; then they are independent.*

Theorem 1.7 (Linear transformation of a Gaussian vector) *Let $[X_1 \ldots X_n]^T$ be a Gaussian vector with a mean vector $\boldsymbol{m}_X$ and a covariance matrix $\boldsymbol{C}_X$. The random vector $\boldsymbol{Y} = \boldsymbol{A}\boldsymbol{X} + \boldsymbol{b}$, where $\boldsymbol{A}$ and $\boldsymbol{b}$ are a matrix and a vector respectively, with the ad hoc length, is Gaussian and we have:*

$$\boldsymbol{m}_Y = \boldsymbol{A}\boldsymbol{m}_X + \boldsymbol{b} \quad \text{and} \quad \boldsymbol{C}_Y = \boldsymbol{A}\boldsymbol{C}_X\boldsymbol{A}^H \quad (1.40)$$

In other words, the Gaussian nature of a vector is untouched by linear transformations.

Equations (1.40) are a direct consequence of definition (1.17) and of property (1.6).

More specifically, if $\boldsymbol{X}$ is a random Gaussian vector $\mathcal{N}(\boldsymbol{m}, \boldsymbol{C})$, then the random variable $\boldsymbol{Z} = \boldsymbol{C}^{-1/2}(\boldsymbol{X} - \boldsymbol{M})$ follows a Gaussian distribution $\mathcal{N}(\boldsymbol{0}, \boldsymbol{I})$. Another way of expressing this is to say that if $\boldsymbol{Z}$ has the distribution $\mathcal{N}(\boldsymbol{0}, \boldsymbol{I})$ then $\boldsymbol{X} = \boldsymbol{M} + \boldsymbol{C}^{1/2}\boldsymbol{Z}$ has the distribution $\mathcal{N}(\boldsymbol{m}, \boldsymbol{C})$.

Note that, if $\boldsymbol{C}$ denotes a positive matrix, a square root of $\boldsymbol{C}$ is a matrix $\boldsymbol{M}$ which verifies:

$$\boldsymbol{C} = \boldsymbol{M}\boldsymbol{M}^H \tag{1.41}$$

Hence, if $\boldsymbol{M}$ is a square root of $\boldsymbol{C}$, then for any unitary matrix $\boldsymbol{U}$, i.e. such that $\boldsymbol{U}\boldsymbol{U}^H = \boldsymbol{I}$, matrix $\boldsymbol{M}\boldsymbol{U}$ is also a square root of $\boldsymbol{C}$. Matrix $\boldsymbol{M}$ is therefore defined to within a unitary matrix. One of the square roots is positive, and is obtained in MATLAB® using the function `sqrtm`.

The Gaussian distribution is defined using the first and second order moments, i.e. the mean and the covariance. Consequently, all moments of an order greater than 2 are expressed as a function of the first two values. The following theorem covers the specific case of a moment of order 4.

Theorem 1.8 (Moment of order 4) *Let X_1, X_2, X_3 and X_4 be four real or complex* centered *Gaussian random variables. Hence,*

$$\begin{aligned}\mathbb{E}\left\{X_1^{\beta_1}X_2^{\beta_2}X_3^{\beta_3}X_4^{\beta_4}\right\} &= \mathbb{E}\left\{X_1^{\beta_1}X_2^{\beta_2}\right\}\mathbb{E}\left\{X_3^{\beta_3}X_4^{\beta_4}\right\} \\ &+\mathbb{E}\left\{X_1^{\beta_1}X_3^{\beta_3}\right\}\mathbb{E}\left\{X_2^{\beta_2}X_4^{\beta_4}\right\} + \mathbb{E}\left\{X_1^{\beta_1}X_4^{\beta_4}\right\}\mathbb{E}\left\{X_2^{\beta_2}X_3^{\beta_3}\right\}\end{aligned} \tag{1.42}$$

where β_i is either "star" (conjugate variable) or "non-star" (non-conjugate variable). Hence:

$$\begin{aligned}&\operatorname{cov}\left(X_1^{\beta_1}X_2^{\beta_2}, X_3^{\beta_3}X_4^{\beta_4}\right) \\ &\quad= \mathbb{E}\left\{X_1^{\beta_1}X_2^{\beta_2}X_3^{\bar\beta_3}X_4^{\bar\beta_4}\right\} - \mathbb{E}\left\{X_1^{\beta_1}X_2^{\beta_2}\right\}\mathbb{E}\left\{X_3^{\bar\beta_3}X_4^{\bar\beta_4}\right\} \\ &\quad= \mathbb{E}\left\{X_1^{\beta_1}X_3^{\bar\beta_3}\right\}\mathbb{E}\left\{X_2^{\beta_2}X_4^{\bar\beta_4}\right\} + \mathbb{E}\left\{X_1^{\beta_1}X_4^{\bar\beta_4}\right\}\mathbb{E}\left\{X_2^{\beta_2}X_3^{\bar\beta_3}\right\}\end{aligned} \tag{1.43}$$

where $\bar\beta_j$ is "star" if β_j is "non-star" and conversely.

Note that, based on (1.39), for *complex* Gaussian variables, the terms $\mathbb{E}\{X_iX_j\} = 0$.

Gaussian conditional distribution

Consider two jointly Gaussian variables $\boldsymbol{X}$ and $\boldsymbol{Y}$, taking their values from $\mathcal{X} \subset \mathbb{C}^p$ and $\mathcal{Y} \subset \mathbb{C}^q$ respectively. The respective means are noted $\mu_{\boldsymbol{X}}$ and $\mu_{\boldsymbol{Y}}$, and

$$\boldsymbol{C} = \begin{bmatrix} \operatorname{cov}(\boldsymbol{X}, \boldsymbol{X}) & \operatorname{cov}(\boldsymbol{X}, \boldsymbol{Y}) \\ \operatorname{cov}(\boldsymbol{Y}, \boldsymbol{X}) & \operatorname{cov}(\boldsymbol{Y}, \boldsymbol{Y}) \end{bmatrix} \tag{1.44}$$

is the joint covariance matrix. This produces the following results:

Property 1.10 *The conditional expectation of $\boldsymbol{X}$ given $\boldsymbol{Y}$ coincides with the orthogonal projection of $\boldsymbol{X}$ onto the affine sub-space spanned by $\mathbf{1}$ and $\boldsymbol{Y}$, written $\boldsymbol{B} + \boldsymbol{A}\boldsymbol{Y}$. Hence:*

- *the conditional expectation is expressed as:*

$$\mathbb{E}\{\boldsymbol{X}|\boldsymbol{Y}\} = \mu_{\boldsymbol{X}} + \operatorname{cov}(\boldsymbol{X}, \boldsymbol{Y}) \left[\operatorname{cov}(\boldsymbol{Y}, \boldsymbol{Y})\right]^{-1} (\boldsymbol{Y} - \mu_{\boldsymbol{Y}}) \tag{1.45}$$

- *the conditional covariance is expressed as:*

$$\operatorname{cov}(\boldsymbol{X}|\boldsymbol{Y}) = \operatorname{cov}(\boldsymbol{X}, \boldsymbol{X}) - \operatorname{cov}(\boldsymbol{X}, \boldsymbol{Y}) \left[\operatorname{cov}(\boldsymbol{Y}, \boldsymbol{Y})\right]^{-1} \operatorname{cov}(\boldsymbol{Y}, \boldsymbol{X}) \tag{1.46}$$

- *the conditional distribution of $p_{\boldsymbol{X}|\boldsymbol{Y}}(\boldsymbol{x}, \boldsymbol{y})$ is Gaussian. The mean is expressed as* (1.45) *and the covariance is given by expression* (1.46).

Let $g(\boldsymbol{Y})$ be the second member of (1.45), and let us demonstrate that $g(\boldsymbol{Y})$ is the conditional expectation of $\boldsymbol{X}$ given $\boldsymbol{Y}$. A rapid calculation shows that $\mathbb{E}\left\{(\boldsymbol{X} - g(\boldsymbol{Y}))\boldsymbol{Y}^H\right\} = 0$. Consequently, the random vectors $\boldsymbol{Z} = (\boldsymbol{X} - g(\boldsymbol{Y}))$ and $\boldsymbol{Y}$ are uncorrelated. As the vectors are jointly Gaussian, following property (1.9), they are independent and hence $\mathbb{E}\{\boldsymbol{Z}|\boldsymbol{Y}\} = \mathbb{E}\{\boldsymbol{Z}\}$. Using the second member of (1.45), we obtain $\mathbb{E}\{\boldsymbol{Z}\} = 0$. On the other hand:

$$\mathbb{E}\{\boldsymbol{X} - g(\boldsymbol{Y})|\boldsymbol{Y}\} = \mathbb{E}\{\boldsymbol{X}|\boldsymbol{Y}\} - g(\boldsymbol{Y})$$

It follows that $\mathbb{E}\{\boldsymbol{X}|\boldsymbol{Y}\} = g(\boldsymbol{Y})$. To demonstrate expression (1.46), let us denote $\boldsymbol{X}^c = \boldsymbol{X} - \mu_{\boldsymbol{X}}$ and $\boldsymbol{X}^c_{\boldsymbol{Y}} = \mathbb{E}\{\boldsymbol{X}|\boldsymbol{Y}\} - \mu_{\boldsymbol{X}}$. Hence, successively:

$$\begin{aligned} \mathbb{E}\{(\boldsymbol{X}^c - \boldsymbol{X}^c_{\boldsymbol{Y}})((\boldsymbol{X}^c - \boldsymbol{X}^c_{\boldsymbol{Y}})^H|\boldsymbol{Y}\} &= \mathbb{E}\{(\boldsymbol{X}^c - \boldsymbol{X}^c_{\boldsymbol{Y}})(\boldsymbol{X}^c - \boldsymbol{X}^c_{\boldsymbol{Y}})^H\} \\ &= \mathbb{E}\left\{(\boldsymbol{X}^c - \boldsymbol{X}^c_{\boldsymbol{Y}})\boldsymbol{X}^{c,H}\right\} \\ &= \operatorname{cov}(\boldsymbol{X}, \boldsymbol{X}) - \operatorname{cov}(\boldsymbol{X}, \boldsymbol{Y}) \left[\operatorname{cov}(\boldsymbol{Y}\boldsymbol{Y})\right]^{-1} \operatorname{cov}(\boldsymbol{Y}, \boldsymbol{X}) \end{aligned}$$

where, in the first equality, we use the fact that $(\boldsymbol{X}^c - \boldsymbol{X}^c_{\boldsymbol{Y}})$ is independent of $\boldsymbol{Y}$. In conclusion, the conditional distribution of $\boldsymbol{X}$, given $\boldsymbol{Y}$, is written:

$$\begin{aligned} p_{\boldsymbol{X}|\boldsymbol{Y}}(\boldsymbol{x}, \boldsymbol{y}) = \mathcal{N}\big(&\mu_X + \operatorname{cov}(\boldsymbol{X}, \boldsymbol{Y}) \left[\operatorname{cov}(\boldsymbol{Y}\boldsymbol{Y})\right]^{-1}(\boldsymbol{Y} - \mu_{\boldsymbol{Y}}), \\ &\operatorname{cov}(\boldsymbol{X}, \boldsymbol{X}) - \operatorname{cov}(\boldsymbol{X}, \boldsymbol{Y}) \left[\operatorname{cov}(\boldsymbol{Y}\boldsymbol{Y})\right]^{-1} \operatorname{cov}(\boldsymbol{Y}, \boldsymbol{X})\big) \end{aligned} \tag{1.47}$$

Note that the distribution for random vector $\mathbb{E}\{\boldsymbol{X}|\boldsymbol{Y}\}$ should not be confused with the conditional distribution $p_{\boldsymbol{X}|\boldsymbol{Y}}(\boldsymbol{x},\boldsymbol{y})$ of $\boldsymbol{X}$ given $\boldsymbol{Y}$. We shall restrict ourselves to the real, scalar case, taking μ_X and μ_Y as the respective means of X and Y, and

$$\boldsymbol{C} = \begin{bmatrix} \sigma_X^2 & \rho\sigma_X\sigma_Y \\ \rho\sigma_X\sigma_Y & \sigma_Y^2 \end{bmatrix}$$

with $-1 \leq \rho \leq 1$ as the covariance matrix. The conditional distribution of X given Y has a probability density $p_{X|Y}(x;y) = \mathcal{N}(\mu_X + \rho\sigma_X(y-\mu_Y)/\sigma_Y, \sigma_X^2(1-\rho^2))$. On the other hand the random variable distribution $\mathbb{E}\{X|Y\}$ has a probability density of $\mathcal{N}(\mu_X, \rho^2\sigma_X^2)$. Indeed based on equation (1.45), $\mathbb{E}\{\mathbb{E}\{X|Y\}\} = \mu_X$ and $\mathbb{E}\left\{(\mathbb{E}\{X|Y\} - \mu_X)^2\right\} = \rho^2\sigma_X^2\sigma_Y^2/\sigma_Y^2 = \rho^2\sigma_X^2$.

1.5 Random variable transformation

1.5.1 Change of variable formula

In many cases, it is necessary to determine the distribution of $Y = g(X)$ from the distribution of X. We shall consider this question in the context of continuous random vectors of dimension 2; the generalization to higher dimensions is straightforward.

Take two random variables X_1 and X_2 with a joint probability density $p_{X_1X_2}(x_1, x_2)$ and two measurable functions $g_1(x_1, x_2)$ and $g_2(x_1, x_2)$. We shall consider the two random variables:

$$\begin{cases} Y_1 = g_1(X_1, X_2) \\ Y_2 = g_2(X_1, X_2) \end{cases}$$

and assume that the transformation defined in this way is bijective. For any pair (y_1, y_2), there is a single solution (x_1, x_2). We may therefore write:

$$\begin{cases} X_1 = h_1(Y_1, Y_2) \\ X_2 = h_2(Y_1, Y_2) \end{cases}$$

In this case, the probability distribution of the random variables (Y_1, Y_2) has a density of:

$$p_{Y_1Y_2}(y_1, y_2) = p_{X_1X_2}(h_1(y_1, y_2), h_2(y_1, y_2)) \left| \det\left\{\frac{\partial \boldsymbol{x}}{\partial \boldsymbol{y}}\right\} \right| \tag{1.48}$$

where $\frac{\partial \boldsymbol{x}}{\partial \boldsymbol{y}}$ denotes the *Jacobian* of $\boldsymbol{h} : \boldsymbol{y} \to \boldsymbol{x}$ which is expressed as:

$$\frac{\partial \boldsymbol{x}}{\partial \boldsymbol{y}} = \begin{bmatrix} \dfrac{\partial h_1(y_1, y_2)}{\partial y_1} & \dfrac{\partial h_2(y_1, y_2)}{\partial y_1} \\ \dfrac{\partial h_1(y_1, y_2)}{\partial y_2} & \dfrac{\partial h_2(y_1, y_2)}{\partial y_2} \end{bmatrix}$$

In cases where the transformation is not bijective, it is necessary to sum all of the solutions giving the pair (x_1, x_2) as a function of the pair (y_1, y_2).

Note that the Jacobian of a bijective function has one particularly useful property. Taking a bijective function $\boldsymbol{x} \in \mathbb{R}^d \leftrightarrow \boldsymbol{y} \in \mathbb{R}^d$, we have:

$$\frac{\partial \boldsymbol{x}}{\partial \boldsymbol{y}} \times \frac{\partial \boldsymbol{y}}{\partial \boldsymbol{x}} = \boldsymbol{I}_d$$

This property allows us to calculate the Jacobian using the expression which is easiest to calculate, and, if necessary, to take the inverse.

Example 1.1 (Law of the sum of two random variables)
As an example, let us consider two random variables X_1 and X_2 with a joint probability density $p_{X_1X_2}(x_1, x_2)$. We wish to determine the joint distribution of the pair (Y_1, Y_2), defined by the following transformation:

$$\begin{cases} Y_1 = X_1 \\ Y_2 = X_1 + X_2 \end{cases} \Leftrightarrow \begin{cases} X_1 = Y_1 \\ X_2 = Y_2 - Y_1 \end{cases}$$

where the determinant of the Jacobian has a value of 1. Applying (1.48), we obtain the following probability density for the pair (Y_1, Y_2):

$$p_{Y_1Y_2}(y_1, y_2) = p_{X_1X_2}(y_1, y_2 - y_1)$$

From this, the probability density of $Y_2 = X_1 + X_2$ may be deduced by identifying the marginal distribution of Y_2. We obtain:

$$p_{Y_2}(y_2) = \int_{\mathbb{R}} p_{X_1X_2}(y_1, y_2 - y_1) dy_1$$

In cases where X_1 and X_2 are independent:

$$p_{X_1X_2}(x_1, x_2) = p_{X_1}(x_1) p_{X_2}(x_2)$$

hence:

$$p_{Y_2}(y_2) = \int_{\mathbb{R}} p_{X_1}(y_1) p_{X_2}(y_2 - y_1) dy_1$$

which is the expression of the convolution product $(p_{X_1} \star p_{X_2})(y_2)$.

1.5.2 δ-method

In cases where only the first two moments are considered, under very general conditions, the δ-method allows us to obtain approximate formulas for the mean and the covariance of:

$$\boldsymbol{Y} = \boldsymbol{g}(\boldsymbol{X}) \tag{1.49}$$

for any function $\boldsymbol{g} : \mathbb{R}^m \mapsto \mathbb{R}^q$. Let $\mu_{\boldsymbol{X}} = \mathbb{E}\{\boldsymbol{X}\} \in \mathbb{R}^m$. Assuming that $\boldsymbol{g}$ is differentiable at point $\mu_{\boldsymbol{X}}$ and using the first order Taylor expansion of $\boldsymbol{g}$ in the neighborhood of $\mu_{\boldsymbol{X}}$, we write

$$\boldsymbol{Y} = \boldsymbol{g}(\boldsymbol{X}) \approx \boldsymbol{g}(\mu_{\boldsymbol{X}}) + \left.\frac{\partial \boldsymbol{y}}{\partial \boldsymbol{x}}\right|_{\boldsymbol{x}=\mu_{\boldsymbol{X}}} (\boldsymbol{X} - \mu_{\boldsymbol{X}}) \tag{1.50}$$

where

$$\left.\frac{\partial \boldsymbol{y}}{\partial \boldsymbol{x}}\right|_{\boldsymbol{x}=\mu_{\boldsymbol{X}}} = \begin{bmatrix} \frac{\partial g_1}{\partial x_1}(\mu_{\boldsymbol{X}}) & \cdots & \frac{\partial g_1}{\partial x_m}(\mu_{\boldsymbol{X}}) \\ \vdots & & \vdots \\ \frac{\partial g_q}{\partial x_1}(\mu_{\boldsymbol{X}}) & \cdots & \frac{\partial g_q}{\partial x_m}(\mu_{\boldsymbol{X}}) \end{bmatrix}$$

is the $q \times m$ Jacobian matrix of $\boldsymbol{g}$ performed at point $\mu_{\boldsymbol{X}}$. For the sake of simplicity, this is noted $\boldsymbol{J}(\mu_{\boldsymbol{X}})$ below. Therefore, taking the expectation of (1.50), we get at first order

$$\mathbb{E}\{\boldsymbol{Y}\} \approx \boldsymbol{g}(\mu_{\boldsymbol{X}}) + \boldsymbol{J}(\mu_{\boldsymbol{X}}) \times \mathbb{E}\{\boldsymbol{X} - \mu_{\boldsymbol{X}}\} = \boldsymbol{g}(\mu_{\boldsymbol{X}}) + 0$$

then

$$\boldsymbol{Y} - \mathbb{E}\{\boldsymbol{Y}\} \approx \boldsymbol{J}(\mu_{\boldsymbol{X}})(\boldsymbol{X} - \mu_{\boldsymbol{X}})$$

Therefore, according to the definition (1.18) of $\text{cov}(\boldsymbol{Y})$, we have

$$\text{cov}(\boldsymbol{g}(\boldsymbol{X})) \approx \boldsymbol{J}(\mu_{\boldsymbol{X}}) \,\text{cov}(\boldsymbol{X})\, \boldsymbol{J}^H(\mu_{\boldsymbol{X}})$$

It is worth noting that $\text{cov}(\boldsymbol{g}(\boldsymbol{X}))$ is a $q \times q$ matrix and $\text{cov}(\boldsymbol{X})$ is a $m \times m$ matrix. In summary we have:

$$\begin{cases} \mathbb{E}\{\boldsymbol{g}(\boldsymbol{X})\} & \approx & \boldsymbol{g}(\mu_{\boldsymbol{X}}) \\ \text{cov}(\boldsymbol{g}(\boldsymbol{X})) & \approx & \boldsymbol{J}(\mu_{\boldsymbol{X}}) \,\text{cov}(\boldsymbol{X})\, \boldsymbol{J}^H(\mu_{\boldsymbol{X}}) \end{cases} \tag{1.51}$$

The δ-method is commonly used when calculating the mean and the covariance of $\boldsymbol{g}(\boldsymbol{X})$ is either intractable or the probability distribution of $\boldsymbol{X}$ is not fully specified.

Exercise 1.1 (δ-method) (see page 235) Consider two random variables (X_1, X_2), Gaussian and independent, with means of μ_1 and μ_2 respectively, and with the same variance σ^2. Using the pair (X_1, X_2), we determine the pair (R, θ) by bijective transformation:

$$(X_1, X_2) = \boldsymbol{h}(R, \theta) : \begin{cases} X_1 & = & R\cos(\theta) \in \mathbb{R} \\ X_2 & = & R\sin(\theta) \in \mathbb{R} \end{cases} \Leftrightarrow$$

$$(R, \theta) = \boldsymbol{g}(X_1, X_2) : \begin{cases} R & = & |X_1 + jX_2| = \sqrt{X_1^2 + X_2^2} \in \mathbb{R}^+ \\ \theta & = & \arg(X_1 + jX_2) \in (0, 2\pi) \end{cases}$$

Use the δ-method to determine the covariance of the pair (R, θ). Use this result to deduce the variance of R. This may be compared with the theoretical value given by:

$$\operatorname{var}(R) = 2\sigma^2 + (\mu_1^2 + \mu_2^2) - \frac{\pi\sigma^2}{2} L_{1/2}^2 \left(\frac{-(\mu_1^2 + \mu_2^2)}{2\sigma^2} \right)$$

where $L_{1/2}(x) = {}_1F_1\left(-\frac{1}{2}; 1; x\right)$ is the hypergeometric function. We see that, when $(\mu_1^2 + \mu_2^2)/\sigma^2$ tends toward infinity, $\operatorname{var}(R)$ tends toward σ^2. Additionally, when $\mu_1 = \mu_2 = 0$, we have $\operatorname{var}(R) = (4 - \pi)\sigma^2/2 \approx 0.43\sigma^2$.

1.6 Fundamental statistical theorems

The following two theorems form the basis of statistical methods, and are essential to the validity of Monte-Carlo methods, which are presented in brief in Chapter 3. In very general conditions, these theorems imply that the empirical average of a series of r.v.s will converge toward the mean. The first theorem, often (erroneously) referred to as a *law*, sets out this convergence; the second theorem states that this convergence is "distributed in a Gaussian manner".

Theorem 1.9 (Law of large numbers) *Let X_n be a series of random vectors of dimension d, independent and identically distributed, with a mean vector $m = \mathbb{E}\{X_1\} \in \mathbb{R}^d$ and finite covariance. In this case,*

$$\frac{1}{N} \sum_{n=1}^{N} X_n \xrightarrow{a.s.}_{N \to +\infty} \mathbb{E}\{X_1\} = m$$

and convergence is almost sure.

One fundamental example is that of empirical frequency, which converges toward the probability. Let X_n be a series of N random variables with values in $a_1, a_2, \ldots, a_J$ and let f_j be the empirical frequency, defined as the relationship between the number of values equal to a_j and the total number N. In this case:

$$f_j = \frac{1}{N} \sum_{n=1}^{N} \mathbb{1}(X_n = a_j) \xrightarrow{a.s.} \mathbb{E}\{\mathbb{1}(X_1 = a_j)\} = \mathbb{P}\{X_1 = a_j\}$$

Theorem 1.10 (Central limit theorem) *Let X_n be a series of random vectors of dimension d, independent and identically distributed, of mean vector $m = \mathbb{E}\{X_1\}$ and covariance matrix $C = \operatorname{cov}(X_1, X_1)$. In this case:*

$$\sqrt{N} \left(\frac{1}{N} \sum_{n=1}^{N} X_n - m \right) \xrightarrow{d}_{N \to +\infty} \mathcal{N}(0, C)$$

with convergence in distribution.

Convergence in distribution is defined as follows:

Definition 1.18 (Convergence in distribution) *A set of r.v. U_N is said to converge in distribution toward an r.v. U if, for any bounded continuous function f, when N tends toward infinity, we have:*

$$\mathbb{E}\{f(U_N)\} \rightarrow_{N\to\infty} \mathbb{E}\{f(U)\} \tag{1.52}$$

Theorem 1.10 is the basis for calculations of confidence intervals (see definition 2.6), and is used as follows: we approximate the probability distribution of the random vector $\sqrt{N}\left(N^{-1}\sum_{n=1}^{N} X_n - m\right)$, for which the expression is often impossible to calculate, by the Gaussian distribution. For illustrative purposes, consider the case where $d = 1$, taking $\widehat{m}_N = N^{-1}\sum_{n=1}^{N} X_n$. Hence, for any $c > 0$:

$$\mathbb{P}\left\{\sqrt{N}\,(\widehat{m}_N - m) \in (-\varepsilon, +\varepsilon)\right\} \approx 2\int_0^{\varepsilon} \frac{1}{\sigma\sqrt{2\pi}} e^{-u^2/2\sigma^2} du$$

Letting $\epsilon = c\sigma$, we have:

$$\mathbb{P}\left\{\widehat{m}_N - \frac{c\sigma}{\sqrt{N}} < m \leq \widehat{m}_N + \frac{c\sigma}{\sqrt{N}}\right\} \approx 2\int_0^{c} \frac{1}{\sqrt{2\pi}} e^{-t^2/2} dt$$

Aim for a probability equal typically to 0.05, $c = 1.96$.

As expected, the smaller σ and/or the higher N, the narrower, i.e. "better", the interval will be.

Exercise 1.2 (Asymptotic confidence interval) (see page 236) Consider a sequence of N independent random Bernoulli variables X_k such that $\mathbb{P}\{X_k = 1\} = p$. To estimate the proportion p, we consider $\widehat{p} = \frac{1}{N}\sum_{k=1}^{N} X_k$.

1. Using the central limit theorem 1.10, determine the asymptotic distribution of $\widehat{p}$.

2. Use the previous result to deduce the approximate expression of the probability that p will lie within the interval between $\widehat{p} - \epsilon/\sqrt{N}$ and $\widehat{p} + \epsilon/\sqrt{N}$.

3. Use this result to deduce an interval which ensures that this probability will be higher than $100\,\alpha\%$, expressed as a function of N and α: typically, $\alpha = 0.95$.

4. Write a program which verifies this asymptotic behavior.

The following theorem, known as the *continuity theorem*, allows us to extend the central limit theorem to more complicated functions:

Theorem 1.11 (Continuity) *Let $\boldsymbol{U}_N$ be a series of random vectors of dimension d such that*

$$\sqrt{N}(\boldsymbol{U}_N - \boldsymbol{m}) \xrightarrow{d}_{N\to+\infty} \mathcal{N}(\boldsymbol{0}_d, \boldsymbol{C})$$

and let $\boldsymbol{g}$ be a function $\mathbb{R}^d \mapsto \mathbb{R}^q$ supposed to be twice continuously differentiable. Thus,

$$\sqrt{N}(\boldsymbol{g}(\boldsymbol{U}_N) - \boldsymbol{g}(\boldsymbol{m})) \xrightarrow{d}_{N\to+\infty} \mathcal{N}(\boldsymbol{0}_q, \boldsymbol{\Gamma})$$

where $\boldsymbol{\Gamma} = \partial\boldsymbol{g}(\boldsymbol{m})\,\boldsymbol{C}\,\partial^T\boldsymbol{g}(m)$ and where

$$\partial\boldsymbol{g} = \begin{bmatrix} \dfrac{\partial g_1(u_1,\ldots,u_d)}{\partial u_1} & \cdots & \dfrac{\partial g_1(u_1,\ldots,u_d)}{\partial u_d} \\ \vdots & & \vdots \\ \dfrac{\partial g_q(u_1,\ldots,u_d)}{\partial u_1} & \cdots & \dfrac{\partial g_q(u_1,\ldots,u_d)}{\partial u_d} \end{bmatrix}$$

is the Jacobian of $\boldsymbol{g}$ and $\partial\boldsymbol{g}(\boldsymbol{m})$ the Jacobian calculated at point $\boldsymbol{m}$.

Applying theorem (1.11), consider the function associating vector $\boldsymbol{U}_N$ with its ℓ-th component, which is written:

$$\boldsymbol{U}_N \mapsto U_{N,\ell} = \boldsymbol{E}_\ell^T \boldsymbol{U}_N$$

where $\boldsymbol{E}_\ell$ is the vector of dimension d of which all components are equal to 0, with the exception of the ℓ-th, equal to 1. Direct application of the theorem gives:

$$\sqrt{N}(U_{N,\ell} - m_\ell) \xrightarrow{d} \mathcal{N}(0, C_{\ell\ell})$$

where m_ℓ is the ℓ-th component of m and $C_{\ell\ell}$ the ℓ-th diagonal element of $\boldsymbol{C}$.

1.7 Other important probability distributions

This section presents a non-exhaustive list of certain other important probability distributions. Some of the associated functions, which are not available in the basic version of MATLAB®, are given in simplified form in the Appendix.

Uniform distribution over (a, b) : noted $\mathcal{U}(a, b)$ of density

$$p_X(x; a, b) = \frac{1}{b-a}\mathbb{1}(x \in (a, b)) \tag{1.53}$$

where $a < b$. The mean is equal to $(b+a)/2$ and the variance to $(b-a)^2/12$.

Exponential distribution : noted $E(\theta)$, of density

$$p_X(x;\theta) = \theta^{-1}e^{-x/\theta}\mathbb{1}(x \geq 0) \tag{1.54}$$

with $\theta > 0$. The mean is equal to θ and the variance to θ^2. We can easily demonstrate that $E(\theta) = \theta E(1)$.

Gamma distribution : noted $G(k,\theta)$, of density

$$p_X(x;(k,\theta)) = \frac{1}{\Gamma(k)\theta^k}e^{-x/\theta}x^{k-1}\mathbb{1}(x > 0) \tag{1.55}$$

where $\theta \in \mathbb{R}^+$ and $k \in \mathbb{R}^+$. The mean is equal to $k\theta$ and the variance to $k\theta^2$. Note that $E(\theta) = G(1,\theta)$.

χ^2 distribution with k d.o.f. : noted χ^2_k. The r.v. $Y = \sum_{i=1}^{k} X_i^2$ where X_i are k Gaussian, independent, centered r.v.s of variance 1 follows a χ^2 distribution with k degrees of freedom (d.o.f.). The mean is equal to k and the variance to $2k$.

Fisher distribution with (k_1, k_2) d.o.f. : noted $F(k_1, k_2)$. Let X and Y be two real, centered Gaussian vectors of respective dimensions k_1 and k_2, with respective covariance matrices I_{k_1} and I_{k_2}, and independent of each other, then the r.v.

$$F_{k_1,k_2} = \frac{k_1^{-1}X^TX}{k_2^{-1}Y^TY} \tag{1.56}$$

follows a Fisher distribution with (k_1, k_2) d.o.f.

Student distribution with k d.o.f. : noted T_k. Let X be a real, centered Gaussian vector, with a covariance matrix I_k, and Y a real, centered Gaussian vector, of variance 1 and independent of X. The r.v.

$$T_k = \frac{Y}{\sqrt{k^{-1}\sum_{i=1}^{k} X_i^2}} \tag{1.57}$$

follows a Student distribution with k d.o.f.

We can show that if Z follows a Student distribution with k degrees of freedom, then Z^2 follows a Fisher distribution with $(1, k)$ degrees of freedom.

Chapter 2

Statistical Inferences

In probability theory, we consider a sample space, which is the set of all possible outcomes, and a collection of its subsets with a structure of σ-algebra, the elements of which are called the events. In what follows, an element of the sample space $\mathcal{X}$ is denoted x, the associated event space is denoted $\mathcal{B}_{\mathcal{X}}$. The pair $\{\mathcal{X}, \mathcal{B}_{\mathcal{X}}\}$ is known as a measurable space [25]. Often, in the following, the sample set will be $\mathbb{R}^n$ and the associated event set the Borel σ-algebra derived from the natural open topology.

2.1 Statistical model

Definition 2.1 (Statistical model) *A statistical model is a family of probability measures defined over the same space $\{\mathcal{X}, \mathcal{B}_{\mathcal{X}}\}$ and indexed by $\theta \in \Theta$, which is written:*

$$\{P_\theta; \theta \in \Theta\} \tag{2.1}$$

When the set $\Theta \subset \mathbb{R}^p$ is of finite dimensions, the model is said to be *parametric*; otherwise, it is said to be *non-parametric*.

In what follows, we shall focus on parametric models, where $\mathcal{X} = \mathbb{R}^n$, for which the probability distribution has either a discrete form or a density $p(x;\theta)$ with respect to the Lebesgue measure. This model is noted:

$$\{p(x;\theta); \theta \in \Theta\} \tag{2.2}$$

θ is said to be identifiable if, and only if:

$$\theta_1 \neq \theta_2 \Leftrightarrow p(x;\theta_1) \neq p(x;\theta_2)$$

We shall only consider models with identifiable parameters. It is essential that two different parameter values will not, statistically, produce the same observations.

Example 2.1 (Gaussian model) In the case of the Gaussian model, the probability measure P_θ is Gaussian over $\mathbb{R}$ with mean m and variance σ^2. Parameter θ is thus $\theta = (m, \sigma^2) \in \Theta = \mathbb{R} \times \mathbb{R}^+$ and the model is parametric. If $\sigma \neq 0$ the probability has a density which is written:

$$p(x;\theta) = \frac{1}{\sigma\sqrt{2\pi}} e^{-(x-m)^2/2\sigma^2}$$

In many applications, the statistical model is connected to a series of observations. For example, consider the observations modeled by n independent Gaussian random variables, with respective means m_1, ..., m_n and respective variances $\sigma_1^2 > 0$, ..., $\sigma_n^2 > 0$. In this case, the statistical model is characterized by a law for which the probability density over $\mathcal{X} = \mathbb{R}^n$ is written:

$$\begin{aligned} p_{X_1 \dots X_n}(x_1, \dots, x_n; \theta) &= \prod_{k=1}^{n} \frac{1}{\sigma_k\sqrt{2\pi}} e^{-(x_k-m_k)^2/2\sigma_k^2} \\ \text{with} \quad & \theta = (m_1, \dots, m_n, \sigma_1^2, \dots, \sigma_n^2) \in \Theta = \mathbb{R}^n \times \mathbb{R}^{+n} \end{aligned}$$

We see that the dimension of the parameter set is dependent on n. In the specific case where the random variables are identically distributed, Θ is no longer dependent on n and is reduced to $\mathbb{R} \times \mathbb{R}^+$. This model, which forms the basis for a significant number of important results, will be noted below as:

$$\{\text{ i.i.d. } \mathcal{N}(n; m, \sigma^2)\} = \left\{ p_X(x_1, \dots, x_n; \theta) = \prod_{k=1}^{n} \frac{1}{\sigma\sqrt{2\pi}} e^{-(x_k-m)^2/2\sigma^2} \right\} \quad (2.3)$$

where parameter $\theta = (m, \sigma^2) \in \mathbb{R} \times \mathbb{R}^+$.

Expression (2.3) may be generalized in cases where the observation is modeled by a series of n random vectors $\boldsymbol{X}_n$ of dimension d which are Gaussian and independent:

$$\begin{aligned} \{\text{ i.i.d. } \mathcal{N}(n; \boldsymbol{m}, \boldsymbol{C})\} &= \dots \quad (2.4) \\ & \left\{ p_{\boldsymbol{X}}(\boldsymbol{x}; \theta) = \prod_{k=1}^{n} \frac{1}{(2\pi)^{d/2}\sqrt{\det\{\boldsymbol{C}\}}} e^{-\frac{1}{2}(\boldsymbol{x}_k - \boldsymbol{m})^T \boldsymbol{C}^{-1}(\boldsymbol{x}_k - \boldsymbol{m})} \right\} \end{aligned}$$

where $\boldsymbol{x}_k = (x_{1,k}, \dots, x_{d,k})$. Parameter $\theta = (\boldsymbol{m}, \boldsymbol{C}) \in \mathbb{R}^d \times \mathcal{M}_d^+$, where $\mathcal{M}_d^+$, denotes the set of positive square matrices of dimension d.

Example 2.2 (Stationary process) Consider a series of n successive observations of a second-order, stationary random process. The statistical model associated with these observations is the set of all probability distributions with a given mean for which the covariance depends solely on the difference between the instants of two observations. The model is clearly non-parametric.

Definition 2.2 (Statistic) *A statistic or estimator is any measurable function of observations.*

The statistical model $\{$ i.i.d. $\mathcal{N}(n; m, \sigma^2)\}$, where σ^2 is presumed to be known, is fundamentally different to model $\{$ i.i.d. $\mathcal{N}(n; m, \sigma^2)\}$ where σ^2 is an unknown parameter. In the first case, an estimator may contain variable σ^2, whereas in the second case this is not possible.

Statistical inference

The aim of *statistical inference* is to obtain conclusions based on observations modeled using random variables or vectors. A number of examples of statistical inference problems are shown below:

- estimation of the value of a parameter,
- estimation of an interval which has a 95% probability of containing the parameter value,
- testing the hypothesis that a parameter belongs to a given region of Θ,
- grouping and/or classifying and/or ordering observations.

We do not aim to provide exhaustive coverage in this brief discussion, but simply to provide some examples of hypothesis and estimator tests.

2.2 Hypothesis tests

Definition 2.3 (Hypothesis) *A hypothesis is a non-empty subset of Θ. A hypothesis is said to be* simple *if it reduces down to a singleton; in all other cases, it is said to be* composite.

Example 2.3 Consider the model $\{$ i.i.d. $\mathcal{N}(n; m, 1)\}$ of mean $m \in \Theta = \mathbb{R}$ and variance 1. The hypothesis $H_0 = \{0\}$ is simple, whilst the alternative hypothesis $H_1 = \Theta - H_0$ is composite.

Example 2.4 Consider the model $\{$ i.i.d. $\mathcal{N}(n; m, \sigma^2)\}$ with$(m, \sigma^2) \in \Theta = \mathbb{R} \times \mathbb{R}^+$. The two hypotheses $H_0 = \{0\} \times \mathbb{R}^+$ and $H_1 = \Theta - H_0$ are composite.

A test of hypothesis H_0 consists of defining a sub-set $\mathcal{X}_1$ of the space $\mathcal{X}$ such that if realization x belongs to $\mathcal{X}_1$, then H_0 is rejected. This is equivalent to defining the statistic $T(X) = \mathbb{1}(X \in \mathcal{X}_1)$ taking values of 0 or 1 and such that:

$$\text{if} \quad T(X) = \begin{cases} 0 & \text{then } H_0 \text{ is accepted,} \\ 1 & \text{then } H_0 \text{ is rejected.} \end{cases} \tag{2.5}$$

Subset $\mathcal{X}_1$ is known as the *critical region* and statistic $T(X)$ as the *critical test function*. Note that $\mathbb{E}_\theta\{T(X)\} = \mathbb{P}_\theta\{T(X) = 1\}$.

In many cases, a statistic $S(X)$ exists with real values and a number η such that the critical function may be written in the form $T(X) = \mathbb{1}(S(X) \geq \eta)$. In this case, the test is said to be *unilateral*. In other cases, we find $\eta_1 \leq \eta_2$ such that $T(X) = \mathbb{1}(S(X) \notin [\eta_1, \eta_2])$. In these cases the test is said to be *bilateral*. $S(X)$ is referred to as the test statistic.

Definition 2.4 (Significance level) *A test associated with the critical function $T(X)$ is said to have a significance level α if*

$$\max_{\theta \in H_0} \mathbb{E}_\theta\{T(X)\} = \alpha \tag{2.6}$$

Note that for a random variable T with values in the interval $\{0, 1\}$, $\mathbb{E}\{T\} = \mathbb{P}\{T = 1\}$. Consequently, $\mathbb{E}_\theta\{T(X)\} = \mathbb{P}_\theta\{T(X) = 1\}$. Thus, in a radar context, the significance level represents the probability of H_1 being accepted when H_0 is true. In the context of radars, the significance level is known as the *false alarm probability*.

Definition 2.5 *A test of a hypothesis H_0 against the alternative $H_1 = \Theta - H_0$, associated with the critical function $T(X)$, is said to be uniformly most powerful (UMP) at level α if, for any $\theta \in H_1$, its power $\mathbb{E}_\theta\{T(X)\}$ is higher than that of any other test of level α. This is written:*

$$\begin{aligned} \exists\, T(X) \in \mathcal{T}_{H_0}(\alpha),\ s.t.\ \forall\, S(X) \in \mathcal{T}_{H_0}(\alpha)\ and\ \forall\, \theta \in H_1 :\\ \mathbb{E}_\theta\{T(X)\} \geq \mathbb{E}_\theta\{S(X)\} \end{aligned} \tag{2.7}$$

where $\mathcal{T}_{H_0}(\alpha)$ denotes the set of tests of hypothesis H_0 of level α.

In the context of radars, the power is interpreted as the probability of *detection*, which consists of accepting H_1 when H_1 is true.

However, UMP tests do not exist for most situations.

2.2.1 Simple hypotheses

Consider the following statistical model:

$$\{p(x;\theta); \theta \in \Theta = \{\theta_0, \theta_1\}\} \tag{2.8}$$

characterized by a set Θ which only contains two elements and by the hypothesis $H_0 = \{\theta_0\}$. Take:

$$\Lambda(x) = \frac{p(x;\theta_1)}{p(x;\theta_0)} \tag{2.9}$$

The following result, obtained by Neyman and Pearson, is fundamental: it gives the expression of the UMP test at level α [20].

Theorem 2.1 *For any value of α, there are two constants η and $f \in (0,1)$, such that the critical function test*

$$T^*(X) = \begin{cases} 1 & \textit{if} \quad \Lambda(X) > \eta \\ f \in (0,1) & \textit{si} \quad \Lambda(X) = \eta \\ 0 & \textit{if} \quad \Lambda(X) < \eta \end{cases} \tag{2.10}$$

is UMP at level α. η and f are chosen in such a way as to satisfy the following constraint:

$$\mathbb{E}_{\theta_0}\{T^*(X)\} = \alpha \tag{2.11}$$

Function $\Lambda(X)$ is known as the likelihood ratio.

This test is said to be "randomized" due to the fact that the decision strategy, when $\Lambda(X) = \eta$, is random. The decision is made using an auxiliary random Bernoulli variable, with a value in the interval $\{0, 1\}$, with a probability f of being equal to 1.

In a large number of practical cases, the probability of the random variable $\Lambda(X)$ being exactly equal to η is null. In this case, $f = 1$ and the critical function is written:

$$T^*(X) = \begin{cases} 1 & \text{if} \quad \Lambda(X) \geq \eta \\ 0 & \text{if} \quad \Lambda(X) < \eta \end{cases} \tag{2.12}$$

The fundamental result is that the optimal test is based on the likelihood ratio.

The inequality $\Lambda(X) \geq \eta$ can often be simplified, as in the case of exponential models. This may be illustrated using an example. Consider the model $\{\text{i.i.d. } \mathcal{N}(n; m, 1)\}$ with $m \in \{m_0, m_1\}$ and $m_1 > m_0$. The likelihood ratio is written:

$$\Lambda(X) = \frac{p(x;\theta_1)}{p(x;\theta_0)} = \exp\left((m_1 - m_0)\sum_{k=1}^{n} X_k - \frac{n}{2}(m_1^2 - m_0^2)\right)$$

However, as the exponential function increases, comparing $\Lambda(X)$ to a threshold is equivalent to comparing the argument to another threshold. This comes down to comparing to a threshold the quantity:

$$\Psi(X) = 2(m_1 - m_0)\sum_{k=1}^{n} X_k - n(m_1^2 - m_0^2)$$

This can be simplified further by comparing the following statistic to a threshold ζ:

$$\Phi(X) = \frac{1}{n}\sum_{k=1}^{n} X_k \tag{2.13}$$

Statistic Φ has a much simpler expression than that used for the likelihood ratio. The critical function is expressed as $T(X) = \mathbb{1}(\Phi(X) \geq \zeta)$. Under H_0 the random variable $\Phi(X)$ is Gaussian, with mean m_0 and variance $1/n$. The threshold value is then determined in such a way as to satisfy level α; this is written:

$$\zeta = m_0 + Q^{[-1]}(\alpha)\sqrt{n}$$

where $Q^{[-1]}(\alpha)$ is the inverse cumulative function of the standard Gaussian distribution.

Note that the UMP test of level α is not dependent on m_1.

Figure 2.1 shows the connection between the significance level (or probability of false alarm), the power (or probability of detection) and the probability densities of the random variable $\Phi(X)$ under the two hypotheses.

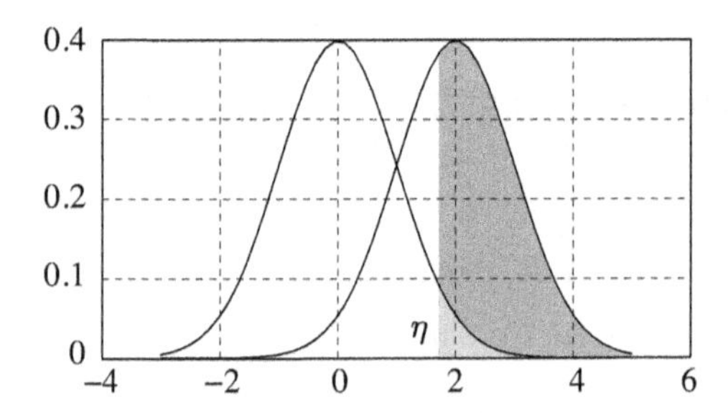

Figure 2.1 – *The two curves represent the respective probability densities of the test function $\Phi(X)$ under hypothesis H_0 ($m_0 = 0$) and under hypothesis H_1 ($m_1 = 2$). For a threshold value η, the light gray area represents the significance level or probability of false alarm, and the dark gray area represents the power or probability of detection.*

Evidently, raising the threshold reduces the probability of a false alarm, but also reduces the probability of detection. A compromise between the two types of error is therefore required, see example 2.5.

For certain problems, costs are assigned to each decision. A cost function is an application of $\Theta \times \Theta \mapsto \mathbb{R}^+$. Let $C_{i,j} \in \mathbb{R}^+$ be the cost of decision i knowing that j. The Bayesian approach consists of determining the threshold which minimizes the mean risk:

$$R_B = \sum_{i=0}^{1}\sum_{j=0}^{1} C_{i,j}\mathbb{P}\{\text{decide } i \mid \text{knowing it is } j\} \tag{2.14}$$

The choice of the cost function depends on the target application, and should be made with assistance from experts in the relevant field.

Example 2.5 (ROC curve) Let us take a statistical model {i.i.d. $\mathcal{N}(n; m, 1)$} where $m \in \{m_0, m_1\}$ with $m_1 > m_0$ and the hypothesis $H_0 = \{m_0\}$. The UMP test of level α compares:

$$\Phi(X) = \frac{1}{n}\sum_{k=1}^{n} X_k$$

with a threshold ζ, calculated based on the value α of the significance level. Starting from this value of threshold ζ, we deduce the power β of the test. The curve giving β as a function of α is known as the Receiving Operational Characteristic (ROC) curve. In our case, its expression as a function of the parameter $\zeta \in \mathbb{R}$ is written:

$$\begin{cases} \alpha = \int_{\zeta}^{+\infty} \frac{1}{\sigma\sqrt{2\pi}} e^{-(t-m_0)^2/2\sigma^2} dt \\ \beta = \int_{\zeta}^{+\infty} \frac{1}{\sigma\sqrt{2\pi}} e^{-(t-m_1)^2/2\sigma^2} dt \end{cases} \quad (2.15)$$

where $\sigma^2 = 1/n$. Figure 2.2 shows the ROC curve for $m_0 = 0$ and $m_1 = 1$ and different values of n. The form of these curves is typical. The ROC curve increases and is concave above the first bisector. Note that the first bisector is the ROC curve associated with a purely random test, which consists of accepting the hypothesis H_1 with probability α. Hence, for a given significance level, the power may not be as high as we would like. The closer the ROC curve comes to the point $(0, 1)$, the more efficient the detector is in discriminating between the two hypotheses. One way of characterizing this efficiency is to calculate the area under the ROC curve, known as the AUC (Area Under the ROC Curve). The expression of the AUC associated to the test function S is given by:

$$\begin{aligned} A &= \int_{-\infty}^{+\infty} \beta(\zeta) d\alpha(\zeta) \\ &= \int_{-\infty}^{+\infty} \int_{-\infty}^{+\infty} \mathbb{1}(u_0 < u_1) p_S(u_0; \theta_0) p_S(u_1; \theta_1)\, du_0 du_1 \end{aligned} \quad (2.16)$$

where $p_S(u_0; \theta_0)$ and $p_S(u_1; \theta_1)$ are the respective distributions of S under the two hypotheses.

The AUC may be seen as the probability that U_0 is lower than U_1 for two independent random variables U_0 and U_1 with respective distributions $p_S(u_0; \theta_0)$ and $p_S(u_1; \theta_1)$.

Experimental ROC curve and AUC

The ROC curve and the AUC are valuable tools for evaluating test functions. However, it is important to note that in problems with composite hypotheses,

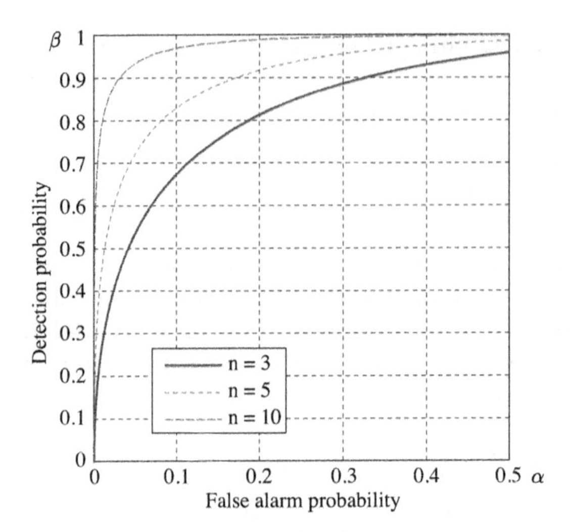

Figure 2.2 – *ROC curve. The statistical model is* $\{i.i.d.\ \mathcal{N}(n; m, 1)\}$*, where* $m \in \{0, 1\}$*. The hypothesis* $H_0 = \{0\}$ *is tested by the likelihood ratio* (2.9)*. The higher the value of* n*, the closer the curve will come to the ideal point, with coordinates* $(0, 1)$*. The significance level* α *is interpreted as the probability of a false alarm. The power* β *is interpreted as the probability of detection.*

the power depends on the choice of the parameter in the hypothesis H_1. There is therefore an infinite number of possible ROC curves, each associated with a value of parameter $\theta \in H_1$. We may therefore choose a sub-set of values of H_1 beforehand, carrying out a random draw in order to obtain a mean curve.

An experimental approach may also be used. To do this, a series of measurements is carried out including N_0 examples under hypothesis H_0 and N_1 examples under H_1. This data is then used to evaluate the test statistic Φ in both cases. We obtain a series of values $\Phi_{n_0|0}$, with n_0 from 1 to N_0 for data labeled H_0 and $\Phi_{n_1|1}$ with n_1 from 1 to N_1 for data labeled H_1. Using these two series of values, it is possible to estimate the ROC curve and the area A under the ROC curve, see exercise 2.1. To obtain an estimation $\widehat{A}$ of the area under the ROC curve, we may use expression (2.16) to deduce the Mann-Whitney statistic, which is written:

$$\widehat{A} = \frac{1}{N_0 N_1} \sum_{n_0=1}^{N_0} \sum_{n_1=1}^{N_1} \mathbb{1}(\Phi_{n_1|1} \geq \Phi_{n_0|0}) \tag{2.17}$$

Note that the choice of the database, containing examples under both H_0 and H_1, is critical: this database must include a sufficient number of examples which are representative of the application in question.

Exercise 2.1 (ROC curve and AUC for two Gaussians) (see page 237)
Consider the statistical model $\{$ i.i.d. $\mathcal{N}(n; m, 1)\}$ where $m \in \{m_0, m_1\}$, with $m_1 > m_0$, and a hypothesis $H_0 = \{m_0\}$ to be tested. We consider that a database is available containing N_0 observations for H_0 and N_1 observations for H_1. In this exercise, these observations will be used to estimate the ROC curve and the AUC associated with statistic (2.13), even though analytical expressions of these two quantities are available in this specific case.

1. Using definition (2.6), propose an estimator, based on the data for H_0, of the significance level. Carry out the same exercise for the estimated power.

2. Write a program:

   - simulating $N_0 = 1000$ draws of length $n = 10$ for H_0 and $N_1 = 1200$ draws of length $n = 10$ for H_1,
   - estimating the ROC curve and comparing it to the theoretical expression given by (2.15),
   - estimating the area under the ROC curve with the Mann-Whitney statistic.

2.2.2 Generalized Likelihood Ratio Test (GLRT)

Let us consider the parametric statistical model $\{P_\theta; \theta \in \Theta \subset \mathbb{R}^p\}$. Below, we shall presume that this family is dominated, and hence P_θ has a probability density which may be simply noted $p(x; \theta)$. The basic hypothesis H_0 and the alternative H_1 are both presumed to be composite.

Generally speaking, no UMP test exists. As an example, consider a situation where the parametric family is dependent on a single scalar parameter θ and where the hypothesis under test is $H_0 = \{\theta \leq \theta_0\}$. If the UMP test of level α of the simple hypothesis $\{\theta_0\}$ against the simple hypothesis $\{\theta_1\}$, with $\theta_1 > \theta_0$, does depend on θ_1, then a UMP test cannot be carried out.

In practice, in the absence of a UMP test, the generalized likelihood ratio test (GLRT) is used, although it has attracted a good deal of criticism.

The GLRT is defined by the critical function $T(X) = \mathbb{1}(\Lambda(X) \geq \eta)$, with the following test statistic:

$$\Lambda(X) = \frac{\max_{\theta \in \Theta} p(X; \theta)}{\max_{\theta \in H_0} p(X; \theta)} \tag{2.18}$$

This situation is illustrated in Figure 2.3. The larger the value of $\Lambda(X)$ is compared to 1, the more it is reasonable to reject hypothesis H_0.

Note that $\Lambda(x)$ is positive. We may therefore take the logarithm and define:

$$\mathcal{L}(X) = 2 \log \Lambda(X) = 2 \left(\max_{\theta \in \Theta} \log p(X; \theta) - \max_{\theta \in H_0} \log p(X; \theta) \right) \tag{2.19}$$

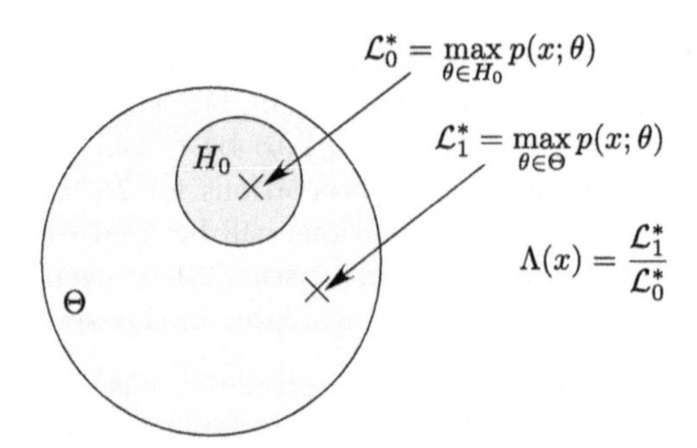

Figure 2.3 – *Diagram showing the calculation of the GLRT*

Subsequently, as the logarithmic function is monotonous and increasing, comparing $\Lambda(X)$ to threshold η is equivalent to comparing $\mathcal{L}(X)$ to threshold $\log \eta$.

Taking $\theta = (\theta_1, \ldots, \theta_q, \theta_{q+1}, \ldots, \theta_p)$ and the hypothesis under test:

$$H_0 = \{\theta : \theta_1 = \ldots = \theta_q = 0, \theta_{q+1}, \ldots, \theta_p\}$$

under relatively general conditions [18], the test function (2.19) for H_0 may be shown to asymptotically follow a χ^2 distribution with q degrees of freedom, written:

$$\mathcal{L}_{H_0}(X) \sim \chi_q^2 \tag{2.20}$$

This result is used in exercises 2.8 and 2.17.

In the specific case where $q = 1$, the hypothesis H_0 to be tested is often:

- of the form $H_0 = \{\theta_1 \leq \mu_0\}$, where μ_0 is a chosen value. The hypothesis is said to be unilateral. It is often tested using a unilateral test; or
- of the form $H_0 = \{\mu_0 \leq \theta_1 \leq \mu_1\}$, where μ_0 and μ_1 are chosen values. In this case, the hypothesis is said to be bilateral. One specific case occurs when $\mu_0 = \mu_1$, which gives the hypothesis $H_0 = \{\mu = \mu_0\}$. Bilateral hypothesis testing is often carried out in these cases. This case is illustrated by the test presented in the following section.

Bilateral mean testing

Consider the statistical model $\{$ i.i.d. $\mathcal{N}(n; m, \sigma^2)\}$ where $(m, \sigma^2) \in \Theta = \mathbb{R} \times \mathbb{R}^+$ and with a hypothesis $H_0 = \{m_0\} \times \mathbb{R}^+$. Let us determine the GLRT at significance level α. The log-likelihood is written:

$$\mathcal{L}(\theta) = \log p(X;\theta) = -\frac{n}{2}\log(2\pi) - \frac{n}{2}\log(\sigma^2) - \frac{1}{2\sigma^2}\sum_{k=1}^{n}(X_k - m)^2$$

Canceling the first derivative with respect to σ^2, we obtain the expression of the maximum:

$$\widetilde{\mathcal{L}}(m) = -\frac{n}{2}\log(2\pi) - \frac{n}{2}\log\sum_{k=1}^{n}(X_k - m)^2$$

For $m = m_0$ (hypothesis H_0), the maximum is expressed as:

$$-\frac{n}{2}\log(2\pi) - \frac{n}{2}\log\sum_{k=1}^{n}(X_k - m_0)^2$$

and for $m \neq m_0$ (hypothesis H_1), the maximum is obtained by canceling the first derivative of $\widetilde{\mathcal{L}}(m)$ with respect to m, and is expressed as:

$$-\frac{n}{2}\log(2\pi) - \frac{n}{2}\log\sum_{k=1}^{n}(X_k - \widehat{m})^2$$

where $\widehat{m} = n^{-1}\sum_{k=1}^{n} X_k$ and, consequently, following (2.18):

$$\begin{aligned}\Lambda^{2/n}(X) &= \frac{\sum_{k=1}^{n}(X_k - \widehat{m})^2}{\sum_{k=1}^{n}(X_k - m_0)^2} = \frac{\sum_{k=1}^{n}(X_k - \widehat{m})^2}{\sum_{k=1}^{n}(X_k - \widehat{m} + (m_0 - \widehat{m}))^2} \\ &= \frac{\sum_{k=1}^{n}(X_k - \widehat{m})^2}{\sum_{k=1}^{n}(X_k - \widehat{m})^2 + n(\widehat{m} - m_0)^2} \\ &= \frac{1}{1 + n(\widehat{m} - m_0)^2/\sum_{k=1}^{n}(X_k - \widehat{m})^2} \end{aligned} \tag{2.21}$$

Due to the monotonous nature of function $1/(1+u^2)$, comparing statistic $\Lambda(X)$ to a threshold is equivalent to comparing statistic $|V(X)|$ to a threshold, where:

$$V(X) = \frac{\sqrt{n}(\widehat{m} - m_0)}{\sqrt{(n-1)^{-1}\sum_{k=1}^{n}(X_k - \widehat{m})^2}} \tag{2.22}$$

Setting $p = 0$ and then $\beta_0 = m_0$ in equation (2.62), we establish that, under H_0, $V(X)$ follows a Student distribution with $(n-1)$ degrees of freedom. The test of H_0 of level α is therefore written:

$$\frac{\sqrt{n}|m_0 - \widehat{m}|}{(n-1)^{-1/2}\sqrt{\sum_{k=1}^{n}(X_k - \widehat{m})^2}} \underset{H_0}{\overset{H_1}{\gtrless}} T_{n-1}^{[-1]}(1 - \alpha/2)$$

Exercise 2.2 (Student distribution for H_0) (see page 238) Take the statistical model {i.i.d. $\mathcal{N}(n; m, \sigma^2)$} and the hypothesis $H_0 = \{m = m_0\}$. Write a program:

– simulating, for H_0, 10000 draws of a sample of size $n = 100$,

- calculating $V(X)$ following expression (2.22),
- comparing the histogram of $V(X)$ to the theoretical curve given by Student's distribution. The `tpdf` function in MATLAB® provides the probability density values of Student's distribution (see also appendix A.2.3).

Exercise 2.3 (Unilateral mean testing) (see page 239) Take the statistical model $\{$ i.i.d. $\mathcal{N}(n;m,\sigma^2)\}$ where $(m,\sigma^2) \in \Theta = \mathbb{R} \times \mathbb{R}^+$ and the hypothesis $H_0 = \{m \geq m_0\} \times \mathbb{R}^+$. Determine the GLRT of H_0 at significance level α.

p-value associated with a test

Let us consider a test of hypothesis H_0, for which the critical function is of unilateral form, $T(X) = \mathbb{1}(S(X) > \eta)$, where $S(X)$ is a function with real values. Based on observation X, it is possible to calculate $T(X)$, which takes a value of either 0 or 1. We wish to determine a confidence level associated with this decision. The first idea would be to give the calculated value of $S(X)$; however, this value is meaningless in absolute terms. However, the probability of observing a value greater than $S(X)$ under the hypothesis H_0 has a clear meaning: the lower the probability, the more reasonable it will be to reject H_0. The probability distribution of $S(X)$ for all values of parameter $\theta \in H_0$ is necessary in order to carry out this calculation. The probability density of this distribution is noted $p_S(s;\theta)$. The following statistic is known as the p-value:

$$p\text{-value} = \min_{\theta \in H_0} \int_{S(X)}^{+\infty} p_S(s;\theta)ds \tag{2.23}$$

The closer this value is to 0, the more reasonable it is to reject H_0.

Clearly, if the critical function is of bilateral form, $T(X) = \mathbb{1}(S(X) \notin (-\eta, +\eta))$, the p-value is defined by:

$$p\text{-value} = 2\min_{\theta \in H_0} \int_{|S(X)|}^{+\infty} p_S(s;\theta)ds \tag{2.24}$$

Note that a common error consists of comparing the p-values of two samples of different sizes. This is irrelevant as, generally, as the size of a sample increases, the distribution of $S(X)$ narrows and thus the p-value decreases. Moreover, note that the p-value is dependent on the chosen test statistic. Consequently, for the same observations, it may take a high value for one test statistic and a low value for another statistic. Its meaning is therefore often debated, but this technique remains widespread. Typically, rejection of H_0 is recommended in cases with a p-value of less than 1%.

Confidence interval associated with a test

Definition 2.6 *Consider a statistical model* $\{P_\theta; \theta \in \theta\}$ *defined over* $\mathcal{X}$ *and a series of observations* $X = \{X_1, \ldots, X_n\}$. *Let* $\alpha \in (0, 1)$. *The confidence region for* θ *at* $100(1-\alpha)\%$ *is a region* $\mathcal{I}_\alpha(X) \subset \Theta$ *such that, for any* θ*:*

$$\mathbb{P}_\theta \{\theta \in \mathcal{I}_\alpha(X)\} \geq 1 - \alpha$$

When θ is of dimension 1, we speak of a confidence interval.

A test may therefore be associated with a region of confidence, as follows. Taking a region of confidence $\mathcal{I}_\alpha(X)$ at $100(1-\alpha)\%$ of θ, then the indicator $\mathbb{1}(\theta_0 \in \mathcal{I}_\alpha(X))$ is a critical test function of the hypothesis $H_0 = \{\theta = \theta_0\}$ at significance level α.

Reciprocally, consider a statistical model, and let $R_\alpha(\theta_0) \subset \mathcal{X}$ be the critical region of a test of significance level α associated with the hypothesis $H_0 : \{\theta = \theta_0\}$. In this case, each value $x \in \mathcal{X}$ may be associated with a set $\mathcal{S}_\alpha(x) \subset \Theta$ defined by:

$$\mathcal{S}_\alpha(x) = \{\theta_0 \in \Theta \quad \text{s.t.} \quad R_\alpha(\theta_0) \ni x\}$$

As the test is of level α, $\mathbb{P}_{\theta_0} \{X \notin R_\alpha(\theta_0)\} \leq \alpha$ and thus:

$$1 - \alpha \leq \mathbb{P}_{\theta_0} \{X \in R_\alpha(\theta_0)\} = \mathbb{P}_{\theta_0} \{\mathcal{S}_\alpha(X) \ni \theta_0\}$$

meaning that $S_\alpha(X)$ is a region of confidence at $100(1-\alpha)\%$ of θ_0.

Example 2.6 (Variance confidence interval) Take the statistical model $\{\text{i.i.d. } \mathcal{N}(n; m, \sigma^2)\}$, with unknown mean and variance. Determine the GLRT of significance level α associated with the hypothesis $H_0 = \{\sigma = \sigma_0\}$. Use this result to deduce a confidence interval at $100(1-\alpha)\%$ of σ^2.

HINTS: based on (2.18), after simplification, we obtain the following test statistic:

$$\Phi(X) = \frac{S(X)}{\sigma_0^2}$$

where $S(X) = \sum_{k=1}^{n}(X_k - \widehat{m})^2$ with $\widehat{m} = n^{-1}\sum_{k=1}^{n} X_k$. Using the results of property 2.2, $\Phi(X)$ may be shown to follow a χ^2_{n-1} distribution with $(n-1)$ degrees of freedom.

We may then identify a critical region at level α:

$$\begin{aligned} R_\alpha(\sigma_0) \quad = \quad & \{X \in \mathbb{R}^n \quad \text{s.t.} \\ & \sigma_0^2 \chi_{n-1}^{2,[-1]}(\alpha/2) \leq S(X) \leq \sigma_0^2 \chi_{n-1}^{2,[-1]}(1 - \alpha/2)\} \end{aligned}$$

where $\chi_{n-1}^{2,[-1]}$ is the inverse of the cumulative function of the χ^2 distribution with $(n-1)$ degrees of freedom. From this, we can deduce a confidence interval at $100(1-\alpha)\%$ of σ^2:

$$\mathcal{I}_\alpha(X) = \left\{\sigma^2 \in \mathbb{R}^+ \quad \text{s.t.} \quad \frac{S(X)}{\chi_{n-1}^{2,[-1]}(1-\alpha/2)} \leq \sigma^2 \leq \frac{S(X)}{\chi_{n-1}^{2,[-1]}(\alpha/2)}\right\}$$

Exercise 2.4 (Mean equality test) (see page 240) Consider two samples, $X_{1,1:n_1}$ and $X_{2,1:n_2}$, i.i.d. Gaussian, independent, with respective means of m_1 and m_2 and a common variance σ^2, and of respective lengths n_1 and n_2. σ is presumed to be unknown. Determine a test for the hypothesis $m_1 = m_2$.

1. Describe the statistical model and the hypothesis H_0.

2. Determine the GLRT of H_0 at significance level α.

3. Show that, for H_0, the test statistic is expressed as the modulus of a Student r.v. with $(n_1 + n_2 - 2)$ degrees of freedom. Use this result to identify the test at level α.

4. Use this result to determine the p-value.

5. Consider Table 2.1, which provides the growth results of a plant, in mm, using two different fertilizers, noted 1 and 2. Give the p-value of the mean equality test and present your conclusions.

F	1	1	1	1	2	2	2
size (mm)	51.0	53.3	55.6	51.0	55.5	53.0	52.1

Table 2.1 – Growth in mm under the action of fertilizers 1 and 2

Exercise 2.5 (Correlation test) (see page 243) Consider two samples $\{X_{1,1}, \ldots, X_{1,n}\}$ and $\{X_{2,1}, \ldots, X_{2,n}\}$, i.i.d. Gaussian, with respective means m_1 and m_2 and respective variances σ_1^2 and σ_2^2, and with a correlation coefficient $\rho \in (-1, 1)$. Consider the hypothesis $H_0 = \{-\rho_0 \leq \rho \leq \rho_0\}$, which tests whether the modulus of the correlation is lower than a given positive value ρ_0.

1. Describe the statistical model and the hypothesis H_0.

2. Determine the GLRT of H_0.

3. Transformation $r \mapsto f = 0.5\log(\frac{1+r}{1-r})$ is known as a Fisher transformation. The Fisher transformations of $\widehat{\rho}$ and ρ_0 are noted $\widehat{f}$ and f_0 respectively. Under the hypothesis H_0, the random variable $\widehat{f}$ can be shown to follow, approximately, a Gaussian distribution of mean f_0 and variance $1/(n-3)$. Use this result to determine the p-value of the test for hypothesis $H_0 = \{|\rho| \leq 0.7\}$ using the data supplied in Table 2.2 and present your conclusions.

H (cm)	162	167	167	159	172	172	168
W (kg)	48.3	50.3	50.8	47.5	51.2	51.7	50.1

Table 2.2 – H: height in cm, W: weight in kg.

Exercise 2.6 (CUSUM) (see page 244) Consider a sequence of n independent random variables. These variables are presumed to have the same probability distribution, of density $p(x;\mu_0)$, between instants 1 and m and the same probability distribution, of density $p(x;\mu_1)$, between instants $m+1$ and n, where m may take any value from 1 to $n-1$ inclusive, or keep the same distribution $p(x;\mu_0)$ over the whole sample. The value of m is unknown.

We wish to test the hypothesis H_0 that no sudden jump exists from μ_0 toward μ_1.

1. Describe the statistical model.

2. Determine the test function T_n of the GLRT of hypothesis H_0, associated with a sample of size n.

3. Taking $s_k = \log p(x_k;\mu_1)/p(x_k;\mu_0)$, $C_n = \sum_{k=1}^{n} s_k$, show that T_n may be calculated recursively using the form $T_k = \max\{0, T_{k-1} + s_k\}$.

4. Write a program implementing this algorithm.

As this test is carried out using the cumulated sum of likelihood ratios, it is generally referred to as the CUSUM. By carrying out maximization of μ_0 and μ_1, it may be extended to cases where these parameters are unknown.

2.2.3 χ^2 goodness of fit test

A *goodness of fit test* is a test using the basic hypothesis H_0 that the observed sample comes from a given probability distribution. As before, hypothesis $H_0 = \{P = P_0\}$ may be simple or composite. Hence, if the hypothesis H_0 is that the observation follows a Gaussian distribution with a given mean and variance, H_0 is simple. However, if the variance is unknown, then H_0 is composite. Evidently, if the set of possible probability distributions under consideration cannot be indexed using a parameter of finite dimensions, the model will be non-parametric.

Consider a series of n real random variables, $\{X_1, \ldots, X_n\}$, i.i.d., and a partition of $\mathbb{R}$ into g disjoint intervals Δ_1, ..., Δ_g. Take:

$$N_j = \sum_{k=1}^{n} \mathbb{1}(X_k \in \Delta_j) \tag{2.25}$$

which represents the number of values in the sample $\{X_1, \ldots, X_n\}$ which are located within the interval Δ_j. Note that $N_1 + \cdots + N_g = n$, and $p_{j,0} = \mathbb{P}_{H_0}\{X_k \in \Delta_j\}$ with $\sum_j p_{j,0} = 1$. In cases where hypothesis H_0 is composite, the quantities $p_{j,0}$ may be dependent on unknown parameters. In this case, these parameters are replaced by consistent estimations. In exercise 2.7, for example, the true value of the variance is replaced by an estimation.

Exercice 2.7 shows that the random variable:

$$X^2 = \sum_{j=1}^{g} \frac{(N_j - np_{j,0})^2}{np_{j,0}} = n\sum_{j=1}^{g} \frac{\left(\frac{N_j}{n} - p_{j,0}\right)^2}{p_{j,0}} \xrightarrow{d} \chi^2_{g-1} \tag{2.26}$$

This result has an obvious meaning: X^2 measures the relative distance between the empirical frequencies N_j/n and the theoretical probabilities. Based on the law of large numbers (theorem 1.9), N_j/n converges under H_0 toward $p_{j,0}$ and under H_1 toward $p_{j,1} \neq p_{j,0}$. Thus, statistic X^2 will take small values for H_0 and large values for H_1; these values will be increasingly large as n increases. This justifies the use of an unilateral test, which compares X^2 to a threshold and proves its consistency.

The choice of the number g and the interval sizes Δ_j is critical. One method consists of selecting g classes of the same probability, or of dividing the set of sample values into classes with the same empirical weight (see exercice 2.8).

Exercise 2.7 (Proof of (2.26)**)** (see page 246) Using the conditions associated with expressions (2.25) and (2.26), take $D = \text{diag}(P)$ with $P = \begin{bmatrix} p_1 & \ldots & p_g \end{bmatrix}^T$, where $p_j = \mathbb{P}\{X_k \in \Delta_j\}$.

1. Determine the expressions of $\mathbb{E}\{N_j\}$ and $\mathbb{E}\{N_j N_m\}$ as a function of p_j. Use these expressions to deduce the fact that the covariance matrix of the random vector $(N_1, \ldots, N_g)$ is written nC, where C is a matrix for which the expression will be given.

2. Show that $D^{-1/2}CD^{-1/2}$ is a projector of rank $(g-1)$.

3. Taking $Y_j = \sqrt{n}(\frac{1}{n}N_j - p_j)$, determine the asymptotic distribution of the random vector $Y = \begin{bmatrix} Y_1 & \ldots & Y_g \end{bmatrix}^T$. Use this result to determine the asymptotic distribution of vector $Z = D^{-1/2}Y$.

4. Finally, identify the asymptotic distribution of variable $X^2 = \sum_{j=1}^{g} Z_j^2$.

Exercise 2.8 (Chi2 fitting test) (see page 247) Write a program simulating the hypothesis $H_0 = \{m = 0\}$ for model $\{\text{i.i.d. } \mathcal{N}(n; m, \sigma^2)\}$. To do this, 3000 drawings of a sample of size $n = 200$ are required, with a Gaussian random variable of mean 0 and unknown variance σ^2. The set of values of each draw should be separated into $g = 8$ blocks containing the same number of samples, i.e. 200/8. Using the function which produces the cumulative function of a Gaussian, calculate the values of $p_j = \mathbb{P}\{X_k \in \Delta_j\}$, in which the true, unknown value of σ is replaced by the estimation given using the `std` function in Matlab. Using this result, deduce the values of X^2 given by expression (2.26), and create a histogram, which will be compared to the theoretical curve of χ^2 with $g - 1 = 7$ degrees of freedom.

2.3 Statistical estimation

2.3.1 General principles

In this section, we shall consider the problem of *parameter estimation.* We shall begin by presenting a number of general results concerning performance evaluation, notably bias and quadratic error. We shall then examine three methods used in constructing estimators: the least squares method, mainly in the case of the linear model, the moment method, and the maximum likelihood approach.

Definition 2.7 (Estimator) *Let $\{P_\theta; \theta \in \Theta\}$ be a statistical model of observation X. An estimator is any (measurable) function of X with values in Θ.*

One fundamental question concerns the evaluation of estimator quality. We might try to choose an estimator $\widehat{\theta}$ such that $\mathbb{P}_\theta\left\{\widehat{\theta} \neq \theta\right\} = 0$. Estimators of this type only exist in exceptional situations, and have no practical interest. In practice, two quantities are used, the bias and the risk: these are defined below.

Definition 2.8 (Bias and risk of an estimator) *Consider an estimator $\widehat{\theta} : \mathcal{X} \mapsto \Theta \subset \mathbb{R}^d$. The bias is the vector of dimension d:*

$$B(\theta, \widehat{\theta}) = \mathbb{E}_\theta\left\{\widehat{\theta}\right\} - \theta$$

The quadratic risk is the square matrix of dimension d:

$$R(\theta, \widehat{\theta}) = \mathbb{E}_\theta\left\{(\theta - \widehat{\theta})(\theta - \widehat{\theta})^T\right\}$$

It is easy to show that

$$R(\theta,\widehat{\theta}) = \operatorname{cov}_\theta\left\{\widehat{\theta}\right\} + B(\theta,\widehat{\theta})B^T(\theta,\widehat{\theta})$$

where

$$\operatorname{cov}_\theta\left\{\widehat{\theta}\right\} = \mathbb{E}_\theta\left\{\left(\widehat{\theta}-\mathbb{E}\left\{\widehat{\theta}\right\}\right)\left(\widehat{\theta}-\mathbb{E}\left\{\widehat{\theta}\right\}\right)^T\right\}$$

is the covariance matrix of $\widehat{\theta}$.

It is worth noticing that an estimator does not depend on the unknown parameter being estimated, but the performance of the estimator does depend on this parameter. Thus, the bias and the risk are generally dependent on θ.

It is pointless to try and find an estimator with the minimum quadratic risk for all values of θ. For this reason, we have to restrict the search class for $\widehat{\theta}$. For example we could limit the search to bias-free estimators, or to the class of linear estimators, or the class of estimators which are invariant by translation, etc.

Another method involves a Bayesian approach, which consists of taking account of available knowledge concerning θ, which takes the form of a probability distribution $p_\Theta(\theta)$ and may be used for minimizing the average risk:

$$R_B(\widehat{\theta}) = \int_\Theta \operatorname{trace}\left\{R(\theta,\widehat{\theta})\right\} p_\Theta(\theta)d\theta \tag{2.27}$$

over the set of all estimators.

Cramer-Rao bound

Quadratic risk has a fundamental lower-bound, known as the Cramer-Rao bound (CRB).

Theorem 2.2 (Cramer-Rao bound (CRB)) *Any estimator $\widehat{\theta}$ of the parameter $\theta \in \Theta \subset \mathbb{R}^d$ verifies*[1]*:*

$$\begin{aligned} R(\theta,\widehat{\theta}) \geq\ & (I_d + \partial_\theta B(\theta,\widehat{\theta}))F^{-1}(\theta)(I_d + \partial_\theta B(\theta,\widehat{\theta}))^T \\ & + B(\theta,\widehat{\theta})\, B^T(\theta,\widehat{\theta}) \end{aligned} \tag{2.28}$$

where $\partial_\theta B(\theta,\widehat{\theta})$ is the Jacobian matrix of the vector $B(\theta,\widehat{\theta})$ with respect to θ, and where

$$F(\theta) = -\mathbb{E}_\theta\left\{\partial^2_\theta \log p(X;\theta)\right\} \tag{2.29}$$

[1] Take two positive square matrices A and B of the same dimension d. We say that $A \geq B$ if, and only if, $A - B \geq 0$, i.e. for any $w \in \mathbb{C}^d$ we have $w^H A w \geq w^H B w$.

where $\partial_\theta^2 \log p(X;\theta)$ *is the Hessian of* $\log p(X;\theta)$ *(square matrix of dimension* d*) for which the element in line* m*, column* ℓ *is written* $\frac{\partial^2 \log p(X;\theta)}{\partial\theta_m \partial\theta_\ell}$.

In the case of an unbiased estimator, formula (2.28) may be simplified, giving:

$$R(\theta,\widehat{\theta}) = \text{cov}_\theta \left\{\widehat{\theta}\right\} \geq F^{-1}(\theta) \tag{2.30}$$

Matrix F is known as the Fisher information matrix. An estimator which reaches the CRB is said to be efficient; however, an efficient estimator does not always exist.

It can be shown that:

$$F(\theta) = \mathbb{E}_\theta \left\{\partial_\theta \log p(X;\theta) \partial_\theta^T \log p(X;\theta)\right\} \tag{2.31}$$

where $\partial_\theta \log p(X;\theta)$ is the Jacobian of $\log p(X;\theta)$ (vector of length d).

2.3.2 Least squares method

Consider a series of observations y_n of the form:

$$y_n = x_n(\beta) + w_n \tag{2.32}$$

where $x_n(\beta)$ is a deterministic model dependent on the parameter of interest $\beta \in \Theta \subset \mathbb{R}^d$ and w_n is a random process representing an additive noise. Take the example of a sinusoid in noise, written:

$$y_n = A\cos(2\pi f_0 n + \phi) + w_n$$

Parameter $\theta = (A, f_0, \phi) \in \Theta = \mathbb{R}^+ \times (0, 1/2) \times (0, 2\pi)$. Note that the model is non-linear with respect to f_0.

Using a series of N observations, the least squares method consists of using the following estimator of β:

$$\widehat{\beta} = \arg\min_{\alpha\in\Theta} \sum_{n=1}^{N} |y_n - x_n(\alpha)|^2 \tag{2.33}$$

This very general method was introduced by Gauss for use in determining the trajectories of the planets. It still plays a key role in estimator construction methods. It may be applied on the condition that we have a deterministic model, dependent on the parameter being estimated. The added noise w_n models the measurement noise, but also the "model" noise, i.e. the fact that we are not completely certain of the presumed model $x_n(\beta)$. As we shall see, the least squares estimator corresponds to the maximum likelihood estimator in cases where w_n is a Gaussian random process.

Example 2.7 (Non linear model) Consider $N = 6$ observations y_n which are written:

$$y_n = \frac{1}{1 + e^{\theta_0 + \theta_1 x_n}} + w_n \tag{2.34}$$

where the parameter of interest is $\theta = (\theta_0, \theta_1)$. Using the function `fminsearch` (one of the components of the `optim` toolbox), write a program to estimate parameter θ using the least squares methods. Use the data provided in Table 2.3.

x_n	21	23	25	27	29	31
y_n	0.02	0.05	0.03	0.08	0.25	0.38
x_n	33	35	37	39	41	43
y_n	0.6	0.7	0.9	0.94	0.99	1

Table 2.3 – Observed pairs (x_n, y_n)

HINTS: type the program:

```
%===== fitLSnonlin.m
clear all
x = [21;23;25;27;29;31;33;35;37;39;41;43];
y = [0.02;0.05;0.03;0.08;0.25;0.38; ...
    0.6;0.7;0.9;0.94;0.99;1.0000];
N = length(x);
Z  = [ones(N,1) x];
theta = fminsearch(@(alpha) ...
    norm(y-(1 ./(1+exp(Z*alpha)))), [1;-1/1000]);
xe    = linspace(20,50,20);
Ze    = [ones(20,1) xe'];
haty = 1 ./ (1+exp(Ze*theta));
plot(x,y,'ok','markerf','k'), hold on;
plot(xe, haty); hold off
```

Figure 2.4 shows the curve obtained based on the estimated value of the pair (θ_0, θ_1).

2.3.3 Least squares method: linear model

Using (2.33) to find an estimator using the least squared method does not generally result in a simple analytical expression, and requires the use of numerical approaches. Moreover, whilst a number of asymptotic results exist, performance often needs to be studied on a case-by-case basis. The exception to this rule is the case of the *linear model*, for which a considerable number

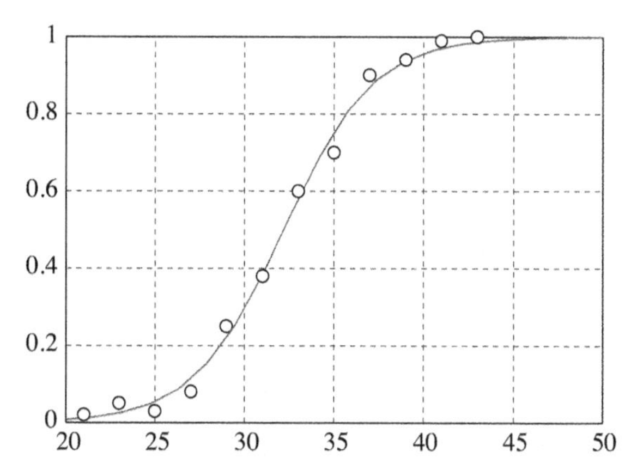

Figure 2.4 – *Least squares method for a non-linear model given by equation* (2.34). *The points represent observed value pairs (Table 2.3)*

of results have been found. This section is based on a detailed study of this model, written, for $n = 1, \ldots, N$:

$$y_n = Z_n^T \beta + \sigma \epsilon_n \tag{2.35}$$

where $\left\{ Z_n = \begin{bmatrix} 1 & X_{n,1} & \ldots & X_{n,p} \end{bmatrix}^T = \begin{bmatrix} 1 & X_n \end{bmatrix}^T \right\}$ is a series of known vectors and ϵ_n is a random sequence of white, zero-mean uncorrelated r.v. with variance 1, and σ is a positive number.

The expression "linear model" refers to linearity of expression (2.35) with respect to β. This is also known as *linear regression*.

The variables y_n are known as responses or dependent variables. The variables $X_{n,j}$ are known as *regressors* or *explanatory variables*, or as *independent variables* (not to be confused with independent random variables). Explanatory variables may be seen in two ways: either as N points in the space $\mathbb{R}^p$, or as p points in the space $\mathbb{R}^N$.

Below, we shall consider the case of real data, but the results are applicable to cases involving complex data values.

Geometric properties

Using matrix notions, (2.35) may be rewritten:

$$y = Z\beta + \sigma\epsilon, \tag{2.36}$$

where

– y is a vector of dimension $N \times 1$;

- $Z = \begin{bmatrix} \mathbf{1}_N & X \end{bmatrix}$ is a matrix of dimension $N \times (p+1)$, in which all of the components in the first column are equal to 1;
- $\beta = (\beta_0, \ldots, \beta_p) \in \mathbb{R}^{p+1}$. Coefficient β_0 is known as the *intercept*;
- $\sigma \in \mathbb{R}^+$.

and where ϵ is a random vector verifying either the hypotheses:

$$\mathbb{E}\{\epsilon\} = \mathbf{0}_N, \quad \mathrm{cov}(\epsilon) = \boldsymbol{I}_N \tag{2.37}$$

or the hypothesis

$$\epsilon \sim \mathcal{N}(\mathbf{0}_N, \boldsymbol{I}_N) \tag{2.38}$$

Model (2.37) is non-parametric, whereas (2.38) is parametric. Clearly, (2.38) entails (2.37), but the reverse is not the case. The parameter for estimation in the proposed statistical model is thus:

$$\theta = (\beta, \sigma^2) \in \mathbb{R}^{p+1} \times \mathbb{R}^+$$

Notation

- r denotes the rank of matrix Z. We can show that $r \leq \min\{N, p+1\}$. If $r = (p+1) \leq N$, Z is said to be of "full column rank", and the square matrix Z^TZ of dimension $p+1$ is invertible.
- Π_Z denotes the orthogonal projector onto the sub-space spanned by the columns of Z. This sub-space is noted $\mathrm{Im}(Z)$ as image of Z. Remember that $\mathbf{0}_N \leq \Pi_Z \leq \boldsymbol{I}_N$, Π_Z has r eigenvalues equal to 1 and that the $(N-r)$ others are null. Hence, $\mathrm{trace}\{\Pi_Z\} = r$.
- $\Pi_Z^\perp = \boldsymbol{I}_N - \Pi_Z$ denotes the projector onto the sub-space orthogonal to $\mathrm{Im}(Z)$. $\Pi_Z^\perp$ is positive. Hence, $\mathrm{trace}\left\{\Pi_Z^\perp\right\} = N - r$.
- $h_{n,n}$ denotes the n-th diagonal element of matrix Π_Z. From this, we can deduce that $\sum_{n=1}^{N} h_{n,n} = \mathrm{trace}\{\Pi_Z\} = r$. If u_n denotes the vector for which all components are null except for the n-th component, which is 1, hence, $h_{n,n} = u_n^T \Pi_Z u_n$. Applying expression (2.51) ($\Pi_1 \leq \Pi_Z \leq \boldsymbol{I}_N$) to vector u_n, we obtain:

$$\frac{1}{N} \leq h_{n,n} \leq 1 \tag{2.39}$$

Note that $h_{n,n}$ is dependent on the X_n variables, but not on the y_n variables. The mean value of the $h_{n,n}$ is r/N. A value for $h_{n,n}$ which is close to 1 indicates that, in the space $\mathbb{R}^p$, point X_n is far from the center of the cloud of points associated with other values. Conversely, a value close to $1/N$ indicates that the point is close to this center. The quantity $h_{n,n}$ is known as the *leverage*, see exercise 2.9.

As $\Pi_Z^{\perp} + \Pi_Z = \boldsymbol{I}_N$, any sum of the diagonal elements of the same rank in $\Pi_Z^{\perp}$ and Π_Z must be equal to 1, i.e. $\pi_Z^{\perp}(i,i) + \pi_Z(i,i) = 1$, and any sum of non-diagonal elements of the same rank will be equal to 0, i.e. $\pi_Z^{\perp}(i,j) + \pi_Z(i,j) = 0$, $i \neq j$.

According to the projection theorem 1.2, the best approximation, in the least square sense, of y in the sub-space $\mathrm{Im}(Z)$ is given by the orthogonal projection:

$$\widehat{y} = \Pi_Z y \tag{2.40}$$

An estimator $\widehat{\beta}$ of β, in the least square sense, therefore verifies:

$$Z\widehat{\beta} = \Pi_Z y \tag{2.41}$$

If Z is of full column rank, then projector $\Pi_Z = Z(Z^T Z)^{-1} Z^T$. There is therefore a single element $\widehat{\beta}$ which verifies equation (2.41), written:

$$\widehat{\beta} = (Z^T Z)^{-1} Z^T y \tag{2.42}$$

In cases where Z is not full rank, there are an infinite number of solutions defined to within an additive factor, an element of the null-space (kernel) of Z.

One expression which is useful in practice can be obtained by replacing y by (2.36) in expression (2.41). We obtain:

$$Z\widehat{\beta} = Z\beta + \sigma \Pi_Z \epsilon \tag{2.43}$$

If Z is full column rank, then (2.43) may be rewritten:

$$\widehat{\beta} = \beta + \sigma (Z^T Z)^{-1} Z^T \epsilon \tag{2.44}$$

This allows us to deduce the properties of $\widehat{\beta}$ from those of ϵ.

Centering variables

In this section, we shall show that the intercept and the other coefficients of β are associated with orthogonal spaces, and can therefore be calculated "separately".

Let $\mathbf{1}_N$ be a vector of length N all the components of which are 1. We can verify that the projector of rank 1 onto the sub-space spanned by $\mathbf{1}_N$ is expressed as:

$$\Pi_{\mathbf{1}_N} = \frac{1}{N} \mathbf{1}_N \mathbf{1}_N^T$$

Let X_c be a matrix of which the column vectors are the centered column vectors of X, written:

$$X_c = X - \Pi_{\mathbf{1}_N} X = (\boldsymbol{I}_N - \Pi_{\mathbf{1}_N})X = \Pi^{\perp}_{\mathbf{1}_N} X \tag{2.45}$$

Finally, let Π_{X_c} be the orthogonal projector over the sub-space spanned by the columns of X_c. Multiplying expression (2.45) by $\Pi_{\mathbf{1}_N}$ on the left, we obtain $\Pi_{\mathbf{1}_N} X_c = 0$, and thus:

$$\Pi_{\mathbf{1}_N} \Pi_{X_c} = \Pi_{X_c} \Pi_{\mathbf{1}_N} = 0 \tag{2.46}$$

This implies that $\mathbf{1}_N$ is orthogonal to X_c and, consequently, $\text{Im}(\begin{bmatrix}\mathbf{1}_N & X_c\end{bmatrix}) = \text{Im}(\mathbf{1}_N) \oplus \text{Im}(X_c)$. Exercise 2.9 demonstrates that:

$$\Pi_Z = \Pi_{\mathbf{1}_N} + \Pi_{X_c} \tag{2.47}$$

From (2.47), we may deduce that the orthogonal projection $\widehat{y}$, given by expression (2.40), is written:

$$\widehat{y} = \Pi_{\mathbf{1}_N} y + \Pi_{X_c} y \tag{2.48}$$

Using (2.46), we have $\Pi_{\mathbf{1}_N} \widehat{y} = \Pi_{\mathbf{1}_N} y$, which can be rewritten:

$$\frac{1}{N} \sum_{n=1}^{N} y_n = \frac{1}{N} \sum_{n=1}^{N} \widehat{y}_n \tag{2.49}$$

Take $y_c = y - \Pi_{\mathbf{1}_N} y$. (2.48) may also be written:

$$\widehat{y} = \Pi_{\mathbf{1}_N} y + \Pi_{X_c} y = \Pi_{\mathbf{1}_N} y + \Pi_{X_c} y_c + \Pi_{X_c} \Pi_{\mathbf{1}_N} y = \Pi_{\mathbf{1}_N} y + \Pi_{X_c} y_c \tag{2.50}$$

Once again, this expression uses (2.46). Expression (2.50) shows that the orthogonal projection may be determined in two separate steps: one for projection onto the space spanned by the centered variables, and the other one for calculating the mean.

Exercise 2.9 (Decomposition of Z) (see page 248)

1. Demonstrate the expression (2.47). To do this, show that $(\Pi_{\mathbf{1}_N} + \Pi_{X_c})Z = Z$.

2. Use this result to show that:

$$\Pi_{\mathbf{1}_N} \leq \Pi_Z \leq \boldsymbol{I}_N \tag{2.51}$$

Remember that, for two positive matrices A and B, $A \geq B$ means that, for any vector v, we have $v^H A v \geq v^H B v$.

3. Multiplying (2.51) by a vector all components of which are null except for the n-th which is 1, demonstrate the double inequality (2.39).

4. Let X_n be the n-th line of matrix X. Write a program which shows that, if $h_{n,n}$ is close to $1/N$, the point of coordinates X_n in $\mathbb{R}^p$ is close to the center of the cloud formed by the other points. Inversely, if $h_{n,n}$ is close to 1, the point is far from the cloud. To allow a display in $\mathbb{R}^p$, set $p = 2$.

Probabilistic properties

Consider the linear model given by expression (2.35) with hypotheses (2.37). The ϵ_n are centered r.v.s, of variance 1, which are not correlated with each other. Suppose that Z is of full column rank, matrix $Z^T Z$ being therefore invertible.

Property 2.1 (Best Linear Unbiased Estimator - BLUE) *The estimator*

$$\widehat{\beta} = (Z^T Z)^{-1} Z^T y$$

is an unbiased estimator of β and has the lowest covariance of all of the linear unbiased estimators. It is said BLUE for Best Linear Unbiased Estimator. This covariance is expressed:

$$\mathrm{cov}\left(\widehat{\beta}\right) = \sigma^2 (Z^T Z)^{-1}$$

Replacing y by (2.35) in $\widehat{\beta}$, we obtain $\widehat{\beta} = \beta + \sigma(Z^T Z)^{-1} Z^T \epsilon$. As $\mathbb{E}\{\epsilon\} = 0$, then $\mathbb{E}\left\{\widehat{\beta}\right\} = \beta$. $\widehat{\beta}$ is unbiased. From this, we deduce that $\mathrm{cov}\left(\widehat{\beta}\right) = \sigma^2 (Z^T Z)^{-1}$.

Now, consider another linear estimator with respect to y of the form $\widehat{b} = Qy$, such that $\mathbb{E}\left\{\widehat{b}\right\} = \beta$. In this case, $\mathbb{E}\left\{\widehat{b}\right\} = QZ\beta = \beta$, implying that $QZ = \boldsymbol{I}_{p+1}$. From this, we deduce that $\mathrm{cov}\left(\widehat{b}\right) = \sigma^2 Q Q^T$.

Additionally, $A = \boldsymbol{I}_N - Z(Z^T Z)^{-1} Z^T$ is a projector; $A = A^T$ and $AA = A$. Hence:

$$Q(\boldsymbol{I}_N - Z(Z^T Z)^{-1} Z^T) Q^T \geq 0$$

By developing the expression and using $QZ = I$, we deduce that:

$$QQ^T - (Z^T Z)^{-1} \geq 0$$

which is the expected result.

In the rest of this chapter, unless indicated otherwise, we shall consider hypothesis (2.38), i.e. that ϵ is white and Gaussian.

Property 2.2 *The solution to problem* (2.36) *using hypothesis* (2.38) *possesses the following properties:*

- $\widehat{\beta} = (Z^T Z)^{-1} Z^T y$,
- $\widehat{y} = \widehat{\beta} y \sim \mathcal{N}(Z\beta; \sigma^2 \Pi_Z)$, *where the rank of* Π_Z *is* r,
- $\widehat{e} = y - \widehat{y} \sim \mathcal{N}(0; \sigma^2 \Pi_Z^{\perp})$ *where the rank of* $\Pi_Z^{\perp}$ *is* $N - r$. *We have* $\mathbf{1}^T \widehat{e} = 0$,
- $\widehat{e}$ *and* $\widehat{y}$ *are independent random variables,*
- *if* Z *is of full rank,* $\widehat{e}$ *and* $\widehat{\beta}$ *are independent random variables.*

Inserting expression (2.36) into expression (2.40), we obtain:

$$\widehat{y} = Z\beta + \sigma \Pi_Z \epsilon \tag{2.52}$$

using $\Pi_Z Z = Z$. From this, we deduce that the prediction residual defined by vector $\widehat{e} = y - \widehat{y} \in \mathbb{R}^N$ is written:

$$\widehat{e} = y - \widehat{y} = \sigma \Pi_Z^{\perp} \epsilon \tag{2.53}$$

This produces the following results:

- based on (2.49), $\mathbf{1}^T \widehat{e} = \sum_{n=1}^{N} e_n = 0$;
- based on (2.52), the distribution of $\widehat{y}$ is Gaussian, and is written:

$$\widehat{y} \sim \mathcal{N}(Z\beta; \sigma^2 \Pi_Z) \quad \Leftrightarrow \quad \frac{\widehat{y} - Z\beta}{\sigma} \sim \mathcal{N}(0; \Pi_Z) \tag{2.54}$$

- based on (2.53), the distribution of error $\widehat{e}$ is Gaussian, and is written:

$$\widehat{e} \sim \mathcal{N}(0; \sigma^2 \Pi_Z^{\perp}) \quad \Leftrightarrow \quad \frac{\widehat{e}}{\sigma} \sim \mathcal{N}(0; \Pi_Z^{\perp}) \tag{2.55}$$

- the random vectors $\widehat{e}$ and $\widehat{y}$ are independent. Indeed, following (2.52) and (2.53), $\text{cov}(\widehat{e}, \widehat{y}) = \sigma^2 \Pi_Z^{\perp} \Pi_Z = 0$, meaning that $\widehat{e}$ and $\widehat{y}$ are not correlated; as they are jointly Gaussian, they are independent;
- if Z is of full rank, then random vectors $\widehat{e}$ and $\widehat{\beta}$ are independent. In this case, $\text{cov}(\widehat{e}, \widehat{y}) = \text{cov}\left(\widehat{e}, \widehat{\beta}\right) Z^T = 0$, and if Z is of full rank, $\text{cov}\left(\widehat{e}, \widehat{\beta}\right) = 0$.

This completes the proof.

Note that the random vector $(\widehat{y} - Z\beta)$ should not be confused with the random vector $\widehat{e} = (y - \widehat{y})$. The former is not observable, as the true value of β is unknown, whereas the latter is observable.

Unbiased estimator of σ^2

Following (2.55), $\mathbb{E}\left\{\widehat{e}\,\widehat{e}^T/\sigma^2\right\} = \Pi_Z^{\perp}$ and thus

$$\mathbb{E}\left\{\widehat{e}^T\widehat{e}\right\} = \text{trace}\left\{\mathbb{E}\left\{\widehat{e}\,\widehat{e}^T\right\}\right\} = \sigma^2(N-r)$$

From this, we deduce that $\widehat{\sigma}^2$, defined as

$$\widehat{\sigma}^2 = (N-r)^{-1}\widehat{e}^T\widehat{e} \tag{2.56}$$

is an unbiased estimator of σ^2.

Calculating a confidence interval for $Z_n\beta$

Using expression (2.54) and multiplying the left hand side by a vector of which all components are null except for the n-th which has a value of 1, we deduce that $(\widehat{y}_n - Z_n\beta)/\sigma \sim \mathcal{N}(0, h_{n,n})$, using theorem (1.7). Applying (2.55) and the definition of Student's law, we deduce that:

$$\frac{\widehat{y}_n - Z_n\beta}{\widehat{\sigma}\sqrt{h_{n,n}}} \sim T_{N-r}$$

where T_{N-r} denotes a Student r.v. with $(N-r)$ degrees of freedom.

Remember that in expression (2.39), $1/N \leq h_{n,n} \leq 1$. If $h_{n,n}$ is small, close to $1/N$, then $\widehat{y}_n$ is weakly dispersed around the true value $Z_n\beta$. However, if $h_{n,n} \approx 1$, $\widehat{y}_n$ is strongly dispersed around the true value $Z_n\beta$ - hence the term leverage.

Calculating a confidence interval for σ^2

The distribution of $\widehat{\sigma}^2/\sigma^2$ is the sum of the squares of $(N-r)$ independent, Gaussian r.v.s of variance 1. $\widehat{\sigma}^2/\sigma^2$ is thus χ^2 with $(N-r)$ degrees of freedom, written:

$$\frac{\widehat{\sigma}^2}{\sigma^2} \sim \chi^2_{N-r} \tag{2.57}$$

From this, we may deduce a confidence interval of level $100(1-\alpha)\%$ of σ, written:

$$\frac{\widehat{\sigma}}{\sqrt{\chi^{2[-1]}_{N-r}(1-\alpha/2)}} \leq \sigma \leq \frac{\widehat{\sigma}}{\sqrt{\chi^{2[-1]}_{N-r}(\alpha/2)}} \tag{2.58}$$

Calculating a confidence interval for new data points

Using the estimated value of β for N learning data points (y_n, Z_n), we can deduce the prediction of y_o (new data) based on an observation z_o, vector of length $(p+1)$. This is written:

$$\widehat{y}_o = z_o^T \widehat{\beta}$$

Hence:

$$y_o - \widehat{y}_o = z_o^T(\beta - \widehat{\beta}) + \sigma\epsilon_o$$

Taking Z as being of full column rank, $\widehat{y}_o$ is a centered Gaussian r.v. of variance:

$$\text{var}\,(\widehat{y}_0) = \sigma^2(z_o^t(Z^tZ)^{-1}z_o + 1) \tag{2.59}$$

Using (2.42) we can write:

$$\frac{y_o - \widehat{y}_o}{\sigma} = -z_o^T(Z^TZ)^{-1}Z^T\epsilon + \epsilon_o$$

Note that, following (2.53), the variance of the n-th component of $\widehat{y}$ is expressed $\sigma^2(1 - h_{n,n})$, where $1/N \leq h_{n,n} \leq 1$ is the n-th diagonal element of Π_Z. Moreover, following (2.59), var $(\widehat{y}_0) \geq \sigma^2(1 - h_{n,n})$. This is legitimate, given that it is harder to make a prediction using new data points than using data taken from the learning set for model β.

Following (2.55), the r.v. $(y_o - \widehat{y}_o)/\sigma$ is independent from $\widehat{e}/\sigma$. Hence:

$$\frac{y_o - \widehat{y}_o}{\widehat{\sigma}\sqrt{1 + z_o^T(Z^TZ)^{-1}z_o}} \sim T_{N-r}$$

From this result, we deduce a confidence interval of $100(1-\alpha)\%$ for y_o:

$$\widehat{y}_o - \widehat{\sigma}\delta(\alpha) \leq y_o \leq \widehat{y}_o + \widehat{\sigma}\delta(\alpha) \tag{2.60}$$

$$\text{with} \quad \delta(\alpha) = T_{N-r}^{[-1]}(1 - \alpha/2)\sqrt{1 + z_o^T(Z^TZ)^{-1}z_o} \tag{2.61}$$

Testing the hypothesis $H_0 = \{\beta_k = 0\}$

We shall begin by showing that:

$$\frac{\widehat{\beta}_k - \beta_k}{\widehat{\sigma}\sqrt{[(Z^TZ)^{-1}]_{k,k}}} \sim T_{N-p-1} \tag{2.62}$$

where T_{N-p-1} denotes a Student r.v. with $(N-p-1)$ degrees of freedom.

Following (2.44), in the case where Z is of full column rank:

$$\frac{\widehat{\beta} - \beta}{\sigma} \sim \mathcal{N}(0; (Z^TZ)^{-1})$$

Multiplying the left hand side by the vector of which all components are null except the k-th which is 1, then dividing by $\sqrt{[(Z^TZ)^{-1}]_{k,k}}$, we obtain:

$$\frac{\widehat{\beta}_k - \beta_k}{\sigma\sqrt{[(Z^TZ)^{-1}]_{k,k}}} \sim \mathcal{N}(0;1) \tag{2.63}$$

Dividing the first member of (2.63) by the square root of the first member of (2.57), we obtain (2.62).

The function defined by (2.62)[2] can be used to deduce a critical function for testing hypothesis $H_0 = \{\beta_k = 0\}$. With H_0, we have:

$$T_k(y) = \frac{\widehat{\beta}_k}{\widehat{\sigma}\sqrt{[(Z^TZ)^{-1}]_{k,k}}} \sim T_{N-p-1} \tag{2.64}$$

The bilateral test with critical function $\mathbb{1}(T_k(y) \notin (-\eta, \eta))$ can therefore be used, where η is determined for a given significance level. Based on expression (2.24), this bilateral test has a p-value of:

$$p\text{-value} = 2\int_{|T_k|}^{+\infty} p_{T_{N-p-1}}(t)dt \tag{2.65}$$

where $p_{T_{N-p-1}}(t)$ is the probability density of Student's distribution with $(N-p-1)$ degrees of freedom. As stated in section 1.7, note that the Student distribution is a specific instance of the Fisher distribution.

Testing the hypothesis $H_0 = \{\beta_1 = \cdots = \beta_p = 0\}$

We wish to test the hypothesis $H_0 = \{\beta_1 = \cdots = \beta_p = 0\}$. To do this, we take:

- the mean $y_c = \frac{1}{N}\sum_{n=1}^{N} y_n$;
- the sum of squares regression SSR $= \|\widehat{y} - y_c\|^2$. Following (2.54), for H_0, the relationship SSR$/\sigma^2$ follows a χ^2 distribution with p degrees of freedom;
- the sum of square errors SSE $= \widehat{e}^T e$. Following (2.55), for H_0, the relationship SSE$/\sigma^2$ follows a χ^2 distribution with $N-(p+1)$ degrees of freedom;
- the sum of square total SST $= \|y - y_c\|^2$. Based on Pythagoras's theorem (see also Figure 1.2), we have SST $=$ SSE $+$ SSR;
- coefficient R such that:

$$R^2 = \frac{\text{SSR}}{\text{SSR} + \text{SSE}} \in (0,1) \tag{2.66}$$

[2]The observation function defined by (2.62) is not a statistic, as it depends on the unknown parameter β. The function is said to be *pivotal*.

The closer R^2 is to 1, the smaller the residuals, and the better the model will explain observations. Inversely, if $R^2 \approx 0$, there will be little connection between the set of explanatory variables and the response y. This may be used to deduce a test for hypothesis H_0. To do this, we note:

$$S(y) = \frac{(N-p-1)R^2}{p(1-R^2)} = \frac{p^{-1}\text{SSR}}{(N-p-1)^{-1}\text{SSE}} \tag{2.67}$$

In accordance with section 1.7, for H_0, $S(X)$ follows a Fisher distribution with $(p, N-p-1)$ degrees of freedom, written:

$$S(y) \sim F_{p,N-p-1} \tag{2.68}$$

We can therefore construct a critical test function $T(X) = \mathbb{1}(S(y) \notin (-\eta, \eta))$ and calculate the threshold η for a given significance level, along with a p-value.

Exercise 2.10 (Car consumption) (see page 248) Use the data file available in MATLAB® with the command `load carsmall`. Among other things, this contains the gas consumption of around one hundred vehicles expressed in miles per gallon (the MPG variable), which we shall note y, along with the weight (Weight variable) noted x_1 and the horsepower (Horsepower variable), noted x_2, of these vehicles. Write a program to test whether variables x_1 and x_2 are explanatory for variable y using a linear model of the form:

$$y_n = \beta_0 + \beta_1 x_{n,1} + \beta_2 x_{n,2} + \sigma\epsilon_n$$

Calculate the p-value of the test of hypothesis $H_0 = \{\beta_k = 0\}$ for the two values $k = 1$ and $k = 2$.

Determine the value of R^2 and the p-value associated with the test of hypothesis $H_0 = \{\beta_1 = \beta_2 = 0\}$.

Example 2.8 (Water fluoridation) The values given in Table 2.4 give the occurrence of dental cavities and the level of fluoride in drinking water observed in 21 American cities for a total of 7257 children [16]. These values are illustrated on the left hand side of Figure 2.5. The linear model does not appear to be suitable. A model of the form $y \approx \gamma(x+1)^\alpha$ with $\alpha < 0$ seems to give a more satisfactory result. Taking the logarithm of the two members, we obtain $\log y \approx \beta_0 + \beta_1 \log(x+1)$, which is linear with respect to $\log(x+1)$. This corresponds well with the illustration in the right hand side of Figure 2.5.

Take $y = \beta_0 + \beta_1 x$, where x denotes the log-percent of the number of cavities and y the log-concentration of fluoride. Test the hypothesis that $\beta_1 = 0$.

fluoride	1.90	2.60	1.80	1.20	1.20	1.20	0
cavities	17.13	17.85	18.29	18.72	20.39	21.99	23.44
fluoride	1.30	0.90	0.60	0.50	0.40	0.30	0
cavities	24.89	29.90	32.22	40.35	47.32	51.02	51.23
fluoride	0.20	0.10	0.20	0.10	0.10	0.10	0
cavities	53.19	56.02	59.73	75.26	58.78	48.84	52.40

Table 2.4 – Cavity figures given as thousands. The fluoride level is given in ppm (parts per million).

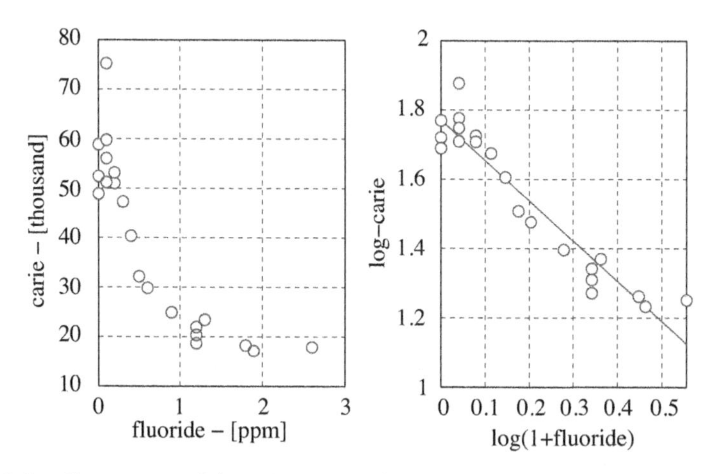

Figure 2.5 – *Percentage of dental cavities observed as a function of fluoride levels in drinking water in ppm (parts per million).*

HINTS: we obtain $R^2 = 0.9092$ and $F = \frac{(N-2)R^2}{(1-R^2)} = 190.20$, and the p-value $\approx 2.4 \times 10^{-11}$. This leads us to reject the hypothesis according to which $\beta_1 = 0$.

Execute the following program:

```
%===== testfluorcaries.m
clear all
caries = [...
    1.90 2.60 1.80 1.20 1.20 1.20 ...
    1.30 0.90 0.60 0.50 0.40 0.30 ...
    0.20 0.10 0.20 0.10 0.10 0.10 0 0 0]';
fluor = [17.13 17.85 18.29 18.72 20.39 21.99 23.44 ...
    24.89 29.90 32.22 40.35 47.32 51.02 51.23 ...
    53.19 56.02 59.73 75.26 58.78 48.84 52.40]';
logf = log10(caries+1); logc = log10(fluor);
p = size(logf,2); N = length(logf);
beta = [ones(N,1) logf]\logc;
```

```
hatlogc = [ones(N,1), logf]*beta;
SSR = sum((hatlogc-mean(logc)) .^2);
SST = sum((logc-mean(logc)) .^2);
R2 = SSR/SST;
S = (N-p-1)*R2/(1-R2)/p;
pvalue = 1-fcdf(S,p,N-p-1);
fprintf('\t*******p-value = %3.1e\n',pvalue)
```

Weighted least squares (WLS)

For model $y = Z\beta + \sigma\epsilon$, the covariance of ϵ is, in certain cases, a known matrix. If this matrix is noted Γ, then the best estimator in terms of least squares is expressed:

$$\widehat{\beta} = (Z^T\Gamma^{-1}Z)^{-1}\, Z^T\Gamma^{-1}Y, \tag{2.69}$$

(2.69) is known as the weighted least squares (WLS) estimator.

Let P be the square matrix of dimension N such that $\Gamma = PP^T$. We have $PY = (PZ)\beta + \sigma\xi$ with $\xi = P\epsilon$. We may verify that $\mathrm{cov}(\xi) = \boldsymbol{I}_N$, as in the case of ordinary least squares (see expression (2.42) of the ordinary least squares (OLS) estimator), with Y replaced by PY and Z by PZ. From this, we deduce that the estimator given by (2.69) is unbiased, and its covariance is expressed:

$$\mathrm{cov}\left(\widehat{\beta}\right) = \sigma^2(Z^T\Gamma^{-1}Z)^{-1} \tag{2.70}$$

It is easy to show that the weighted least squares estimator minimizes the following function:

$$J(\beta) = (Y - Z\beta)^T\Gamma^{-1}(Y - Z\beta) = \|y - \beta Z\|^2_{\Gamma^{-1}} \tag{2.71}$$

2.3.4 Method of moments

Let us begin with an example: consider N observations $X_1, \ldots, X_N$ distributed following a Gamma distribution of parameter $\theta = (k, \lambda) \in \mathbb{R}^+ \times \mathbb{R}^+$, with a probability density of:

$$p_X(x;\theta) = \frac{1}{\lambda^k\Gamma(k)}\, x^{k-1}\, e^{-x/\lambda}\mathbb{1}(x \geq 0) \tag{2.72}$$

where $\Gamma(k) = \int_0^{+\infty} t^{k-1}e^{-t}dt$. We can show that $\mathbb{E}\{X_n\} = k\lambda$ and $\mathbb{E}\{X_n^2\} = k(1+k)\lambda^2$.

The basic idea behind the method of moments is to relate statistical and empirical moments written:

$$\begin{cases} k\lambda & \approx & \frac{1}{N}\sum_{n=1}^{N} X_n \\ k(1+k)/\lambda^2 & \approx & \frac{1}{N}\sum_{n=1}^{N} X_n^2 \end{cases}$$

To obtain estimators, we must simply solve these two equations, with two unknowns, with respect to (k, λ). Hence:

$$\begin{cases} \widehat{k} & = & \dfrac{\widehat{m}^2}{\widehat{\sigma}^2} \\ \widehat{\lambda} & = & \dfrac{\widehat{\sigma}^2}{\widehat{m}} \end{cases} \tag{2.73}$$

where $\widehat{m} = N^{-1}\sum_{n=1}^{N} X_n$ and $\widehat{\sigma}^2 = N^{-1}\sum_{n=1}^{N}(X_n - \widehat{m})^2$.

Exercise 2.11 (Central limit for a moment estimator) (see 250) For the moment estimator given by (2.73), determine the asymptotic behavior deduced from the central limit theorem 1.10 and the continuity theorem 1.11.

Generalized method of moments

Clearly, multiple estimators may be obtained by choosing moments of the form $\mathbb{E}_\theta\{g(X_n)\}$. It is even possible to choose more moments (i.e. equations) than there are parameters (i.e. unknowns) to estimate. Starting with M moments of the form $\mathbb{E}\{g_m(X_1)\} = S_m(\theta)$ (X_1 may be used with no loss of generality, as the r.v.s are taken to be i.i.d.), the generalized method of moments consists of taking the following estimator:

$$\begin{aligned} \widehat{\theta} &= \arg\min_\theta \sum_{m=1}^{M}\left(\frac{1}{N}\sum_{n=1}^{N} g_m(X_n) - S_m(\theta)\right)^2 \\ &= \arg\min_\theta \|\widehat{S} - S(\theta)\|^2 \end{aligned} \tag{2.74}$$

where $\widehat{S}$ is a vector of length M, written:

$$\widehat{S} = \begin{bmatrix} \frac{1}{N}\sum_{n=1}^{N} g_1(X_n) \\ \vdots \\ \frac{1}{N}\sum_{n=1}^{N} g_M(X_n) \end{bmatrix} \quad \text{and} \quad S(\theta) = \begin{bmatrix} \mathbb{E}_\theta\{g_1(X_1)\} \\ \vdots \\ \mathbb{E}_\theta\{g_M(X_1)\} \end{bmatrix} \tag{2.75}$$

The estimator given by expression (2.74) may be improved by using the covariance matrix of $\widehat{S}$ and by using the weighted least squares approach shown in expression (2.71). This gives the following estimator:

$$\widehat{\theta} = \arg\min_{\theta}(\widehat{S} - S(\theta))^T C^{-1}(\theta)(\widehat{S} - S(\theta)) \tag{2.76}$$

where the covariance matrix is given by:

$$C(\theta) = \mathbb{E}_\theta\left\{(\widehat{S} - \mu(\theta))(\widehat{S} - \mu(\theta))^T\right\} \quad \text{where} \quad \mu(\theta) = \mathbb{E}_\theta\left\{\widehat{S}\right\} \tag{2.77}$$

Using the fact that the r.v.s X_n are i.i.d., we have:

$$\begin{aligned} C(\theta) &= \frac{1}{N}\left[\mathbb{E}_\theta\left\{(g_m(X_1) - \mu_m(\theta))\,(g_{m'}(X_1) - \mu_{m'}(\theta))\right\}\right]_{1\leq m,m'\leq M} \\ &= \frac{1}{N}\left[\mathbb{E}_\theta\left\{g_m(X_1)\,g_{m'}(X_1)\right\} - \mu_m(\theta)\mu_{m'}(\theta)\right]_{1\leq m,m'\leq M} \end{aligned}$$

(2.76) is generally minimized using a computational approach via a minimization program such as `fminsearch` in MATLAB®, see exercise 2.12.

Another possibility is to replace matrix $C(\theta)$ in expression (2.76) by a consistent estimation, giving:

$$\widehat{\theta} = \arg\min_{\theta}(\widehat{S} - S(\theta))^T \widehat{C}^{-1}(\widehat{S} - S(\theta)) \tag{2.78}$$

then applying a general minimization program.

However, a major issue with the generalized method of moments is that there is nothing to indicate, *a priori*, how many and which moments we have to choose, except in cases where a "sufficient" statistic exists (see the Fisher factorization criterion in [20, 27]); however, in these cases, the maximum likelihood approach is preferable.

In conclusion, the method of moments is easy to apply, but a case-by-case study is required in order to obtain satisfactory performances. However, in the case of large samples, the law of large numbers 1.9 and the central limit theorem 1.10 can be used for performance evaluation.

Exercise 2.12 (Estimation of the mixture proportions) (see page 251)
Consider a series of N random variables X_n, i.i.d., where the commun distribution writes:

$$p_X(x) = \frac{\alpha}{\sigma_1\sqrt{2\pi}}e^{-(x-m_1)^2/2\sigma_1^2} + \frac{1-\alpha}{\sigma_2\sqrt{2\pi}}e^{-(x-m_2)^2/2\sigma_2^2}$$

where $\alpha \in (0, 1)$.

Suppose that α is unknown and that m_1, m_2, σ_1 and σ_2 are known:

1. Give the expression of the statistical model.
2. Determine a moment estimator for α based on the statistic $\widehat{S}(X) = N^{-1}\sum_{n=1}^{N} X_n$. Use this result to deduce the expressions of $\mathbb{E}\{X_n\}$ and $\mathbb{E}\{X_n^2\}$.
3. Determine a moment estimator for α based on the two statistics $\widehat{S}_1(X) = N^{-1}\sum_{n=1}^{N} X_n$ and $\widehat{S}_2(X) = N^{-1}\sum_{n=1}^{N} X_n^2$ and using the expression of $S(\alpha)$ given (2.75).
4. Determine the expression of the covariance $C(\alpha)$, defined by (2.77). Use expression (2.76) to find a moment estimator for α.
5. Write a program to simulate and compare the behavior of the three estimators.

2.3.5 Maximum likelihood

The maximum likelihood estimation method is based on the fact that a good estimator of θ may be reasonably thought to maximize the probability of what has been observed. In the case of "continuous" r.v.s, the probability is replaced by the probability density.

Definition 2.9 (Likelihood) *Let $X = (X_1, \ldots, X_N)$ be a set of N observations, with a joint probability density $p_X(x_1, \ldots, x_N; \theta)$ where $\theta \in \Theta$. The likelihood is the function of Θ in $\mathbb{R}^+$:*

$$\theta \mapsto p(X_1, \ldots, X_N; \theta) \tag{2.79}$$

The maximum likelihood estimator (MLE) is the estimator defined by:

$$\widehat{\theta}(X_1, \ldots, X_N) = \arg\max_{\theta \in \Theta} p_X(X_1, \ldots, X_N; \theta) \tag{2.80}$$

The logarithm of the likelihood function is known as the log-likelihood. As the logarithm is an increasing function, likelihood may be replaced by log-likelihood in expression (2.80). Note that:

- one sufficient condition for the existence of a maximum is that the set Θ is compact and that function $\ell(x; \theta)$ is continuous over Θ;
- the MLE is not generally unique;
- the MLE is invariable by re-parameterization. This means that if $\theta \mapsto g(\theta)$ is a function defined over Θ and if $\widehat{\theta}$ is an MLE of θ, then $g(\widehat{\theta})$ is an MLE of $g(\theta)$.

In most practical situations, the maximum likelihood estimator converges toward the true value of the parameter as the number of samples tends toward infinity. More precisely, under very general conditions, it can be shown that:

$$\sqrt{N}\left(\widehat{\theta}_{\mathrm{MLE}}-\theta\right) \xrightarrow{d} \mathcal{N}(0, F^{-1}) \tag{2.81}$$

where

$$F = -\lim_{N\to+\infty} \frac{1}{N}\mathbb{E}_\theta\left\{\partial_\theta^2 \log p(X_{1:N};\theta)\right\} \tag{2.82}$$

This result is fundamental, showing that maximum likelihood estimators are asymptotically unbiased and asymptotically efficient in that their asymptotic covariance matrix is the limit of the Cramer-Rao bound, expression (2.30), normalized by N.

This is the reason why this estimator is generally preferred, even when the problem (2.80) can only be solved using a computational approach. Note, however, that the maximum likelihood approach can fail: an example is shown in exercise 2.14.

MLE of the i.i.d. Gaussian model

Consider a series of N observations {i.i.d. $\mathcal{N}(N; m, C)$} where m is a vector of dimension d and C is a square matrix of dimension d, taken to be strictly positive. C is therefore invertible with $\det\{C\} \neq 0$. The log-likelihood is written:

$$\begin{aligned}\ell(\theta) &= -\frac{Nd}{2}\log(2\pi) - \frac{N}{2}\log\det\{C\} \\ &\quad -\frac{1}{2}\sum_{k=1}^{N}(X_k - m)^T C^{-1}(X_k - m)\end{aligned} \tag{2.83}$$

where $\theta = (m, C) \in \mathbb{R}^d \times \mathcal{M}_d^+$.

Maximization of ℓ with respect to m is carried out by canceling the gradient of ℓ:

$$\partial_m \ell(\theta) = \sum_{k=1}^{N} C^{-1}(X_k - m) = 0$$

Hence, assuming C is of full rank, we have:

$$\widehat{m}_{\mathrm{MLE}} = \frac{1}{N}\sum_{k=1}^{n} X_k$$

Let:

$$\widehat{C} = \frac{1}{N}\sum_{k=1}^{N}(X_k - \widehat{m}_{\text{MLE}})(X_k - \widehat{m}_{\text{MLE}})^T$$

Inserting $\widehat{m}_{\text{MLE}}$ and $\widehat{C}$ into ℓ, using the identity $v^T A v = \text{trace}\left\{Avv^T\right\}$ and the linearity of the trace, we obtain the log-likelihood as a function of C:

$$\widetilde{\ell}(C) = -\frac{Nd}{2}\log(2\pi) - \frac{N}{2}\log\det\{C\} - \frac{N}{2}\text{trace}\left\{C^{-1}\widehat{C}\right\} \tag{2.84}$$

We now need to maximize $\widetilde{\ell}(C)$ with respect to C. To do this, note that the trace and the determinant (for matrices of ad hoc dimensions) verify:

$$\text{trace}\{AB\} = \text{trace}\{BA\} \text{ and } \det\{AB\} = \det\{BA\} = \det\{A\}\det\{B\}$$

Let:

$$S \;=\; \widehat{C}^{-1/2}C\widehat{C}^{-1/2} \tag{2.85}$$

S is positive and can therefore be diagonalized. Carrying (2.85) into (2.84), we obtain, successively:

$$\begin{aligned}\widetilde{\ell}(C) \;&=\; -\frac{Nd}{2}\log(2\pi) - \frac{N}{2}\log\det\left\{\widehat{C}^{1/2}S\widehat{C}^{1/2}\right\} - \frac{N}{2}\text{trace}\left\{S^{-1}\right\} \\ &= -\frac{Nd}{2}\log(2\pi) - \frac{N}{2}\log\det\left\{\widehat{C}\right\} - \frac{N}{2}\log\det\{S\} - \frac{N}{2}\text{trace}\left\{S^{-1}\right\}\end{aligned} \tag{2.86}$$

Let λ_s be the eigenvalues of S. Expression (2.86) may be rewritten:

$$\widetilde{\ell}(C) = -\frac{Nd}{2}\log(2\pi) - \frac{N}{2}\log\det\left\{\widehat{C}\right\} - \frac{N}{2}\sum_{s=1}^{d}\left(\log\lambda_s + \lambda_s^{-1}\right) \tag{2.87}$$

Canceling the derivative with respect to λ_s we obtain $\lambda_s = 1$ and thus $S = \boldsymbol{I}_d$ which is clearly a positive matrix. Using $S = \boldsymbol{I}_d$ in expression (2.85), we deduce that the matrix C which maximizes the likelihood is, in fact, the matrix $\widehat{C}$. In conclusion, the maximum likelihood estimators of the i.i.d. Gaussian model are:

$$\widehat{m}_{\text{MLE}} \;=\; \frac{1}{N}\sum_{k=1}^{N}X_k \tag{2.88}$$

$$\widehat{C}_{\text{MLE}} \;=\; \frac{1}{N}\sum_{k=1}^{N}(X_k - \widehat{m}_{\text{MLE}})(X_k - \widehat{m}_{\text{MLE}})^T \tag{2.89}$$

The maximum is expressed as:

$$\ell_{\max} = -\frac{Nd}{2}\log(2\pi) - \frac{N}{2}\log\det\left\{\widehat{C}_{\text{MLE}}\right\} - \frac{Nd}{2} \tag{2.90}$$

In the specific case where $d = 1$, we obtain:

$$\begin{aligned}
\widehat{m}_{\text{MLE}} &= \frac{1}{N}\sum_{k=1}^{N} X_k \\
\widehat{\sigma}^2_{\text{MLE}} &= \frac{1}{N}\sum_{k=1}^{N}(X_k - \widehat{m}_{\text{MLE}})^2 \\
\ell_{\max} &= -\frac{N}{2}\log(2\pi) - \frac{N}{2}\log\det\left\{\widehat{\sigma}^2_{\text{MLE}}\right\} - \frac{N}{2}
\end{aligned}$$

Example 2.9 (Poisson distribution) Consider N observations, i.i.d., with values in $\mathbb{N}$ and a Poisson distribution:

$$\mathbb{P}\{X_n = x\} = \frac{\lambda^x}{x!}e^{-\lambda}$$

where $\lambda \in \mathbb{R}^+$. Determine an MLE for λ.

HINTS: The log-likelihood is written:

$$\ell(\lambda) = -N\lambda + \log\lambda\sum_{n=1}^{N} X_n - \sum_{n=1}^{N} X_n!$$

Canceling the derivative with respect to λ, we obtain $\widehat{\lambda} = N^{-1}\sum_{n=1}^{N} X_n$.

Exercise 2.13 ($\Gamma(k,\lambda)$ distribution) (see page 253) Consider N observations X_n, i.i.d. with

1. distribution $\Gamma(1,\lambda)$ of density $p_X(x;\lambda) = \lambda^{-1}e^{-x/\lambda}\mathbb{1}(x \geq 0)$. Determine the expression of the MLE of λ. Carry out a simulation. Compare the obtained dispersions to the Cramer-Rao bound, expression (2.82);

2. distribution $\Gamma(k,\lambda)$; the expression of the distribution is given in (2.72). Determine the expression of the MLE of $\theta = (k,\lambda)$.

Exercise 2.14 (Singularity in the MLE approach) (see page 255) Consider N observations, i.i.d., with values in $\mathbb{R}$, with the probability distribution:

$$p_{X_n}(x_n;\theta) = \frac{1}{2\sqrt{2\pi}\sigma_1}e^{-(x_n-m_1)^2/2\sigma_1^2} + \frac{1}{2\sqrt{2\pi}\sigma_2}e^{-(x_n-m_2)^2/2\sigma_2^2}$$

where $\theta = (m_1, m_2, \sigma_1, \sigma_2) \in \Theta = \mathbb{R}\times\mathbb{R}\times\mathbb{R}^+\times\mathbb{R}^+$. Show that the likelihood $\ell(\theta)$ may tend toward infinity. More precisely, for any $A > 0$, it exists $\theta \in \Theta$ such that $\ell(\theta) \geq A$.

Exercise 2.15 (Parameters of a homogeneous Markov chain) (see page 255) Consider a series of N Markovian r.v.s $X_1, \ldots, X_N$ with values in the finite set $\mathcal{S} = \{1, \ldots, S\}$, with an initial distribution $\mathbb{P}\{X_1 = s\} = \alpha_s \geq 0$ and a transition law:

$$\mathbb{P}\{X_n = s | X_{n-1} = s'\} = p_{s|s'} \geq 0$$

Hence, $\sum_{s=1}^{S} \alpha_s = 1$ and for any $s' \in \{1, \ldots, S\}$, $\sum_n p_{s|s'} = 1$. For ease of calculation, the following notation may be used:

$$\begin{aligned} \mathbb{P}\{X_1 = x_1\} &= \sum_{s=1}^{S} \alpha_s \mathbb{1}(x_1 = s) \\ \mathbb{P}\{X_n = x_n | X_{n-1} = x_{n-1}\} &= \sum_{s=1}^{S} \sum_{s'=1}^{S} p_{s|s'} \mathbb{1}(x_n = s, x_{n-1} = s') \end{aligned}$$

where x_n belong to $\mathcal{S}$.

Note (this may be verified by identification) that, if the function $g(x) = \sum_{s=1}^{S} g_s \mathbb{1}(s = x)$, then for any function f:

$$f(g(x)) = \sum_{s=1}^{S} f(g_s) \mathbb{1}(x = s)$$

1. Determine the expression of the likelihood of the series of observations $X_1, \ldots, X_N$ as a function of $p_{s|s'}$ and α_s.

2. Use this result to deduce a maximum likelihood estimator for $p_{s|s'}$.

3. Write a program creating a series of N Markov r.v.s with values in $\{1, \ldots, S\}$ for a transition law P (drawn at random) and for an initial distribution α (drawn at random), then estimate the matrix P.

Logistic regression

Consider a series of N observations (X_n, Y_n), taken to be i.i.d. Variable $X_n \in \mathcal{X}$ is said to be explanatory and variable $Y_n \in \mathcal{Y}$ is said to be the explained or response variable.

Variables X_n and Y_n may be quantitative, categorical (qualitative) with no notion of order, or ordered categorical. Unordered categorical values include, for example, gender (man, woman) or place of residence. Ordered *categorical* values include size (very tall, tall, small, very small) or product satisfaction (good, bad, no opinion). In the case where response Y_n is quantitative, we speak of *regression*. When Y_n is qualitative without order, we speak of classification.

Finally, when Y_n is qualitative and ordered, we speak of *ranking*.

In this section, we consider that the value of the explanatory variable falls within $\mathcal{X} = \mathbb{R}^p$ and that the categorical response has values within $\mathcal{Y} = \{0, 1\}$. Logistic regression consists of modeling the probability of response Y_n, conditionally to the explanatory variable X_n, in the following manner:

$$\mathbb{P}\{Y_n = y_n | X_n = x_n; \beta_0, \beta\} = \begin{cases} \dfrac{e^{\beta_0 + \beta^T x_n}}{1 + e^{\beta_0 + \beta^T x_n}} & \text{if} \quad y_n = 0 \\ \dfrac{1}{1 + e^{\beta_0 + \beta^T x_n}} & \text{if} \quad y_n = 1 \end{cases} \tag{2.91}$$

where $\beta_0 \in \mathbb{R}$ and $\beta \in \mathbb{R}^p$.

Writing the log-likelihood of N observations, assumed to be independent, we obtain $p_{X,Y}(x, y; \beta) = \prod_{n=1}^N p(y_n | x_n; \beta) p_{X_n}(x_n)$. Assuming that the marginal distribution of X does not depend on β, the log-likelihood may be written, to within an additive constant (which does not depend on β):

$$\ell(\alpha) = \sum_{n=1}^N \alpha^T Z_n \, \mathbb{1}(Y_n = 0) - \sum_{n=1}^N \log\left(1 + e^{\alpha^T Z_n}\right) \tag{2.92}$$

where $\alpha = \begin{bmatrix}\beta_0 & \beta\end{bmatrix}^T$ and $Z_n = \begin{bmatrix}1 & X_n^T\end{bmatrix}^T$.

The search for a maximum likelihood estimator is a non-linear maximization problem. It may be numerically solved, for example using the Newton-Raphson method.

The Newton-Raphson algorithm is an iterative algorithm which aims to minimize a function $\ell(\alpha)$ with values in $\mathbb{R}$. Let α_p be the value calculated in step p. Thus the value at step $p + 1$ writes:

$$\alpha_{p+1} = \alpha_p - \left[\frac{\partial^2 \ell}{\partial^2 \alpha}\right]_{\alpha=\alpha_p}^{-1} \times \left.\frac{\partial \ell}{\partial \alpha}\right|_{\alpha=\alpha_p} \tag{2.93}$$

When maximizing expression (2.92), the first and second derivatives are written:

$$\begin{aligned} \frac{\partial \ell}{\partial \alpha} &= \sum_{n=1}^N Z_n \mathbb{1}(Y_n = 0) - \sum_{n=1}^N Z_n \frac{e^{\alpha^T Z_n}}{1 + e^{\alpha^T Z_n}} \\ &= -\sum_{n=1}^N Z_n \mathbb{1}(Y_n = 1) + \sum_{n=1}^N Z_n \frac{1}{1 + e^{\alpha^T Z_n}} \end{aligned} \tag{2.94}$$

and

$$\frac{\partial^2 \ell}{\partial^2 \alpha} = -\sum_{n=1}^N Z_n Z_n^T \frac{e^{\alpha^T Z_n}}{(1 + e^{\alpha^T Z_n})^2} \tag{2.95}$$

This algorithm is implemented in exercise 2.16.

Whilst the logistic model is not identically distributed, property (2.81) may still be used, where

$$\begin{aligned} F(\alpha) &= \lim_{N\to+\infty}\frac{1}{N}\sum_{n=1}^{N} Z_n Z_n^T \frac{e^{\alpha^T Z_n}}{(1+e^{\alpha^T Z_n})^2} \\ &\approx \frac{1}{N}\sum_{n=1}^{N} Z_n Z_n^T \frac{e^{\alpha^T Z_n}}{(1+e^{\alpha^T Z_n})^2} \end{aligned}$$

$\widetilde{Z}$ is the matrix of dimension $N \times (p+1)$, for which line n has the following expression:

$$\widetilde{Z}_n = \frac{e^{Z_n^T \alpha/2}}{1+e^{Z_n^T \alpha}} Z_n$$

Thus, $F(\alpha) \approx N^{-1}\widetilde{Z}^T\widetilde{Z}$. From this, we see that:

$$(\widehat{\alpha}_{\text{MLE}} - \alpha) \xrightarrow{d} \mathcal{N}(0, (\widetilde{Z}^T\widetilde{Z})^{-1}) \quad (2.96)$$

This expression allows the calculation of asymptotic confidence intervals for the components of α, along with test statistics following the construction presented in the sub-section on page 37. These results are applied in exercise 2.16.

Exercise 2.16 (Logistic regression) (see page 257) Write a function which estimates parameter α of the logistic model, starting from iteration (2.93) of the Newton-Raphson algorithm.

Apply the result to the data series shown in Table 2.5 which gives the state of a joint as a function of temperature:

1. using (2.96), calculate a confidence interval at 95% of α;
2. using (2.20), calculate the p-value of the log-GLRT of hypothesis $H_0 = \{\alpha_2 = 0\}$;
3. compare the two previous results.

Exercise 2.17 (GLRT for the logistic model) (see page 259) The function determined in exercise 2.16 calculates the log-likelihood of a logistic model based on a set of observations. It may therefore be used to determine the log-GLRT (2.19) of the hypothesis that one of the coefficients of model (2.91) is null.

Consider the logistic model where parameter $\alpha = \begin{bmatrix}0.3 & 0.5 & 1 & 0 & 0\end{bmatrix}$. Write a program to verify that the log-GLRT associated with hypothesis $H_0 = \{\alpha : \alpha_4 = \alpha_5 = 0\}$ follows, in accordance with (2.20), a χ^2 distribution with 2 degrees of freedom.

K	53	56	57	63	66	67	67	67	68	69	70	70
S	1	1	1	0	0	0	0	0	0	0	0	1
K	70	70	72	73	75	75	75	76	78	79	80	81
S	1	1	0	0	0	1	0	0	0	0	0	0

Table 2.5 – Temperature in degrees Fahrenheit and state of the joint: 1 signifies the existence of a fault and 0 the absence of faults

EM algorithm

Consider an observation Y with likelihood function $p_Y(y;\theta)$ the maximization of which is intractable. On the other hand we assume that a joint probability law $p_{X,Y}(x,y;\theta)$ exists such that:

$$\mathcal{L}(\theta) = p_Y(y;\theta) = \int_{\mathcal{X}} p_{X,Y}(x,y;\theta)dx$$

In this context, the pair (X,Y) is known as *complete data* and X as *incomplete data.* Because $\log p_{X,Y}(x,y;\theta)$ cannot be maximized directly, since X has not been observed, the idea consists to maximize the conditional expectation $\mathbb{E}\left\{\log p(X,Y;\theta)|Y\right\}$ which is, by definition, a function of Y and is therefore observable. The EM (Expectation-Maximization) algorithm is used to solve this problem. Each iteration involves the two following steps:

Expectation

$$Q(\theta,\theta^{(p)}) = \mathbb{E}_{\theta^{(p)}}\left\{\log p_{X,Y}(X,Y;\theta)|Y\right\} \tag{2.97}$$

Maximization

$$\theta^{(p+1)} = \arg\max_{\theta} Q(\theta,\theta^{(p)}) \tag{2.98}$$

where $\theta^{(p)}$ denotes the estimated value of the parameter at the p-th iteration of the algorithm. This two step process is repeated until convergence. Because the EM algorithm is only able to reach a local maximum, the choice of the initial value of θ is crucial. It is commonly advised to choose random starting points and keep the value giving the highest maximized likelihood.

The following property is fundamental:

Property 2.3 *For each iteration of the EM algorithm, the likelihood of the observations increases; this is written:*

$$p_Y(Y;\theta^{(p+1)}) \geq p_Y(Y;\theta^{(p)}) \tag{2.99}$$

Indeed, if θ is such that $Q(\theta, \theta^{(p)}) \geq Q(\theta^{(p)}, \theta^{(p)})$, then using Bayes' rule, $p_{X,Y}(X, Y; \theta) = p_{X|Y}(X, Y; \theta) p_Y(y; \theta)$, we have:

$$\begin{aligned} 0 \leq & \; Q(\theta, \theta^{(p)}) - Q(\theta^{(p)}, \theta^{(p)}) \qquad (2.100) \\ = & \; \mathbb{E}_{\theta^{(p)}} \left\{ \log p_{X|Y}(X, Y; \theta) | Y \right\} + \log p_Y(Y; \theta) \\ & \; - \mathbb{E}_{\theta^{(p)}} \left\{ \log p_{X|Y}(X, Y; \theta^{(p)}) | Y \right\} - \log p_Y(Y; \theta^{(p)}) \\ = & \; \mathbb{E}_{\theta^{(p)}} \left\{ \log \frac{p_{X|Y}(X, Y; \theta)}{p_{X|Y}(X, Y; \theta^{(p)})} | Y \right\} + \log p_Y(Y; \theta) - \log p_Y(Y; \theta^{(p)}) \end{aligned}$$

Making use of the concavity of the log function and the Jensen inequality, we obtain:

$$\mathbb{E}_{\theta^{(p)}} \left\{ \log \frac{p_{X|Y}(X, Y; \theta)}{p_{X|Y}(X, Y; \theta^{(p)})} | Y \right\} \leq \log \mathbb{E}_{\theta^{(p)}} \left\{ \frac{p_{X|Y}(X, Y; \theta)}{p_{X|Y}(X, Y; \theta^{(p)})} | Y \right\} = 0$$

Carrying this result in (2.100) shows that:

$$\log p_Y(Y; \theta^{(p)}) - \log p_Y(Y; \theta) \geq Q(\theta, \theta^{(p)}) - Q(\theta^{(p)}, \theta^{(p)})$$

Therefore choosing θ such that $Q(\theta, \theta^{(p)}) - Q(\theta^{(p)}, \theta^{(p)}) \geq 0$ leads to the expression (2.99). From the demonstration, we also see that the likelihood of the observations increases as long as $Q(\theta, \theta^{(p)})$ increases. It is not necessary to take the maximum of $Q(\theta, \theta^{(p)})$ as it is expressed in (2.98). In this case, the algorithm is known as the Generalized Expectation Maximization (GEM) algorithm.

Two examples of applications of the EM algorithm are shown below, the first for mixture model, and the other for model with censured data.

Mixture model

Consider a series of N observations, i.i.d., with a probability density written

$$p_{Y_n}(y_n; \theta) = \sum_{k=1}^{K} \alpha_k f_k(y_n; \mu_k) \qquad (2.101)$$

where, for example

$$f_k(y_n; \mu_k) = \frac{1}{\sigma_k \sqrt{2\pi}} e^{-(y_n - m_k)^2 / 2\sigma_k^2} \quad \text{where } \mu_k = (m_k, \sigma_k) \qquad (2.102)$$

The parameter for estimation is therefore $\theta = (m_1, \sigma_1, \alpha_1, \ldots, m_K, \sigma_K, \alpha_K)$ with $\sum_k \alpha_k = 1$ and where $\sigma_k \in \mathbb{R}^+$. This model is known as the Gaussian Mixture Model (GMM). It is used in a variety of fields, particularly speaker recognition and population mixing. A distribution of this type is shown in

Figure 2.6. There are three discernible "states" or "modes": one around a value of 1, the second around 5 and the third around 8. In certain cases, there is a clear explanation for the presence of these "modes". For example, using a problem concerning the size of adult individuals, if two modes occur, these will clearly correspond to the male and female populations.

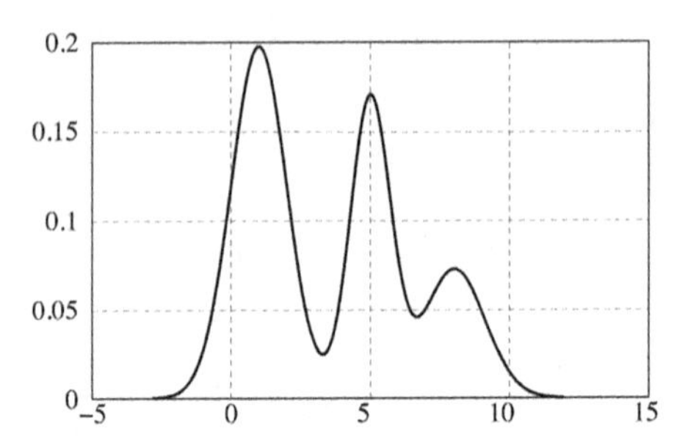

Figure 2.6 – *Multimodal distribution*

From (2.101), the analytical expression of maximum likelihood is given by:

$$\widehat{\theta} = \arg\max_{\theta} \sum_{n=1}^{N} \log \sum_{k=1}^{K} \alpha_k f_k(y_n; \mu_k) \tag{2.103}$$

Unfortunately, in this case, the calculation is intractable. This leads us to use the EM algorithm, which consists of an iterative search process for local maximums of the likelihood function $\log p_{Y_1 \dots, Y_N}(y_1, \dots, y_N; \theta)$.

Consider a series of random variables S_n, i.i.d., with values in $\{1, \dots, K\}$ and such that $\mathbb{P}\{S_n = k\} = \alpha_k$. We note:

$$p_{Y_n, S_n}(y_n, S_n = s_n; \theta) = \sum_{k=1}^{K} \alpha_k f_k(y_n; \mu_k) \mathbb{1}(s_n = k) \tag{2.104}$$

We then verify that $\sum_{s_n=1}^{K} p_{Y_n, S_n}(y_n, S_n = s_n; \theta) = \sum_{k=1}^{K} \alpha_k f_k(y_n; \mu_k)$, which is expression (2.101) of the mixture model. We may therefore consider the pair (S_n, Y_n) as complete data. Taking the logarithm of the complete law (2.104), we obtain:

$$\begin{aligned} \ell(s, y; \theta) &= \sum_{n=1}^{N} \sum_{k=1}^{K} \log(\alpha_k f_k(y_n; \mu_k)) \mathbb{1}(s_n = k) \\ &= \sum_{n=1}^{N} \sum_{k=1}^{K} \log(f_k(y_n; \mu_k)) \mathbb{1}(s_n = k) + \sum_{n=1}^{N} \sum_{k=1}^{K} \log(\alpha_j) \mathbb{1}(s_n = k) \end{aligned}$$

We shall now determine the expression of the auxiliary function Q. Note that as the pairs (S_n, Y_n) are independent, the conditional expectation of S_n conditionally to $Y_{1:N}$ is only dependent on Y_n. Using the expression of f_k given by (2.102) and taking $v_k = \sigma_k^2$, we obtain:

$$\begin{aligned} Q(\theta, \theta^{(p)}) = &= \mathbb{E}_\theta \left\{ \ell(S, Y; \theta) | Y \right\} \\ &= -\sum_{n=1}^{N} \sum_{k=1}^{K} \frac{1}{2} \left(\log(v_k) + \frac{(y_n - m_k)^2}{v_k} \right) \mathbb{E}_{\theta^{(p)}} \left\{ \mathbb{1}(S_n = k) | Y_n \right\} \\ &\quad + \sum_{n=1}^{N} \sum_{k=1}^{K} \log(\alpha_k) \mathbb{E}_{\theta^{(p)}} \left\{ \mathbb{1}(S_n = k) | Y_n \right\} - \frac{1}{2} NK \log(2\pi) \end{aligned}$$

The expression of $\mathbb{E}_{\theta^{(p)}} \left\{ \mathbb{1}(S_n = k) | Y_n \right\}$ is obtained using Bayes' rule:

$$\begin{aligned} \mathbb{E}_{\theta^{(p)}} \left\{ \mathbb{1}(S_n = k) | Y_n \right\} &= \mathbb{P}_{\theta^{(p)}} \left\{ S_n = k | Y_n \right\} \\ &= \frac{p_{Y_n|S_n=k}(y_n; \theta^{(p)}) \, \mathbb{P}\left\{ S_n = k; \theta^{(p)} \right\}}{p_{Y_n}(y_n; \theta^{(p)})} \\ &= \frac{\alpha_k^{(p)} f_k(y_n; \mu_k^{(p)})}{\sum_{j=1}^{K} \alpha_j^{(p)} f_j(y_n; \mu_j^{(p)})} \end{aligned}$$

We must therefore simply calculate $\alpha_k^{(p)} f_k(y_n; \mu_k^{(p)})$ then normalize using

$$c_n^{(p)} = p_{Y_n}(y_n; \theta^{(p)}) = \sum_{j=1}^{K} \alpha_j^{(p)} f_j(y_n; \mu_j^{(p)}) \tag{2.105}$$

In what follows, we shall use:

$$\gamma_{k,n}^{(p)} = \mathbb{P}_{\theta^{(p)}} \left\{ S_n = k | Y_n \right\} \tag{2.106}$$

which verifies, for any n, $\sum_{k=1}^{K} \gamma_{k,n}^{(p)} = 1$ and thus $\sum_{n=1}^{N} \sum_{k=1}^{K} \gamma_{k,n}^{(p)} = N$.

We may now determine the value of the parameter which maximizes function $Q(\theta, \theta^{(p)})$, canceling the derivatives with respect to θ.

For a maximization with respect to α, we consider the Lagrange multiplier η, associated to the constraint $\sum_k \alpha_k = 1$. Canceling the derivative with respect to α_k of the Lagrangian:

$$\mathcal{L} = Q(\theta, \theta^{(p)}) + \eta \left(1 - \sum_{j=1}^{K} \alpha_j \right)$$

we obtain:

$$\frac{1}{\alpha_k} \sum_{n=1}^{N} \mathbb{E}_{\theta^{(p)}} \left\{ \mathbb{1}(S_n = k) | Y_n \right\} = \eta$$

hence, after normalization:

$$\alpha_k^{(p+1)} = \frac{1}{N}\sum_{n=1}^{N}\gamma_{n,k}^{(p)}$$

Placing this value into function Q, we obtain the function $\widetilde{Q}$. Canceling the derivative of $\widetilde{Q}$ with respect to m_k, we have:

$$m_k^{(p+1)} = \frac{\sum_{n=1}^{N}\gamma_{k,n}^{(p)}\, y_n}{\sum_{n=1}^{N}\gamma_{k,n}^{(p)}}$$

then, canceling the derivative with respect to v_k, we obtain:

$$v_k^{(p+1)} = \frac{\sum_{n=1}^{N}\gamma_{k,n}^{(p)}\left(y_n - m_k^{(p+1)}\right)^2}{\sum_{n=1}^{N}\gamma_{k,n}^{(p)}}$$

Finally, note that, following expression (2.105), the log-likelihood of the observations for each iteration is written:

$$\ell^{(p)} = \sum_{n=1}^{N}\log p_{Y_n}(y_n;\theta) = \sum_{n=1}^{N}c_n^{(p)} \tag{2.107}$$

It is important to verify that $\ell^{(p)}$ increases with each iteration of the EM algorithm.

The previous algorithm may be easily generalized to cases with a mixture of K Gaussians of dimension d, with respective means $m_1, \ldots, m_K$ and covariances $C_1, \ldots, C_K$. We note:

$$p(X;m,C) = \frac{1}{(2\pi)^{d/2}\sqrt{\det\{C\}}}e^{-\frac{1}{2}(X-m)^T C^{-1}(X-m)}$$

The algorithm is written:

Data: K and $X_n \in \mathbb{R}^d$ for $n = 1$ to N
Result: $\widehat{\alpha}$, $\widehat{m}$, $\widehat{C}$, ℓ
Initialization: $\alpha_{1:K}^{(1)}, m_{1:K}^{(1)}, C_{1:K}^{(1)}$;
while *stopping condition* **do**
  for $n = 1$ *to* N **do**
    for $k = 1$ *to* K **do**
      $g_{k,n} = p(X_n; m_{1:K}^{(p)}, C_{1:K}^{(p)})$;
    end
    $c_n = \sum_{k=1}^{K} \alpha_k^{(p)} g_{k,n}$;
    for $k = 1$ *to* K **do**
      $\gamma_{k,n} = \alpha_k^{(p)} g_{k,n}/c_n$;
    end
  end
  $\alpha_k^{(p+1)} = \sum_{n=1}^{N} \gamma_{k,n}/N$;
  $\ell^{(p+1)} = \sum_{n=1}^{N} c_n$;
  for $k = 1$ *to* K **do**
    $S_k = \sum_{n=1}^{N} \gamma_{k,n}$
    $m_k^{(p+1)} = \frac{1}{S_k} \sum_{n=1}^{N} \gamma_{k,n} X_n$
    $C_k^{(p+1)} = \frac{1}{S_k} \sum_{n=1}^{N} \gamma_{k,n} (X_n - m_k^{(p+1)})(X_n - m_k^{(p+1)})^T$
  end
end

Algorithm 1: EM algorithm for GMM

REMARKS:

- initialization may be carried out either at random or using a rough estimation, for example obtained by dividing the data into K blocks and calculating the empirical means and covariances for each block;

- the stopping condition may concern the relative increase of $\ell^{(p)}$ in the form

$$\rho = \frac{|\ell^{(p+1)} - \ell^{(p)}|}{|\ell^{(p)}|}$$

  and/or a maximum number of iterations;

- the EM algorithm is similar to the K-means algorithm [11].

Exercise 2.18 (GMM) (see page 260) Write:

- a function which creates N data points of a mixture of K Gaussians with means and variances m_k, σ_k^2, with proportions α_k,
- a function which estimates the parameters of a GMM using the EM algorithm. Initialize this function by dividing ordered observations into K groups and estimating the mean and the variance for each group.
- a program to test the EM algorithm.

Exercise 2.19 (Estimation of states of a GMM) (see page 262) Consider the mixture model used in exercise 2.18. We presume that the parameters of the model have been estimated, for example using the program written in exercise 2.18.

1. Using equation (2.106), propose an estimator of state S.
2. Write a program which estimates the state of a series of observations and compares this state with the true value of obtained data. Begin by applying a learning process for θ using a sample of 10000 values.

Censored data model

In this section, we shall consider the distribution of survival times of patients receiving treatment. The survival period for each patient is collected at a given moment. There are two possibilities: either the patient is deceased and the survival period is therefore known, or the patient is still living, in which case we know only that the survival period is greater than the observed treatment time. Observations of this type are said to be right-censored.

In more general terms, samples are either the values taken by the r.v. over different outcomes, or an interval which the outcome value belongs to.

More precisely, consider a series of N random variables $\{X_n\}$, scalar and independent, with a probability density noted $p_X(x;\theta)$ and a cumulative function $F_X(x;\theta)$.

Take the pair (y_n, c_n) where, when $c_n = 0$, the realization of the r.v. X_n is y_n and, when $c_n = 1$, we only know that X_n is greater than y_n, i.e. $X_n \in (y_n, +\infty)$. Note that, when $c_n = 0$, the likelihood of the observation may be calculated, i.e. $p_X(y_n;\theta)$; however, when $c_n = 1$, this is no longer possible, and we only know that $X_n \in (y_n, +\infty)$, which has a probability of $(1 - F_X(y_n;\theta))$. An approach similar to the maximum likelihood method consists of maximizing the following expression:

$$\mathcal{L}(\theta) = \sum_{n=1}^{N} \log p_X(y_n;\theta)\mathbb{1}(c_n = 0) + \sum_{n=1}^{N} \log(1 - F_X(y_n;\theta))\mathbb{1}(c_n = 1) \quad (2.108)$$

Other forms of censoring model exist, for example the clipping mechanism, where, when $c_n = 1$, then X_n has a known value a.

Generally speaking, it is impossible to maximize expression (2.108); this leads us to use the EM algorithm. To do this, we consider that the full model is the survival value X_n, and thus:

$$Q(\theta, \theta') = \mathbb{E}_{\theta'} \left\{ \log\, p_X(X_1, \ldots, X_N; \theta) | Y_1, \ldots, Y_N, c_1, \ldots, c_N \right\}$$

where the X_n are complete data points and Y_n are incomplete data points. Using the independence hypothesis, we obtain:

$$Q(\theta, \theta') = \sum_{n=1}^{N} \mathbb{E}_{\theta'} \left\{ \log\, p_{X_n}(X_n; \theta) | Y_n, c_n \right\}$$

- when $c_n = 0$, as $X_n = y_n$, the conditional expectation of $\log p_X(X_n; \theta)$ conditionally to Y_n is, based on the definition of conditional expectation itself, equal to $\log p_X(y_n; \theta)$,
- when $c_n = 1$, the conditional expectation of $p_X(X_n; \theta)$ conditionally to the fact that $X_n > y_n$ is written:

  $$\mathbb{E}_{\theta'} \left\{ \log p_X(X_n; \theta) | X_n > y_n \right\} = g(y_n)$$

  where

$$g(y) = \frac{\int_y^{+\infty} \log p_X(x; \theta) p_X(x; \theta') dx}{1 - F_X(y; \theta')} \tag{2.109}$$

Applying Bayes' rule, we obtain:

$$\begin{aligned} \mathbb{P}_{\theta'} \left\{ X_n \leq x | X_n > y \right\} &= \frac{\mathbb{P}_{\theta'} \left\{ X_n \leq x, X_n > y \right\}}{\mathbb{P}_{\theta'} \left\{ X_n > y \right\}} \\ &= \frac{\mathbb{P}_{\theta'} \left\{ y < X_n \leq x \right\}}{\mathbb{P}_{\theta'} \left\{ X_n > y \right\}} \mathbb{1}(y < x) \\ &= \frac{\int_y^x p_X(u; \theta') du}{1 - F_X(y; \theta')} \mathbb{1}(y < x) \end{aligned}$$

Consequently, applying a derivation with respect to x, we obtain the density of X_n conditionally to $X_n > y_n$, giving:

$$p_{X_n | X_n > y_n}(x, y; \theta') = \frac{p_X(x; \theta')}{1 - F_X(y; \theta')} \mathbb{1}(x > y)$$

Finally, using the fact that for any function $h(x)$ independent of y:

$$\mathbb{E}_{\theta'} \left\{ h(X) | X \geq y \right\} = \int_{-\infty}^{+\infty} h(x) p_{X|X>y}(x; \theta') dx$$

we obtain, using $h(x) = \log p_X(x;\theta)$ (NB: we use θ and not θ'), the result set out in (2.109). Finally, we have:

$$\begin{aligned} Q(\theta,\theta') &= \sum_{n=1}^{N} \log\, p_{X_n}(y_n;\theta)\, \mathbb{1}(c_n=0) \\ &+ \sum_{n=1}^{N} \frac{\int_{y_n}^{+\infty} \log p_X(x;\theta) p_X(x;\theta')dx}{1-F_X(y_n;\theta')}\, \mathbb{1}(c_n=1) \end{aligned} \tag{2.110}$$

In cases where the cancellation of the derivative with respect to θ is hard to express, we may choose to use the GEM algorithm instead of the EM algorithm; this consists of calculating an increase in function Q. This may be done using a number of steps from the gradient algorithm. The method of moments may be used for initialization (see section 2.3.4).

Exercise 2.20 (MLE on censored data) (see page 264) This exercise considers an unusual case where the recurrence of the EM algorithm has an analytical solution. We suppose that, following medical treatment, patient survival times X_n follow an exponential distribution of parameter θ, i.e. $p_X(x;\theta) = \theta e^{-x\theta}\mathbb{1}(x \geq 0)$ and thus $\log p(x;\theta) = \log\theta - \theta x$. During the evaluation process, certain survival times are censored as the patients are still living at the moment of evaluation.

1. Using expression (2.110), determine function $Q(\theta,\theta')$.

2. Determine the recurrence over θ associated with the EM algorithm.

3. Carry out a simulation to compare the estimator obtained in the previous question with an estimation leaving aside the indication of censored data and an estimation which only uses uncensored data, using results from exercise 2.13, question 1.

4. Table 2.6 shows the survival period and state of patients for a given treatment. Write a program to estimate θ.

Y	1	30	7	4	8	5	10	2	9	36	3	9
c	0	1	0	0	0	0	1	0	0	1	0	0
Y	3	35	8	1	5	11	56	2	3	15	1	10
c	0	1	0	0	0	1	0	0	0	1	0	1

Table 2.6 – Survival period Y in number of months and state c of the patient. A value of $c = 1$ indicates that the patient is still living, hence their survival period is greater than value Y. A value of $c = 0$ indicates that the patient is deceased

2.3.6 Estimating a distribution

This section provides a simplified discussion of non-parametric estimations of probability densities and cumulative functions. We shall not consider kernel-based methods, which are essential when searching for a consistent estimation of probability density. The probability density estimation problem is similar to the estimation of spectral density, presented in section 4.3. The variance of an estimator may be reduced by carrying out smoothing using a kernel function.

Estimating probability density from a histogram

Consider a sample of N r.v.s $X_1, \ldots, X_N$, taken to be i.i.d. The distribution is taken to have a density noted $p(x)$. The approach used here to estimate $p(x)$ is very similar to that used in section 2.2.3 for goodness tests, and the same notation will be used. Note that the range of observed values is divided into g intervals; this is equivalent to applying a rectangular smoothing kernel to the data.

The total observation interval is split into g intervals Δ_j of respective length $\ell(\Delta_j)$. We can therefore write:

$$\mathbb{P}\{X_n \in \Delta_j\} = \int_{\Delta_j} p(x)dx \approx \ell(\Delta_j)\, p(c_j)$$

Based on the empirical distribution, we obtain an estimator of $\mathbb{P}\{X_n \in \Delta_j\}$ which is written:

$$\widehat{P}_j = \frac{1}{N}\sum_{k=1}^{N} \mathbb{1}(X_k \in \Delta_j)$$

For a more detailed definition of the empirical distribution, see section 2.3.7, page 79.

Equalizing the last two expressions, we obtain an estimator for $p(c_j)$ written:

$$\widehat{p}(c_j) = \frac{1}{\ell(\Delta_j)}\widehat{P}_j = \frac{\sum_{k=1}^{N} \mathbb{1}(X_k \in \Delta_j)}{N\ell(\Delta_j)}$$

Exercise 2.7 shows that the asymptotic distribution of a vector $\widehat{P}$, with g components $\widehat{P}_j$, converges in law toward vector P, with g components of values $\mathbb{P}\{X_n \in \Delta_j\}$:

$$\sqrt{N}(\widehat{P} - P) \xrightarrow{d} \mathcal{N}(0, C)$$

where $C = \text{diag}\,(P) - PP^T$. Hence:

$$\sqrt{N}(\widehat{p}(c_j) - p(c_j)) \xrightarrow{d} \mathcal{N}(0, \gamma_j)$$

where $\gamma_j = C_{jj}/\ell^2(\Delta_j)$. This expression allows us to calculate the confidence intervals of $p(c_j)$ of the form:

$$\widehat{p}(c_j) - \frac{c(\alpha)\sqrt{\gamma_j}}{\sqrt{N}} \leq p(c_j) \leq \widehat{p}(c_j) + \frac{c(\alpha)\sqrt{\gamma_j}}{\sqrt{N}}$$

where α is given by $\int_{-\infty}^{c}(2\pi)^{-1/2}e^{-u^2/2}du = 1-\alpha/2$. In practice, γ_j is replaced by an estimate obtained from $\widehat{P}$.

The following program provides an example:

```
%===== estimdsproba.m
clear all
N=10000;g=50; alpha=0.05; c=norminv(1-alpha/2);
x=linspace(-5,5,g)'; Deltax=(x(2)-x(1));
X=randn(N,1); hx=hist(X,x);
hatP=hx'/N; hatpdf=hatP/Deltax;
%bar(x,hatpdf)
pdf0=normpdf(x,0,1);
hatC=diag(hatP)-hatP*hatP';
CI95=c*sqrt(diag(hatC))/sqrt(N)/Deltax;
pfd0I=hatpdf-CI95; pfd0S=hatpdf+CI95;
%=====
plot(x,pdf0,'b'); hold on, plot(x,[pfd0I,pfd0S],'.-r');
hold off, grid on
```

Estimation of the cumulative function

Consider a sample of N i.i.d. r.v.s $X_1, \ldots, X_N$. The cumulative function is noted $F(x)$. In cases where the distribution has a probability density, this is the integral of the density. There are thus two possible approaches: (i) estimate $F(x) = \mathbb{P}\{X_n \leq x\}$ from the empirical law, or (ii) estimate the probability density using the method proposed in section 2.3.6 and carry out a cumulative sum.

The empirical distribution method consists of assigning a probability of $1/N$ to each observed value. The empirical distribution of the cumulative function is therefore expressed:

$$\widehat{F}_N(x) = \frac{1}{N}\sum_{n=1}^{N} \mathbb{1}(X_n \leq x) \tag{2.111}$$

Taking $\{X_{(n)}\}$ to be the series of values arranged in increasing order, known as the order statistic, $\widehat{F}_N(s)$ may be written:

$$\widehat{F}_N(s) = \frac{1}{N}\mathbb{1}(X_{(1)} \leq s < X_{(2)}) + \frac{2}{N}\mathbb{1}(X_{(2)} \leq s < X_{(3)}) + \cdots \tag{2.112}$$

$\widehat{F}_N(s)$ is therefore a stepwise function, with steps equal to $1/N$ and located in the ordered value series. From this, we deduce the algorithm for performing

$\widehat{F}_N(s)$: (i) ordering the values in ascending order and (ii) assign a step $1/N$ to the ordered values.

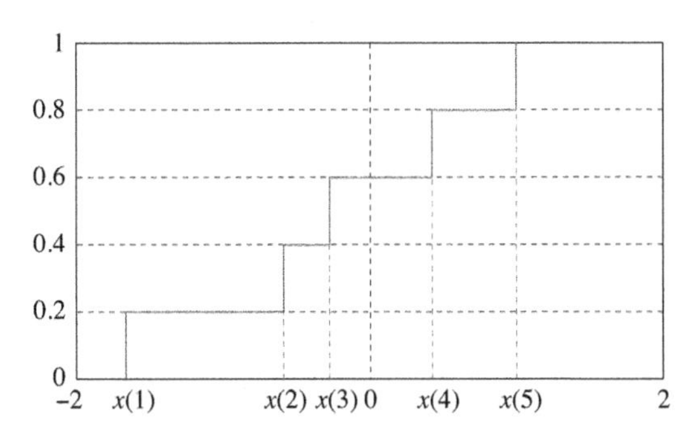

Figure 2.7 – *Step equal to* $1/N$ *for the five values of the series, ranked in increasing order*

Following (2.111), $\widehat{F}_N(x)$ appears as the mean of the sequence of i.i.d. r.v.s $Y_i = \mathbb{1}(X_i \leq x)$. Noting that $\mathbb{E}\{Y_1\} = F(x)$ and that $\text{var}(Y_1) = F(x)(1 - F(x))$, the central limit theorem is written:

$$\sqrt{N}\left(\widehat{F}_N(x) - F(x)\right) \to N(0, \gamma) \tag{2.113}$$

where $\gamma = F(x)(1 - F(x))$. Hence, an approximate confidence interval at 95% may be obtained using:

$$\widehat{F}_N(x) - \frac{1.96\sqrt{\widehat{\gamma}}}{\sqrt{N}} \leq F(x) \leq \widehat{F}_N(x) + \frac{1.96\sqrt{\widehat{\gamma}}}{\sqrt{N}}$$

where $\widehat{\gamma} = \widehat{F}_N(x)(1 - \widehat{F}_N(x))$.

In the following program, we consider two cases, depending on whether a sample of size N follows a multinomial distribution (`select='M'`) or a Gaussian distribution (`select='G'`).

```
%===== cumulEstimate.m
% Uses normcdf from the stats toolbox
clear all
N=500; mchoice='G'; %'G': gaussian, 'M': multinomial
switch mchoice
    case 'M'
        X   = zeros(N,1);
        pi0 = [1/2 1/4 1/8 1/8]; trueCF = cumsum(pi0);
        for n=1:N
            U    = rand;
            X(n) = find(U<trueCF,1,'first')-1;
        end
    case 'G'
```

```
        X=randn(N,1);
end
Xsort = sort(X);
%=====
stairline0 = (0:N)/N;
stairs([0 Xsort'], stairline0);
set(gca,'xlim',[-2 2],'ylim',[0 1])
set(gca,'xtick',[-2 0 2],'xticklab',[-2 0 2])
grid on, hold on;
switch mchoice
    case 'M'
        stairs((0:length(pi0)-1),trueCF,'r')
    case 'G'
        plot(Xsort,normcdf(Xsort,0,1),'r')
end
hold off
```

Exercise 2.21 (Estimation of a quantile) (see page 265) Consider an r.v. with a cumulative function noted $F(x)$. The quantile at $100c\%$ associated with $F(x)$ is defined by:

$$c \mapsto s = \min\{t : F(t) > c\} \tag{2.114}$$

The notion of a quantile may be interpreted as the inverse function of $(1 - F(x))$. The quantile situated at 50% is the median.

1. Use definition (2.114) to deduce an estimator $\widehat{s}_N$ of s.

2. Applying the δ-method (see section 1.5.2) to the expression (2.113), deduce the asymptotic distribution of $\widehat{s}_N$:

3. Use this result to deduce an approximate confidence interval at 95%.

4. Write a program which verifies these results.

Exercise 2.22 illustrates the estimation of the cumulative function in the context of image processing.

Exercise 2.22 (Image equalization) (see page 266) Image equalization in grayscale consists of transforming the value of pixels so that the pixel value distribution of the transformed image is uniform. In section 3.3.1, we saw that the application of function $F(x)$ to an r.v. of cumulative function $F(x)$ gives an r.v. with a uniform distribution over $(0, 1)$. This result can be applied in equalizing images. To do this, we begin by estimating the cumulative function of the original image, then we apply the estimated distribution to the image.

Consider a grayscale image where the value $I_{n,m}$ of pixel (n, m) is modeled as an r.v. with values in the set $\mathcal{I} = \{0, \ldots, S - 1\}$. Let p_s be the probability that a pixel will be equal to s, and $F(x)$ the cumulative function. This is a piecewise linear function with steps located in $\mathcal{I}$ of height $p_s = \mathbb{P}\{I_{n,m} = s\}$ (see Figure 2.7). Thus, to estimate the cumulative function of the image, we estimate the probability p_s using the empirical law, written, for s from 0 to $S-1$:

$$\widehat{p}_s = \frac{1}{NM}\sum_{n=1}^{N}\sum_{m=1}^{M}\mathbb{1}(I_{n,m} = s)$$

This gives us an estimate of the cumulative function:

$$\widehat{F}(x) = \sum_{s=0}^{S-1}\widehat{p}_s\mathbb{1}(s \leq x)$$

Write a program to equalize levels of gray in an image. Verify the result by estimating the cumulative function of the transformed image. Note that the cumulative function of the uniform distribution over $(0,1)$ is written $F(x) = x\mathbb{1}(x \in (0,1))$.

2.3.7 Bootstrap and others

In this section, we shall give a brief introduction to non-parametric estimation of estimator variance. In practice, three approaches are generally used: the *bootstrap approach*, the *jackknife approach*, and *cross-validation*. In all cases, once the variance has been estimated and based on the assumption of Gaussian behavior, it becomes possible to estimate a confidence interval.

Before considering the way in which estimator variance is determined, let us define the empirical distribution and expectation.

Definition 2.10 (Empirical distribution) *Consider a series of N observations $X_1, \ldots, X_N$, taken to be i.i.d. and with values in $\mathbb{R}^d$, with a Borel set $\mathcal{B}^{\otimes d}$. The empirical distribution associated with observations $X_1, \ldots, X_N$ is the probability distribution defined over $(\mathbb{R}^d, \mathcal{B}^{\otimes d})$, which associates each Borelian $b \in \mathcal{B}^{\otimes d}$ with the probability:*

$$\widehat{\mathbb{P}}_N(b) = \sum_{n=1}^{N}\frac{1}{N}\mathbb{1}(X_n \in b) \tag{2.115}$$

In short, each observation has an assigned probability of $1/N$. Moreover, based on the hypothesis that the X_n are i.i.d., the empirical distribution of the sequence $X_1, \ldots, X_N$ can be taken to be the product of the identical empirical laws, written:

$$\widehat{\mathbb{P}}_N^{(N)}(b_1, \ldots, b_N) = \prod_{n'=1}^{N}\sum_{n=1}^{N}\frac{1}{N^N}\mathbb{1}(X_n \in b_{n'})$$

where $b_{n'} \in \mathcal{B}^{\otimes d}$ are N Borelians in $\mathbb{R}^d$.

Note that, strictly speaking, the empirical distribution is random due to the use of series X_n: each outcome ω is associated with a realization of the series $X_1, \ldots, X_N$, and thus with a realization of the empirical distribution.

The empirical expectation is obtained from the empirical law:

Definition 2.11 (Empirical expectation) *Consider a series of N observations $X_1, \ldots, X_N$, taken to be i.i.d. with values in $\mathbb{R}^d$ and a Borel set $\mathcal{B}^{\otimes d}$. The empirical distribution of this series is noted $\widehat{\mathbb{P}}_N^{(N)}$. The empirical expectation of the function $g(X_1, \ldots, X_N)$ with values in $\mathbb{R}^q$ is the vector:*

$$\widehat{G}_N = \mathbb{E}_{\widehat{\mathbb{P}}_N^{(N)}} \{g(X_1, \ldots, X_N)\}$$

Using (2.116), *it is easy to show that:*

$$\mathbb{E}_{\widehat{\mathbb{P}}_N^{(N)}} \{g(X_1, \ldots, X_N)\} = \frac{1}{N^N} \sum_{1 \leq i_1, \ldots, i_N \leq N} g(X_{i_1}, \ldots, X_{i_N}) \tag{2.116}$$

Note that this is equivalent to assigning a probability of $1/N^N$ to all of the N-uplets constructed, with duplication, using these observations. In the specific case where $g(X_1, \ldots, X_N) = \prod_{n=1}^{N} g_n(X_n)$, we have:

$$\widehat{G}_N = \frac{1}{N^N} \prod_{n=1}^{N} \sum_{n=1}^{N} g_n(X_n)$$

In the specific case where $g(X_1, \ldots, X_N) = g(X_1)$, we have:

$$\widehat{G}_N = \frac{1}{N} \sum_{n=1}^{N} g(X_n)$$

Now, consider a series of N observations, i.i.d., of unknown distribution, and let θ be the parameter of interest and $\widehat{\theta}(X_1, \ldots, X_N)$ an estimator of θ. To estimate the variance of $\widehat{\theta}$, expression (2.116) could be applied to calculate the first two moments of $\widehat{\theta}(X_1, \ldots, X_N)$; however, this calculation is often impossible, even for simple estimators. Note that, in general, N^N terms are involved (a very large number of terms for $N = 100$, for example). This leads us to consider a Monte-Carlo type approach, which consists of carrying out a draw from these N^N possible cases; the bootstrap approach, introduced by B. Efron [7], may be seen in this way.

Bootstrap

Consider a series of observations X_1, ..., X_N, and a parameter of interest θ of dimension p. The bootstrap technique consists of carrying out B random draws, with replacement, of N samples from the set of observations. B is always very small compared to N^N, typically, $B = 300$. The samples of the b-th drawing are noted $X_1^{(b)}$, ..., $X_N^{(b)}$, and the associated estimation value is noted $\widehat{\theta}(X_1^{(b)}, \ldots, X_N^{(b)})$. For simplicity's sake, we note $\widehat{\theta}(X^{(b)}) = \widehat{\theta}(X_1^{(b)}, \ldots, X_N^{(b)})$. The bootstrap technique may then be used to obtain the empirical covariance:

$$\widehat{\sigma}_B^2(\widehat{\theta}) = \frac{1}{B-1} \sum_{b=1}^{B} (\widehat{\theta}(X^{(b)}) - \widehat{\mu}_B(\widehat{\theta}))(\widehat{\theta}(X^{(b)}) - \widehat{\mu}_B(\widehat{\theta}))^T \tag{2.117}$$

$$\text{with} \qquad \widehat{\mu}_B(\widehat{\theta}) = \frac{1}{B} \sum_{b=1}^{B} \widehat{\theta}(X^{(b)})$$

Note that the term $(B-1)$ in (2.117) leads to the creation of an unbiased estimator of the covariance matrix.

Example 2.10 Consider a series of N Gaussian r.v.s of mean μ and variance 1, and consider the estimation of the mean $\widehat{\mu} = N^{-1} \sum_{n=1}^{N} X_n$. Create a program comparing the variance via the bootstrap technique and the theoretical variance of $\widehat{\mu}$, with a value of $1/N$, over L simulations.

HINTS: Type:

```
%===== bootstraponmean
clear all
N        = 100;
mu       = 3;
B        = 300;
Lruns    = 100;
sigma2b = zeros(Lruns,1); mub = zeros(B,1);
for irun = 1:Lruns
    X    = randn(N,1)+mu;
    U    = fix(rand(N,B)*N)+1;
    Xb   = X(U); mub = mean(Xb,1);
    sigma2b(irun) = std(mub) .^2;
end
boxplot(sigma2b), hold on
xx = get(gca,'xlim');
plot(xx, ones(2,1)/N), hold off
mean(sigma2b)
```

Exercise 2.23 (Bootstrap for a regression model) (see page 267) Consider a series of N observations of the linear model:

$$X = Z\beta + \sigma\epsilon$$

where

$$Z = \begin{bmatrix} 1 & 1 \\ 1 & 2 \\ \vdots & \vdots \\ 1 & N \end{bmatrix} \quad \text{and} \quad \beta = \begin{bmatrix} 3 & 2 \end{bmatrix}^T$$

and where ϵ is a centered Gaussian vector with covariance $\boldsymbol{I}_N$. Remember that, according to property 2.1, the covariance matrix of $\widehat{\beta} = (Z^T Z)^{-1} Z^T X$ is expressed as $(Z^T Z)^{-1}$. Write a program comparing the covariance obtained using the bootstrap technique and the theoretical covariance over L simulations.

Jackknife

The Jackknife - multi-usage - technique was introduced by M. H. Quenouille [23, 17] to reduce the bias of an estimator. J. W. Tukey [28] extended the technique for estimating estimator variance.

The fundamental idea involves calculating $\widehat{\theta}$ several times, removing a sample each time. More precisely, if $X^{(j)}$ denotes the series of observations from which sample X_j is removed, we obtain the N following values:

$$\widehat{\theta}^{(j)} = \widehat{\theta}(X^{(j)}) \tag{2.118}$$

From this, we obtain the empirical covariance by jackknife associated with the estimator $\widehat{\theta}$:

$$\widehat{\sigma}_J^2 = \sum_{j=1}^{N} (\widehat{\theta}^{(j)} - \widehat{\mu}_J)(\widehat{\theta}^{(j)} - \widehat{\mu}_J)^T \tag{2.119}$$

$$\text{with } \widehat{\mu}_J(\widehat{\theta}) = \frac{1}{N} \sum_{j=1}^{N} \widehat{\theta}^{(j)} \tag{2.120}$$

Cross-validation

Cross-validation may be seen as a generalization of the jackknife approach. Instead of removing a single value (Leave One Out, LOO), we remove several values. Typically, a series of observations is divided into K blocks of the same length, and each of the K blocks is used successively as a test database, with the remaining $(K-1)$ blocks acting as a learning base. Let block k be our test data. Over the remaining $(K-1)$ blocks, we estimate θ, obtaining a value of $\widehat{\theta}^{(-k)}$. We then calculate the error for the test data using a parameter value of $\widehat{\theta}^{(-k)}$. We then calculate an average over the K blocks, see Figure 2.8.

Without losing generality, we may suppose that $N = LK$, where L is the size of a block. The block of test data k is noted $X_\ell^{(k)} = X_{(k-1)L+\ell}$ and the remaining data is noted $X_\ell^{(-k)}$. The covariance of the estimator $\widehat{\theta}(X)$ is written:

$$\widehat{\sigma}^2_{CV} = \frac{1}{K}\sum_{k=1}^{K}\frac{1}{L-1}\sum_{\ell=1}^{L}(\widehat{\theta}(X_\ell^{(k)}) - \widehat{\theta}(X_\ell^{(-k)}))(\widehat{\theta}(X_\ell^{(k)}) - \widehat{\theta}(X_\ell^{(-k)}))^T \quad (2.121)$$

learning	learning	learning	testing	learning	learning

Figure 2.8 – *Cross-validation: a block is selected as the test base, for example block no. 4 in this illustration, and the remaining blocks are used as a learning base, before switching over.*

Consider an example using the linear model:

$$y = Z\theta + \sigma\epsilon$$

where ϵ is a centered random vector of covariance $\boldsymbol{I}_N$ and matrix Z, of dimension $N \times r$ with $r < N$, is taken to be full column rank. The series is divided into L blocks of K data points. Using $(L-1)$ blocks for learning and one block for testing, we calculate the prediction variance using expression (2.59). We then use each block in turn for testing purposes and calculate the mean.

This approach may be used to estimate the order of the model. If we add columns to Z, the projection theorem guarantees that the average prediction error for the learning data can only decrease or, at worst, remain constant if the additional vector is contained in the space created by the columns of Z. In simple terms, the more columns are added, the easier it is to "explain" the noise ϵ. However, the more columns are added, the more the error over the test data will increase; this is illustrated in exercise 2.24.

Exercise 2.24 (Model estimation by cross-validation) (see page 268)
Write a program:

- which produces $N = 3000$ data points following a linear model of the form $y = Z\theta + \sigma\epsilon$, where Z is a matrix, which may be created using the function `randn`, including $p = 10$ columns (regressors). It is best to create a matrix X with N lines and 20 columns and to extract the first 10 columns in order to obtain matrix Z. Use $\sigma = 2$;

- which applies a cross validation of order $K = 10$;

- which varies the number $\widehat{p}$ of regressors, taken to be from 1 to 20;
- which calculates the prediction errors over the learning base and over the test base as a function of p;
- which displays the error curves as a function of p.

Chapter 3

Monte-Carlo Simulation

3.1 Fundamental theorems

As we saw in section 1.6, theorems 1.9 and 1.10 form the basis for statistical methods, and are crucial to the validity of Monte-Carlo methods. These theorems set out the way in which empirical means converge toward statistical moments. Noting that a statistical moment is defined as the integral of a certain function, this statement says, in some ways, that you can approximate this integral using a mean based on a random (or pseudo-random) sequence. Using these two theorems, we see that the convergence as a function of the number N of samples is of the order of $N^{-1/2}$.

It is therefore interesting to compare the value $N^{-1/2}$ to those obtained using deterministic numerical methods, such as the trapezoid method or Simpson's method. Figure 3.1 illustrates the calculation of the volume of a unit sphere using a direct method (I_{d}) and the Monte-Carlo method (I_{MC}) as a function of sample size.

The deterministic method can be seen to have a convergence speed of the order of $N^{-2/d}$, where d is the dimension of the space over which the function to integrate is defined. Consequently, when the dimension increases, the speed decreases. On the other hand, Monte-Carlo methods present two advantages compared to deterministic methods: (i) the convergence speed does not depend on the dimension d and (ii) their use does not depend on the regularity of the function being integrated.

In a less formal manner, the trapezoidal method can be seen as using a grid with a large number of points, many of which have a negligible effect on the calculated value of the integral; following the Monte-Carlo method, only the significant values are used. There is, however, one major drawback, in that the numerical result depends on the realization: the error is therefore random.

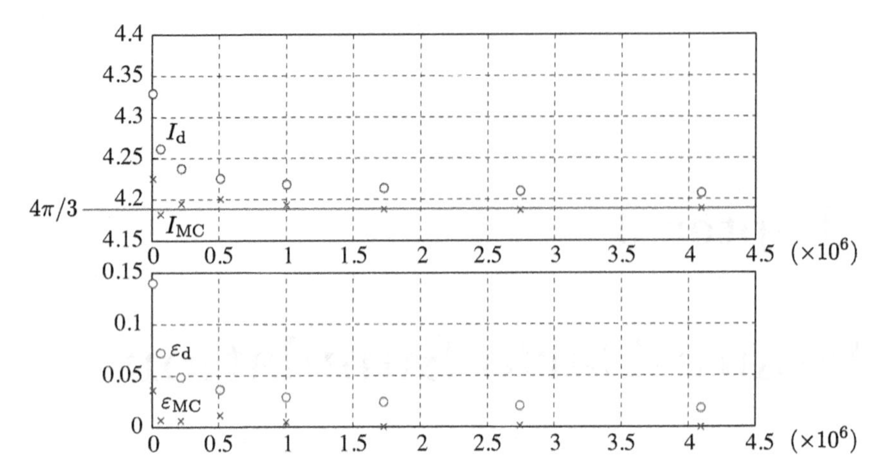

Figure 3.1 – *Direct calculation (circles) and calculation using a Monte-Carlo method (crosses), showing that the Monte-Carlo method presents faster convergence in terms of sample size. The error ε_{MC} represented in the lower diagram is random in nature. In the case of direct calculation, the error is monotone and decreasing*

3.2 Stating the problem

The aim of Monte-Carlo methods is to calculate an integral using a random generator rather than a deterministic value set. The term Monte-Carlo refers to the role of chance in games played in the world-famous casino in Monaco. Let us consider an integral over $\mathbb{R}$, which is written:

$$I(g) = \int_{-\infty}^{+\infty} g(x)dx \tag{3.1}$$

If the integral is defined over $\mathcal{S} \subset \mathbb{R}$, it may be written using the indicative function of $\mathcal{S}$ in the form:

$$\int_{\mathcal{S}} f(x)dx = \int_{-\infty}^{+\infty} \underbrace{f(x)\mathbb{1}_{\mathcal{S}}(x)}_{g(x)} dx$$

where $g(x) = f(x)\mathbb{1}_{\mathcal{S}}(x)$ and where $\mathbb{1}_{\mathcal{S}}(x)$ has a value of 1 if $x \in \mathcal{S}$ and 0 in all other cases. It therefore takes the form of expression (3.1).

In many applications, the integral for calculation is associated with the mathematical expectation of a function f, i.e. an integral of the form:

$$I(g) = \int_{-\infty}^{+\infty} \underbrace{f(x)\, p_X(x)}_{g(x)}\, dx \tag{3.2}$$

where $p_X(x)$ is the density of a probability distribution. Sometimes, the function to integrate does not have an explicit form, and can only be calculated using an algorithm.

The central idea behind the Monte-Carlo method is to use a random generator, with a distribution characterized by a probability density μ, then to use the law of large numbers to calculate the integral. We may write:

$$I(g) = \int_{-\infty}^{+\infty} g(x)dx = \int_{-\infty}^{+\infty} \frac{g(x)}{\mu(x)}\mu(x)dx = \int_{-\infty}^{+\infty} h(x)\mu(x)dx \tag{3.3}$$

where $h(x) = g(x)/\mu(x)$. If we have a realization of N random variables $X_1, \ldots, X_N$, independent and identically distributed following distribution μ, we may write:

$$I(g) \approx \frac{1}{N}\sum_{n=1}^{N} h(X_n) = \frac{1}{N}\sum_{n=1}^{N} \frac{g(X_n)}{\mu(X_n)}$$

More precisely, the law of large numbers states that

$$\frac{1}{N}\sum_{n=1}^{N} \frac{g(X_n)}{\mu(X_n)} \underset{N\to+\infty}{\longrightarrow} I(g)$$

where the convergence is in probability. Note that, when the integral $I(g)$ is associated with a mathematical expectation as in expression (3.2), it is not, *a priori*, necessary to draw the sample using the distribution $p_X(x)$. As we shall see, the choice of a different distribution can even lead to a better approximation in these cases.

Note that a real random variable is a measurable function from a sample space Ω into $\mathbb{R}$, written $X : \omega \in \Omega \mapsto x \in \mathbb{R}$. Each outcome (or experiment) is associated with a real value x known as the realization. In the context of a statistical simulation, stating that N independent draws will be used signifies that N distinct random variables $X_1, \ldots, X_N$ will be considered, and that for each outcome $\omega \in \Omega$, we obtain N realizations $x_1, \ldots, x_N$, and not N realizations of a single random variable! In more general terms, we consider an infinite series of random variables, i.e. a family $\{X_n\}$ of r.v. indexed by $\mathbb{N}$. The practical applications of statistical methods are thus essentially linked to the properties obtained when N tends toward infinity. These properties are often easier to establish when the random variables in the series are taken to be independent.

This approach is easily extended to the integral of a function defined over $\mathbb{R}^d$ and with values in $\mathbb{R}$. Let $g : (x^1, \ldots, x^d) \in \mathbb{R}^d \mapsto \mathbb{R}$ be a function of this type, and consider the integral:

$$I(g) = \int_{-\infty}^{+\infty} \cdots \int_{-\infty}^{+\infty} g(x^1, \ldots, x^d) dx^1, \ldots, dx^d$$

To calculate an approximate value for $I(g)$, we carry out N independent random drawings, for which the probability distribution defined over $\mathbb{R}^d$ has a probability density $\mu(x^1, \ldots, x^d)$, and we write:

$$I(g) \approx \frac{1}{N} \sum_{n=1}^{N} h(X_n^1, \ldots, X_n^d)$$

where $h(x^1, \ldots, x^d) = g(x^1, \ldots, x^d)/\mu(x^1, \ldots, x^d)$.

Two types of problems should be considered when using the Monte-Carlo method to calculate an integral:

1. how to determine the "optimum" way of choosing the drawing distribution μ in order to calculate a given interval;

2. how to create samples following a given distribution.

We shall begin by considering the second problem, the generation of random variables. In this context, we shall begin by presenting distribution transformation methods, followed by sequential methods based on Markov chains. We shall then consider the first issue in a section on variance reduction.

Before that, a bit of history. The first method for calculating integrals using a Monte-Carlo type technique was proposed by N. Metropolis in 1947, in the context of a statistical physics problem. In 1970, K. Hastings published an article establishing the underlying principle for general random variable generation methods, known as the Metropolis-Hastings sampler and based on Markov chains; in this context, we speak of MCMC, Monte-Carlo Markov Chains. In 1984, S. Geman and D. Geman proposed the "Gibbs" sampler, a specific form of the Metropolis-Hastings sampler, used by the authors in the context of image restoration.

3.3 Generating random variables

In this section, we shall consider that we have access to a generator using uniform distribution over the interval $(0, 1)$, able to supply a given number of independent draws. Without going into detail, note that a variety of algorithms propose generators of this type. One example, which is no longer particularly widespread, is the *Mersenne Twister* algorithm (MT) based on the Mersenne prime 19937 [14]. MATLAB® uses a variant, named MT19937ar ("ar" stands for ARray). This variant has a period of $2^{19937} - 1$ of the order of 10^{6600} (a period of the order of 10^{170} would be sufficient for the majority of simulations). Calculated using 32-bit integers, it produces values which are uniformly distributed across the hypercube $(0,1)^d$ with $d > 600$. A 64-bit version of this algorithm is also available in MATLAB®.

We shall now consider the generation of sequences following a given probability distribution, starting with a generator giving a uniform distribution over $(0, 1)$.

3.3.1 The cumulative function inversion method

Taking a real valued random variable, with a cumulative function $F(x)$, this method is based on the following result. Letting

$$F^{(-1)}(u) = \inf\{t,\ F(t) \geq u\} \tag{3.4}$$

the inverse of the cumulative function F, the random variable $X = F^{(-1)}(U)$ follows a distribution with the cumulative function F if, and only if, U is a random variable which is uniformly distributed over the interval $(0, 1)$.

Firstly, note that, by definition, the cumulative function of a real r.v. is the probability that it belongs to the interval $(-\infty, x]$. A cumulative function is a monotone increasing function which may contain jumps and may be constant over certain intervals. A typical form is shown in Figure 3.2.

Take $Y = F(X)$. Using the fact that $F(x)$ is monotone, we may write, successively:

$$F(x) = \mathbb{P}\{X \leq x\} = \mathbb{P}\left\{F^{-1}(Y) \leq x\right\} = \mathbb{P}\{Y \leq F(x)\}$$

Finally, denoting $F(x) = y$, we have $\mathbb{P}\{Y \leq y\} = y \times \mathbb{1}(y \in (0,1))$ which is, by definition, the cumulative function of the uniform law.

Example 3.1 (Rayleigh law) A probability distribution is said to follow the Rayleigh law if it has a density expressed as:

$$p(x) = \frac{x}{\sigma^2} e^{-x^2/2\sigma^2} \mathbb{1}(x \geq 0)$$

where $\sigma^2 > 0$. From this, we may deduce the cumulative function and its inverse:

$$u = F(x) = 1 - e^{-x^2/2\sigma^2} \Leftrightarrow F^{(-1)}(u) = \sigma\sqrt{-2\log(1-u)}$$

Write a program to create a sequence of length N following a Rayleigh law with parameter $\sigma = 1$. Compare the histogram obtained in this way with the theoretical probability density.

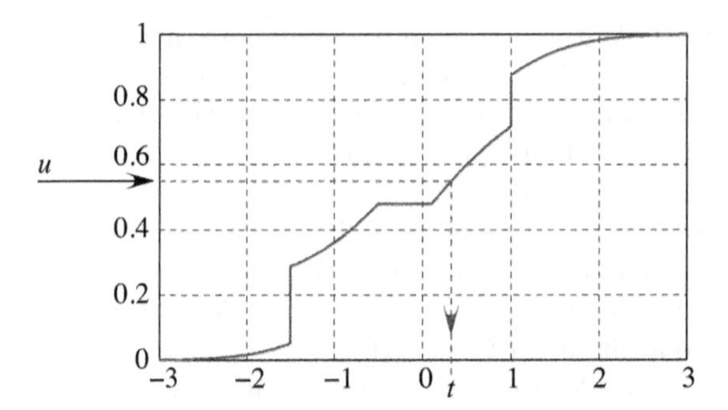

Figure 3.2 – *Typical form of a cumulative function. We choose a value in a uniform manner between 0 and 1, then deduce the realization* $t = F^{(-1)}(u)$

HINTS: Type the program:

```
%===== rayleighsimul.m
clear
N = 100000;
sigma = 1; sigma2 = sigma ^2;
U = rand(N,1); X = sigma*sqrt(-2*log(1-U));
dx = 0:0.1:5;
%=====
[hx,dx] = hist(X,dx);
px = hx/N/(dx(2)-dx(1));
rtheo = dx .* exp(-(dx .^ 2)/(2*sigma2))/sigma2;
bar(dx,px,'y')
hold on, plot(dx,rtheo,'k','linew',1.5), hold off
```

Exercise 3.1 (Multinomial law) (see page 269) A multinomial random variable is a random variable with values in $\mathcal{A} = \{a_1, \ldots, a_P\}$ such that $\mathbb{P}\{X = a_i\} = \mu_i$, where $\mu_i \geq 0$ and $\sum_i \mu_i = 1$. The associated cumulative function is written:

$$F(x) = \sum_{i=1}^{P} \mu_i \mathbb{1}(a_i \leq x)$$

and the inverse is written:

$$F^{(-1)}(u) = \min\{a_j \in \mathcal{A} : \textstyle\sum_{i=1}^{j} \mu_i \geq u\}$$

Write a program which creates a sequence of length N following a multinomial law with values in $\{1, 2, 3, 4, 5\}$, for which the probabilities are 0.1, 0.2, 0.3, 0.2, and 0.2 respectively.

Exercise 3.2 (Homogeneous Markov chain) (see page 269) Consider a sequence $\{X_n\}$ where $n \in \mathbb{N}$ and which has values in the finite discrete set $\mathcal{S} = \{1, \ldots, K\}$. By definition (for more details, see the section page 96) a sequence $\{X_n\}$ is said to be a *Markov chain* if $\mathbb{P}\{X_{n+1} = x_{n+1} | \{X_s = x_s, s \leq n\}\}$ coincides with $\mathbb{P}\{X_{n+1} = x_{n+1} | X_n = x_n\}$. The chain is said to be *homogeneous* if, in addition, the transition probabilities $\mathbb{P}\{X_{n+1} = x_{n+1} | X_n = x_n\}$ do not depend on n.

Consider a Markov chain with $K = 3$ states, the initial distribution and the transition distribution of which are shown in Table 3.1.

$\mathbb{P}\{X_0 = 1\} = 0.5$
$\mathbb{P}\{X_0 = 2\} = 0.2$
$\mathbb{P}\{X_0 = 3\} = 0.3$

p_{ij}	$i = 1$	$i = 2$	$i = 3$
$j = 1$	0.3	0	0.7
$j = 2$	0.1	0.4	0.5
$j = 3$	0.4	0.2	0.4

Table 3.1 – The initial probabilities $\mathbb{P}\{X_0 = i\}$ and the transition probabilities $\mathbb{P}\{X_{n+1} = j | X_n = i\}$ of a three-state homogeneous Markov chain. We verify that, for all i, $\sum_{j=1}^{3} \mathbb{P}\{X_{n+1} = j | X_n = i\} = 1$

Write a program to create a sequence of N values from this Markov chain. Hint: exercise 3.1 may be used. Check that the estimated transition probabilities correspond to those given in Table 3.1.

3.3.2 The variable transformation method

Linear transformation

Let X be a linear vector, the probability distribution of which has a density of $p_X(x)$. The random vector $Y = AX + B$, where A is a square matrix taken to be invertible and B is a vector of *ad hoc* dimension, follows a probability distribution of density

$$p_Y(y) = \frac{p_X(A^{-1}(y - B))}{|\det\{A\}|} \tag{3.5}$$

Thus, to simulate a random vector with a vector mean M and covariance matrix C, we must simply apply:

$$X = AW + M$$

where W is a random vector of vector mean 0 and covariance matrix I and where A is a square root of C. Remember that if W is Gaussian, then X will also be Gaussian.

Exercise 3.3 (2D Gaussian) (see page 270) Write a program:

1. which uses a centered Gaussian generator of variance 1 to generate a sequence of length N of a centered Gaussian random variable, of dimension 2, with the following covariance matrix:

$$R = \begin{bmatrix} 2 & 0.95 \\ 0.95 & 0.5 \end{bmatrix}$$

2. which displays the obtained points and the principal directions associated with the covariance matrix in the plane.

The non-linear case

If X follows a distribution of density $p_X(x)$ and if f is a derivable monotone function, then the random variable $Y = f(X)$ follows a distribution of density:

$$p_Y(y) = \frac{1}{|f'(f^{(-1)}(y))|} p_X\left(f^{(-1)}(y)\right)$$

where $f^{(-1)}(y)$ denotes the inverse function of f, and f' its derivative. This result can be extended to a bijective transformation for variables with multiple dimensions, replacing the derivative by the determinant of the Jacobian. The example shown below concerns a function with two variables.

Example 3.2 (Distribution of the modulus of a Gaussian vector) Consider two independent, centered Gaussian random variables (X, Y) of the same variance σ^2. We are concerned with the bijective variable substitution $(r, \theta) \mapsto (x, y)$ defined by:

$$\begin{cases} r = \sqrt{x^2 + y^2} & r \in \mathbb{R}^+ \\ \theta = \arg(x + jy) & \theta \in (0, 2\pi) \end{cases} \Leftrightarrow \begin{cases} x = r\cos(\theta) & X \in \mathbb{R} \\ y = r\sin(\theta) & Y \in \mathbb{R} \end{cases}$$

As this substitution is bijective and its Jacobian is equal to r, which is positive, the joint distribution of the pair (R, Θ) has a density of:

$$\begin{aligned} p_{R\Theta}(r, \theta) &= r\, p_{XY}(r\cos(\theta), r\sin(\theta))\mathbb{1}(r \geq 0)\mathbb{1}(\theta \in (0, 2\pi)) \\ &= \frac{r}{2\pi\sigma^2} e^{-r^2/2\sigma^2}\mathbb{1}(r \geq 0)\mathbb{1}(\theta \in (0, 2\pi)) \end{aligned}$$

Exercise 3.4 (Box-Muller method) (see page 270) Using the results of example 3.2, determine an algorithm to create two centered, Gaussian random variables, of the same variance σ^2, from two independent uniform random variables in $(0, 1)$. Create a program which uses this algorithm, which is known as the *Box-Muller algorithm*.

Exercise 3.5 (The Cauchy distribution) (see page 271) A Cauchy random variable may be obtained in two ways:

1. consider a random variable U which is uniform over $(0,1)$. Note $Z = z_0 + a\tan(\pi(U - 1/2))$. Determine the distribution of Z.

   The distribution of Z is a Cauchy distribution of parameters (a, z_0) and is noted $\mathcal{C}(z_0, a)$. It has no moment;

2. consider two centered, independent Gaussian variables X and Y of variance 1. We then construct $Z = z_0 + aY/X$ where $a > 0$;

3. write a program which creates a sample following the Cauchy distribution of parameters $a = 0.8$ and $z_0 = 10$, using both methods. Compare the obtained histograms to the probability density of the Cauchy distribution.

Sequence of correlated variables

It may be useful to produce a sequence of correlated random variables. ARMA(P,Q) processes, see [2], are a key example. They are defined by the following recurrence equation:

$$X_n + a_1 X_{n-1} + \cdots + a_P X_{n-P} = W_n + \cdots + b_Q W_{n-Q}$$

where W_n is a centered white noise of variance σ^2 and where all of the roots of the polynomial

$$A(z) = z^P + a_1 z^{P-1} + \cdots + a_P$$

are strictly within the unit circle. We know that the spectral density is expressed as:

$$S(f) = \sigma^2 \left| \frac{B(e^{2j\pi f})}{A(e^{2j\pi f})} \right|^2$$
$$\text{with } B(z) = z^Q + b_1 z^{Q-1} + \cdots + b_Q$$

In theory, it is easy to determine the series of covariances using $S(f)$, but the calculation itself can prove difficult.

3.3.3 Acceptance-rejection method

In the *acceptance-rejection method*, we use an auxiliary distribution $q(x)$ which is known to be "easy" to create in order to construct samples with a distribution $p(x)$ considered "difficult". Moreover, we consider that a value M exists such that, for any x, $Mq(x) \geq p(x)$ (see Figure 3.3). We use the following algorithm:

– draw a random variable X for which the distribution has a density of q;

– independently, draw a uniform random variable U with values in $(0, 1)$;

– if $UMq(X) \leq p(X)$, apply $Y = X$, otherwise reject the result.

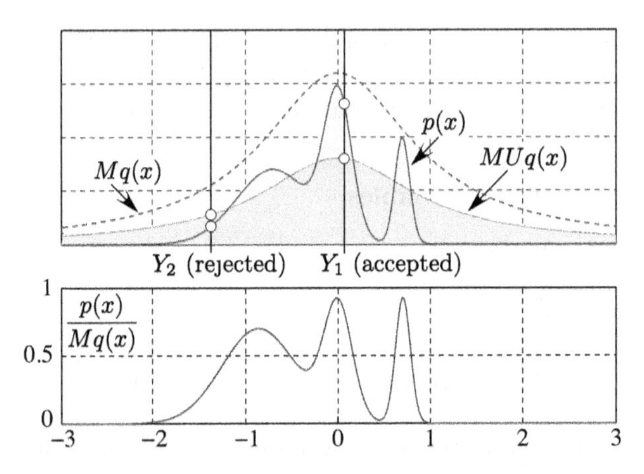

Figure 3.3 – *The greater the value of M, the more samples need to be drawn before accepting a value. The value chosen for M will therefore be the smallest possible value such that, for any x, $Mq(x) \geq p(x)$. The curve noted $MUq(x)$ in the figure corresponds to a draw of the r.v. U from the interval 0 to 1. For this draw, the only accepted values of Y are those for which $p(x) \geq MUq(x)$*

The distribution of Y can be shown to have a probability density p. Let Z be the random variable which takes a value of 0 if the draw is rejected and a value of 1 otherwise. The probability that $Z = 1$ is written:

$$\mathbb{P}\{Z = 1\} = \mathbb{P}\left\{U \leq \frac{p(X)}{Mq(X)}\right\}$$

Taking into account the fact that U and X are independent, we can write:

$$\begin{aligned}\mathbb{P}\{Z = 1\} &= \iint_{\{(u,t):u\leq\frac{p(t)}{Mq(t)}\}} \mathbb{1}(u \in (0, 1))du \times q(t)dt \\ &= \int_{-\infty}^{+\infty} \frac{p(t)}{Mq(t)} \times q(t)dt = \frac{1}{M}\end{aligned}$$

To obtain this equation, we begin by integrating with respect to u then with respect to t. Now, let us calculate the conditional probability:

$$\begin{aligned}
\mathbb{P}\{Y \leq x|Z=1\} &= \frac{\mathbb{P}\{Y \leq x, Z=1\}}{\mathbb{P}\{Z=1\}} = \frac{\mathbb{P}\left\{X \leq x, U \leq \frac{p(X)}{Mq(X)}\right\}}{\mathbb{P}\{Z=1\}} \\
&= M \iint_{\{t \leq x, u \leq \frac{p(t)}{Mq(t)}\}} \mathbb{1}(u \in (0,1)) du \times q(t) dt \\
&= M \int_{-\infty}^{x} \int_{0}^{\frac{p(t)}{Mq(t)}} du \times q(t) dt \\
&= M \int_{-\infty}^{x} \frac{p(t)}{Mq(t)} \times q(t) dt = \int_{-\infty}^{x} p(t) dt
\end{aligned}$$

Carrying out a derivation with respect to x, we see that the density of Y conditional on $Z=1$ is expressed as:

$$p_{Y|Z=1}(x) = p(x)$$

which concludes the proof.

This method has two main drawbacks:

- when M is large, the acceptation rate is low;
- it is not possible to predict how many draws are required to get N samples.

To illustrate, if $p(x)$ is the Gaussian distribution $\mathcal{N}(0, \sigma^2)$ and $q(x)$ is the Gaussian distribution $\mathcal{N}(0,1)$, then M is equal to $1/\sigma$. If $\sigma = 0.01$ then $M = 100$ hence the mean acceptance rate will be $1/100$.

3.3.4 Sequential methods

Sequential methods implement algorithms with value distributions which converge toward the distribution which we wish to simulate. Two of the most commonly used approaches are the *Metropolis-Hastings sampler* and the *Gibbs sampler*. These techniques, which are both based on the construction of a Markov chain, will be presented below. The acronym MCMC, *Monte-Carlo Markov Chain*, is used to refer to both samplers.

The structure of these algorithms means that they require a *burn-in period*, leading to the removal of a certain number of initial values. However, it is often difficult to evaluate this burn-in period. We expect that, asymptotically, the random variables are identically distributed and simultaneously approximately independent. The main interest of these methods lies in the fact that the samples they produce may be used for high dimensional random vectors.

To begin with, let us introduce a number of Markov chain properties.

Markov chain

A Markov chain is a discrete-time random process with values in $\mathcal{X}$ and such that the conditional distribution of X_n with respect to the algebra spanned by all past events $\{X_s; s < n\}$ coincides with the conditional distribution of X_n given only $X_{n-1} = x'$. The conditional probability measure of X_n knowing X_{n-1} is denoted $Q_n(dx; x')$. $Q_n(dx; x')$ is known as the transition law. A Markov chain is said to be homogeneous if the transition law $Q_n(dx; x')$ does not depend on n.

In cases where the set of possible values of X_n is finite and denoted $\mathcal{X} = \{1, \dots, K\}$, the transition law Q is characterized by the probabilities $q(x|x') = \mathbb{P}\{X_n = x|X_{n-1} = x'\}$ for any pair $(x, x') \in \mathcal{X} \times \mathcal{X}$. In this case, for any $x' \in \mathcal{X}$, we have:

$$\sum_{x\in\mathcal{X}} q(x|x') = 1$$

In cases where transition law Q has a density, noted $q(x|x')$, we can write that, for any $A \subset \mathcal{X}$ and for any $x' \in \mathcal{X}$:

$$\mathbb{P}\{X_n \in A|X_{n-1} = x'\} = \int_A q(x|x')dx$$

In this case, for any $x' \in \mathcal{X}$, we have:

$$\int_{\mathcal{X}} q(x|x')dx = 1$$

In what follows, unless stated otherwise, we will only use the notation associated with cases where the transition law possesses a density.

Let us determine a recurrence equation for the probability distribution of X_n. Using Bayes' rule $p_{X_n X_{n-1}}(x, x') = p_{X_n|X_{n-1}}(x, x')p_{X_{n-1}}(x') = q(x|x')p_{X_{n-1}}(x')$, we deduce that:

$$p_{X_n}(x) = \int q(x|x')p_{X_{n-1}}(x')dx' \tag{3.6}$$

A chain is said to be stationary if $p_{X_n}(x)$ is not dependent on n. This means that a density $p(x)$ exists such that:

$$p(x) = \int q(x|x')p(x')dx' \tag{3.7}$$

Equation (3.7) may be seen as an equation with eigenvectors $p(x)$ for the transition kernel $q(x|x')$. The condition

$$p(x)q(x'|x) = p(x')q(x|x') \tag{3.8}$$

on $p(x)$ is sufficient to satisfy (3.7). Indeed integrating the two members of (3.8) with respect to x', we obtain (3.7). Expression (3.8) is known as the *detailed balance equation.*

Metropolis-Hastings algorithm

We wish to carry out a draw using a distribution $p(x)$ known as being "difficult" to produce samples. Firstly, we select two conditional probability densities $r(x|x')$ and $a(x|x')$. The draw using $r(x|x')$ is taken to be "easy". The conditional distribution $a(x|x')$ is implemented in an acceptance/rejection mechanism, and is chosen so that the following condition is verified:

$$p(x)a(x'|x)r(x'|x) = p(x')a(x|x')r(x|x') \tag{3.9}$$

At stage n, we have x_{n-1} and the algorithm entails the application of the following two steps:

1. draw a sample X following the distribution $r(x|x_{n-1})$;

2. independently, draw a Bernoulli variable B_n with a value in the set $\{0, 1\}$ and such that $\mathbb{P}\{B_n = 1\} = a(X|X_{n-1})$. B is then used in the following acceptance/rejection mechanism:

$$X_n = \begin{cases} X & \text{if } B_n = 1 \\ X_{n-1} & \text{if } B_n = 0 \end{cases}$$

Condition (3.9) implies that $p(x)$ verifies condition (3.8). Indeed let us denote:

$$P_R(x) = 1 - \int a(u'|x)r(u'|x)du'$$

The two stages of the algorithm mean that the conditional distribution of X_n given X_{n-1} has a density of $q(x|x') = a(x|x')r(x|x') + P_R(x')\mathbb{1}(x = x')$. We can thus write, successively, that:

$$\begin{aligned} p(x')\left(q(x|x') - a(x|x')r(x|x')\right) &= p(x')\left(1 - P_R(x')\right)\mathbb{1}(x = x') \\ &= p(x)(1 - P_R(x))\mathbb{1}(x = x') \\ &= p(x)(q(x'|x) - \rho(x'|x)r(x'|x)) \end{aligned}$$

Using (3.9), we obtain (3.8). In conclusion, $p(x)$ is the stationary distribution associated with $q(x|x')$.

In the Metropolis-Hastings algorithm, the acceptance/rejection law has the following specific expression:

$$a(x|x') = \min\left(\frac{p(x)r(x'|x)}{p(x')r(x|x')}, 1\right) \tag{3.10}$$

This supposes that $r(x|x') \neq 0$ for any pair (x, x'). Expression (3.10) must be shown to verify condition (3.9). One important property is that the target distribution $p(x)$ only needs to be known to within a multiplicative constant. Two specific cases may arise:

1. the chosen distribution $r(x|x')$ is symmetrical, i.e. such that $r(x|x') = r(x'|x)$. The algorithm can therefore be simplified, and is written:

   (a) draw a sample X using the distribution $r(x|x_{n-1})$,

   (b) independently, draw a Bernoulli variable B_n with a value in the set $\{0, 1\}$ such that $\mathbb{P}\{B_n = 1\} = p(x)/p(x_{n-1})$, on the condition that $p(x) < p(x_{n-1})$. B is then used in the following acceptance/rejection mechanism:

   $$X_n = \begin{cases} X & \text{if } p(X) > p(X_{n-1}) \text{ or } B = 1 \\ X_{n-1} & \text{if } p(X) \le p(X_{n-1}) \text{ and } B = 0 \end{cases}$$

   Put simply, if X is more likely than X_{n-1}, it will always be accepted;

2. the chosen distribution $r(x|x') = r(x)$ is independent of x'. In this case, the acceptance/rejection mechanism is written:

   $$a(x|x') = \min\left(\frac{p(x)r(x_{t-1})}{p(x_{t-1})r(x)}, 1\right)$$

Exercise 3.6 (Metropolis-Hastings algorithm) (see page 272) We wish to calculate $I = \int x^2 p(x)dx$ where $p(x) \propto e^{-x^2/2\sigma^2}$. It is worth noticing that $p(x)$ is given up to an unknown multiplicative factor (even if we know that this factor is $1/\sigma\sqrt{2\pi}$). For $r(x|x')$, we select a uniform distribution over the interval $[-5\sigma, 5\sigma]$, hence $r(x_{t-1})/r(x) = 1$. Write a program implementing the Metropolis-Hastings algorithm to calculate the value of the integral I, which has a theoretical value of σ^2.

Gibbs sampler

When simulating multivariate samples, it is sometimes interesting to change single components individually. The *Gibbs sampler*, a specific case of the Metropolis-Hastings algorithm, fulfills this function. Consider a distribution with a joint probability noted $p_{X_1,\ldots,X_d}(x_1, \ldots, x_d)$; let us use the Metropolis-Hastings algorithm, where the proposition law $r(x|x')$ is the conditional distribution $p_{X_k|X_{-k}}(x_1, \ldots, x_d)$, with X_{-k} the set of variables excluding X_k. The probability of acceptance/rejection, given by expression (3.10), is therefore equal to 1 for each pair (x, x'), meaning that the sample is always accepted.

Exercise 3.7 (Gibbs sampler) (see page 272) Consider a bivariate Gaussian distribution of mean $\mu = \begin{bmatrix}\mu_1 & \mu_2\end{bmatrix}^T$ and covariance matrix:

$$C = \begin{bmatrix} \sigma_1^2 & \rho\sigma_1\sigma_2 \\ \rho\sigma_1\sigma_2 & \sigma_2^2 \end{bmatrix}$$

1. Determine the expression of the conditional distribution $p_{X_2|X_1}(x_1, x_2)$.
2. Using the Gibbs sampler, write a program which simulates a bivariate Gaussian distribution of mean $(0, 0)$ and covariance matrix C.

Gibbs sampler in a Bayesian context

The Gibbs sampler is used in a Bayesian context. Using a statistical model characterized by a probability density $p_X(x; \theta)$, we consider that $\theta \in T$ is a random vector, of which the distribution has a probability density $p_\Theta(\theta)$. Considering $p_X(x; \theta)$ as the conditional probability of X given Θ, it is possible to deduce the conditional distribution of Θ given X:

$$p_{\Theta|X}(x, \theta) = \frac{p_X(x; \theta) p_\Theta(\theta)}{\int_T p_X(x; t) p_\Theta(t) dt} \propto p_X(x; \theta) p_\Theta(\theta) \tag{3.11}$$

This expression can be used to generate samples of θ, given X, with the Gibbs sampler.

3.4 Variance reduction

3.4.1 Importance sampling

Consider the calculation of the following integral:

$$I = \int_{-\infty}^{+\infty} f(x) p(x) dx \tag{3.12}$$

With no loss of generality, we may consider that $f(x) \geq 0$. This is possible as $f(x) = f^+(x) - f^-(x)$, where $f^+(x) = \max(f(x), 0)$ and $f^-(x) = \max(-f(x), 0)$ are both positive, and to note that, by linearity, $\int f(x)dx = \int f^+(x)dx - \int f^-(x)dx$.

Direct application of the Monte-Carlo method consists of drawing N independent samples X_1, ..., X_N following distribution $p(x)$ and approximating the integral using:

$$\widehat{I}_1 = \frac{1}{N} \sum_{n=1}^{N} f(X_n) \tag{3.13}$$

Based on the hypothesis that the N variables are independent and identically distributed, the law of large numbers ensures that $\widehat{I}_1$ converges in probability toward I. As the mean is $\mathbb{E}\left\{\widehat{I}_1\right\} = \mathbb{E}\left\{f(X_1)\right\} = I$, the estimator is unbiased. Its variance is given by:

$$\text{var}\left(\widehat{I}_1\right) = \frac{1}{N} \text{var}\left(f(X_1)\right) = \frac{1}{N}\left(\int_{-\infty}^{+\infty} f^2(x) p(x) dx - I^2\right) \tag{3.14}$$

Now, consider a situation where instead of drawing N samples following the distribution associated with $p(x)$, the draw is carried out using an auxiliary distribution of density μ, known as the *instrumental* or *proposition distribution.* This technique is known as *importance sampling.* The use of an auxiliary distribution may appear surprising, but, as we shall see, improves the results of the calculation. Let us return to expression (3.12), which may be rewritten:

$$I = \int_{-\infty}^{+\infty} f(x)\frac{p(x)}{\mu(x)}\mu(x)dx \tag{3.15}$$

We may now apply the previous approach to function $f(x)p(x)/\mu(x)$ and take the following estimation of I:

$$\widehat{I}_2 = \frac{1}{N}\sum_{n=1}^{N} f(X_n)h(X_n) \tag{3.16}$$

where the values of function $h(x) = p(x)/\mu(x)$ for the values of $x = X_n$ are known as importance weights. The law of large numbers implies that $\widehat{I}_2$ converges in probability toward I. Using the fact that the random variables X_1, ..., X_N are independent and identically distributed, we may deduce the mean:

$$\mathbb{E}\left\{\widehat{I}_2\right\} = \mathbb{E}\left\{f(X_1)h(X_1)\right\} = \int_{-\infty}^{+\infty} f(x)h(x)\mu(x)dx = I$$

meaning that the estimator is unbiased and the variance:

$$\text{var}\left(\widehat{I}_2\right) = \mathbb{E}\left\{\widehat{I}_2^2\right\} - I^2 = \mathbb{E}\left\{(f(X_1)h(X_1) - I)^2\right\}$$

From (3.15), the term I^2 does not depend on the choice of μ. Consider the first term $\mathbb{E}\left\{\widehat{I}_2^2\right\}$. Based on the Schwarz inequality, we have:

$$\mathbb{E}\left\{\widehat{I}_2^2\right\} = \int f^2(x)\frac{p^2(x)}{\mu(x)}dx \times \underbrace{\int \mu(x)dx}_{=1} \geq \left(\int f(x)p(x)dx\right)^2$$

in which equality occurs if and only if $\mu(x) \propto f(x)p(x)$ and, by normalization,

$$\mu(x) = \frac{1}{\int f(u)p(u)du} \times f(x)p(x)$$

Notably, the variance of $\widehat{I}_2$ has a value of 0! However, this result is pointless, as it presumes knowledge of $\int f(u)p(u)du$, the quantity which we wish to calculate. Nevertheless, it shows that the "optimum" distribution needs to be as close as possible to the function being integrated. Specifically, if f is the indicator of the interval $(\alpha, +\infty)$, it is better to use an auxiliary law with a large number of values greater than α; this would not be the case using a centered Gaussian distribution of variance 1 for high values of α.

The importance sampling method may be modified by using the following quantity as the estimator of the integral I:

$$\widehat{I}_3 = \frac{N^{-1}\sum_{n=1}^{N} f(X_n)h(X_n)}{N^{-1}\sum_{n=1}^{N} h(X_n)} \tag{3.17}$$

The law of large numbers states that:

$$N^{-1}\sum_{n=1}^{N} h(X_n) \longrightarrow \mathbb{E}\{h(X_1)\} = 1$$

using the fact that $\mathbb{E}\{h(X_1)\} = \int \frac{p(x)}{\mu(x)}\mu(x)dx = 1$. From this, we see that $\widehat{I}_3$ tends in probability toward I. The interest of expression (3.17) in relation to expression (3.15) may therefore be questioned. The response to this query is that, in certain situations, we only know distribution p to within an unknown multiplicative coefficient λ. Let the available expression be noted $\widetilde{p}(x) = \lambda p(x)$. Placing this expression into $h(x)$, we obtain $h(x) = \widetilde{p}(x)/\lambda\mu(x)$ then, applying the result to (3.17), the unknown constant λ disappears: knowledge of this constant is therefore not required.

Introducing the normalized importance weights:

$$w(X_n) = \frac{h(X_n)}{\sum_{j=1}^{N} h(X_j)} \tag{3.18}$$

expression (3.17) may be rewritten:

$$\widehat{I}_3 = \sum_{n=1}^{N} f(X_n)w(X_n) \tag{3.19}$$

Note that, following expression (3.18), the $w(X_n)$ are positive and their sum has a value of 1. We may therefore begin by calculating $h(X_n)$ for all values of X_n, then calculate the values $w(X_n)$ by normalization.

Example 3.3 Consider a centered Gaussian random variable of variance 1. We wish to calculate the probability that this variable will be greater than a given value α. This probability is written:

$$I = \mathbb{P}\{X > \alpha\} = \int_{\alpha}^{+\infty} p(x)dx = \int_{-\infty}^{+\infty} \mathbb{1}(x > \alpha)p(x)dx \tag{3.20}$$

where $p(x) = \dfrac{1}{\sqrt{2\pi}}e^{-x^2/2}$

HINTS: The direct approach consists of drawing N independent, centered Gaussian samples of variance 1 and using the following value for the integral:

$$\widehat{I}_1 = N^{-1}\sum_{n=1}^{N} Y_n, \text{ where } Y_n = \mathbb{1}(X_n > \alpha)$$

The random variables Y_n have values in $\{0,1\}$ with $\mathbb{P}\{Y_n\} = \mathbb{P}\{\mathbb{1}(X_n > \alpha)\} = I$. Consequently, $\mathbb{E}\{Y_n\} = I$ and $\text{var}(Y_n) = I(1-I)$. Moreover, the independence of the variables X_n implies that the variables Y_n will also be independent. Hence, $\widehat{I}_1$ is a random variable of mean I and variance $N^{-1}\text{var}(Y_1) = N^{-1}I(1-I)$. Applying the central limit theorem to Y_n, we obtain:

$$\sqrt{N}(\widehat{I}_1 - I) \rightarrow \mathcal{N}(0, I(1-I))$$

Hence, we deduce a confidence interval of 95% of the value of I using this method:

$$I_{95\%} = \left[\widehat{I}_1 - 1.96\sqrt{\widehat{I}_1(\widehat{I}_1 - 1)/N}, \widehat{I}_1 + 1.96\sqrt{\widehat{I}_1(\widehat{I}_1 - 1)/N}\right]$$

Consider an approach using the importance sampling technique and the Cauchy distribution $\mathcal{C}(0,1)$ as the proposition distribution. The integral is written:

$$\widehat{I}_2 = N^{-1}\sum_{n=1}^{N} \mathbb{1}(X_n > \alpha)h(X_n)$$

$$\text{where } h(x) = \frac{\pi}{\sqrt{2\pi}}e^{-x^2/2}(1+x^2)$$

From this, we see that $\mathbb{E}\left\{\widehat{I}_2\right\} = I$ and that:

$$\text{var}\left(\widehat{I}_2\right) = \mathbb{E}\left\{\widehat{I}_2^2\right\} - I^2 = \frac{1}{2}\int_{\alpha}^{+\infty} e^{-x^2}(1+x^2)dx - I^2$$

which may be compared to the variance $\text{var}\left(\widehat{I}_1\right) = I(1-I)$.

Numerically we can show that, for large values of α, $\widehat{I}_2$ will be better than $\widehat{I}_1$.

Exercise 3.8 (Direct draws and importance sampling) (see page 273) Write a program which estimates $\mathbb{P}\{X > \alpha\}$ when X is Gaussian, centered, of variance 1. Consider three cases. In the first case, samples are produced following the Gaussian distribution, and in the second following the Cauchy distribution. In the third case, samples are produced following the Cauchy distribution, but considering that we only know that the probability density of X is proportional to $e^{-x^2/2}$.

3.4.2 Stratification

Returning to the calculation of the integral:

$$I = \int_{\mathbb{R}} f(x) p_X(x) dx \tag{3.21}$$

To reduce the variance of the estimator associated with a drawing of N i.i.d. samples, following a distribution characterized by the probability density $p_X(x)$, the stratification method involves splitting $\mathbb{R}$ into "strata", optimizing the number of points to draw in each of these strata. As we shall see, this technique requires:

- the ability to calculate $\mathbb{P}\{X \in A\}$ for any segment A,
- the ability to carry out a random draw conditional on $X \in A$.

Let $A_1, \ldots, A_S$ be a partition of $\mathbb{R}$ and

$$p_s = \mathbb{P}\{X \in A_s\} = \int_{A_s} p_X(x) dx \tag{3.22}$$

with $\sum_{s=1}^{S} p_s = 1$

We deduce the conditional probability density of X given $X \in A_s$:

$$p_{X|X \in A_s}(x) = \frac{1}{p_s} p_X(x) \mathbb{1}_{A_s}(x) \tag{3.23}$$

It follows that the integral I may be rewritten:

$$I = \sum_{s=1}^{S} p_s \int_{\mathbb{R}} f(x) p_{X|X \in A_s}(x) dx \tag{3.24}$$

Inserting (3.23), we obtain:

$$\int_{\mathbb{R}} f(x) p_X(x) dx = \sum_{s=1}^{S} p_s \int_{\mathbb{R}} f(x) \underbrace{\frac{1}{p_s} p_X(x) \mathbb{1}_{A_s}(x)}_{p_{X|X \in A_s}(x)} dx$$

Hence the idea of carrying out independent draws from the S strata under conditional distributions. Let $X_{k_s}^{(s)}$ be the N_s variables of stratum s distributed following $p_{X|X\in A_s}(x)$. An approximate value of I may be obtained using:

$$\widehat{I}_{\text{st}} = \sum_{s=1}^{S} p_s \frac{1}{N_s} \sum_{k_s=1}^{N_s} f(X_{k_s}^{(s)})$$

From this result we see that $\mathbb{E}\left\{\widehat{I}_{\text{st}}\right\} = I$, and estimator $\widehat{I}_{\text{st}}$ is therefore unbiased. Let us now determine the variance. Using the hypothesis of independent drawings in the S strata, we have:

$$\operatorname{var}\left(\widehat{I}_{\text{st}}\right) = \sum_{s=1}^{S} \frac{p_s^2}{N_s^2} \sum_{k_s}^{N_s} \operatorname{var}\left(f(X)|X \in A_s\right) = \sum_{s=1}^{S} \frac{1}{N_s} p_s^2 \sigma_s^2 \tag{3.25}$$

where

$$\begin{aligned}\sigma_s^2 &= \operatorname{var}\left(f(X)|X \in A_s\right) = \int_{\mathbb{R}} (f(x) - I_s)^2 p_{X|X\in A_s}(x)dx \\ &= \frac{1}{p_s} \int_{A_s} f^2(x) p_X(x)dx - I_s^2\end{aligned} \tag{3.26}$$

It is interesting to note that, in expression (3.24), the integral

$$\int_{\mathbb{R}} f(x) p_{X|X\in A_s}(x)dx \tag{3.27}$$

is interpreted as the conditional expectation of $f(X)$ given $Y_s = \mathbb{1}(X \in A_s)$. Following property 5 of the properties 1.9, $\operatorname{var}\left(\mathbb{E}\left\{f(X)|Y_s\right\}\right) \le \operatorname{var}\left(f(X)\right)$ and therefore $\operatorname{var}\left(\widehat{I}\right) \le \operatorname{var}\left(\widehat{I}_1\right)$ where $\widehat{I}_1$ is given by (3.13), i.e. an approximation of I using a single stratum.

It is now possible to minimize (3.25) under the constraint $N = \sum_{s=1}^{S} N_s$. Canceling the derivative of the Lagrangian $\mathscr{L}(N_s) = \sum_{s=1}^{S} N_s^{-1} p_s^2 \sigma_s^2 + \lambda(N - \sum_{s=1}^{S} N_s)$, we obtain:

$$N_s = \left\lfloor N \frac{p_s \sigma_s}{\sum_{s=1}^{S} p_s \sigma_s} \right\rfloor$$

This solution is unusable, as it requires knowledge of σ_s and thus of the value of the integral I. One solution (not ideal) is to take $N_s = \lfloor N p_s \rfloor$, giving a variance

$$\operatorname{var}\left(\widehat{I}_{\text{st}}\right) = \frac{1}{N} \sum_{s=1}^{S} p_s \sigma_s^2$$

To use this method, we now need to carry out a random draw following the conditional distribution $p_{X|X\in A_s}(x)$. One simple approach would be to carry out a random draw using the distribution $p_X(x)$, then to conserve only those values situated within the interval A_s. The problem with this approach is that the number of values to draw is random. This drawback may be overcome if (i) we use the property of inversion of the cumulative function (see section 3.3.1), which uses the uniform distribution over $(0,1)$, and if (ii) we note that for a uniform random variable U, the conditional distribution of U given that U belongs to an interval B is itself uniform, written $p_{U|U\in B}(u) = \mathbb{1}(u \in B)/\ell$, where ℓ denotes the length of B.

In summary, to calculate (3.21):

- we choose S intervals $(a_0, a_1]$, $]a_1, a_2]$, ..., $]a_{S-1}, a_S)$, where $a_0 = -\infty$ and $a_S = +\infty$;
- we calculate the S integer values

$$\begin{aligned} N_1 &= \lfloor N p_1 \rfloor, \\ &\vdots \\ N_s &= \left\lfloor N \textstyle\sum_{j=1}^{s} p_j \right\rfloor - \left\lfloor N \textstyle\sum_{j=1}^{s-1} p_j \right\rfloor, \\ &\vdots \\ N_S &= \left\lfloor N \textstyle\sum_{j=1}^{S} p_j \right\rfloor - \left\lfloor N \textstyle\sum_{j=1}^{S-1} p_j \right\rfloor \end{aligned}$$

  where $p_s = F_X(a_s) - F_X(a_{s-1})$ and $F_X(x)$ denotes the cumulative function associated with $p_X(x)$;
- we calculate $[b_0, b_1]$, $]b_1, b_2]$, ..., $]b_{S-1}, b_S]$ where $b_s = F_X(a_s)$ with $b_0 = 0$ and $b_S = 1$;
- for each value of s, we draw N_s values U_1^s, ..., $U_{N_s}^s$, independently and following a uniform distribution $\mathcal{U}(b_{s-1}, b_s)/(b_s - b_{s-1})$;
- for each value of s, we calculate $X_1^{(s)} = F_X^{[-1]}(U_1^s), \ldots, X_{N_s}^{(s)} = F_X^{[-1]}(U_{N_s}^s)$;
- we calculate

$$\widehat{I} = \sum_{s=1}^{S} p_s \frac{1}{N_s} \sum_{k_s=1}^{N_s} f(X_{k_s}^{(s)})$$

Exercise 3.9 (Stratification) (see page 275) In this exercise, we shall use a Monte-Carlo method with and without stratification to compare the calculation of the following integral:

$$I = \int_{-\infty}^{\infty} \cos(ux) \frac{1}{\sqrt{2\pi}} e^{-x^2/2} dx$$

1. determine the analytical expression of I (use the characteristic function $\mathbb{E}\left\{e^{juX}\right\}$);
2. write a program which compares the calculation with and without stratification.

3.4.3 Antithetic variates

We wish to estimate the integral:

$$I = \int_{\mathbb{R}} f(x) p_X(x) dx \tag{3.28}$$

The *antithetic variates method* consists of finding a pair of random variables $(X, \widetilde{X})$ such that functions $f(X)$ and $f(\widetilde{X})$ have the same expectation, the same variance, and satisfy the condition:

$$\operatorname{cov}\left(f(X), f(\widetilde{X})\right) < 0 \tag{3.29}$$

Using an i.i.d. series of random variables X_n distributed following the law $p_X(x)$, we take the following estimator of I:

$$\widehat{I}_a = \frac{1}{N} \sum_{n=1}^{N/2} \left(f(X_n) + f(\widetilde{X}_n)\right) \tag{3.30}$$

Hence

$$\begin{aligned} \operatorname{var}\left(\widehat{I}_a\right) &= \frac{1}{N}\left(\operatorname{var}\left(f(X)\right) + \operatorname{cov}\left(f(X), f(\widetilde{X})\right)\right) \\ &< \frac{1}{N} \operatorname{var}\left(f(X)\right) \end{aligned} \tag{3.31}$$

Exercise 3.10 (Antithetic variates approach) (see page 276) We wish to calculate the integral:

$$I = \int_0^1 \frac{1}{1+x} dx = \log(2)$$

Integral I may be seen as the expectation, following the uniform distribution $\mathcal{U}(0,1)$ of $f(X) = 1/(1+X)$. Take $\widetilde{X} = 1 - X$. Hence, if $X \sim \mathcal{U}(0,1)$, then $\widetilde{X} \sim \mathcal{U}(0,1)$.

Let $\widehat{I}$ and $\widehat{I}_a$ be the estimators given by (3.12) and (3.30) respectively:

1. determine the values of $\operatorname{var}\left(f(X)\right)$; $\operatorname{cov}\left(f(X), f(\widetilde{X})\right)$; $\operatorname{var}\left(\widehat{I}\right)$; and $\operatorname{var}\left(\widehat{I}_a\right)$ for $N = 100$;
2. write a program to compare the direct calculation with the antithetic variate method.

Chapter 4

Second Order Stationary Process

In this chapter, we shall consider *second order wide sense stationary* (WSS) processes. At first, we shall begin with the estimation of the mean and covariance function, before introducing the notion of linear prediction. We will then consider non-parametric spectral estimation.

4.1 Statistics for empirical correlation

In this section, we shall present results relating to the estimation of the mean, of the covariance function and of the correlation function of a WSS random process. A number of properties of WSS processes are presented hereafter; for further details, see [2].

Second order stationarity

A second order WSS random process is a sequence of random variables X_n defined on the same probability space and such that:

- $\mathbb{E}\{X_n\} = \mu \in \mathbb{C}$ is independent of n,
- $\mathbb{E}\{X_n X_n^*\} < +\infty$
- $\mathbb{E}\{(X_{n+h} - \mu)(X_n - \mu)^*\} = R(h)$ is independent of n.

The sequence $R(h)$ is known as the (auto-)covariance function.

It follows that $R(0) \geq 0$ and $R(h) = R^*(-h)$. In the following, we shall assume the existence[1] of the function:

$$\gamma(f) = \sum_{h=-\infty}^{+\infty} R(h) e^{-2j\pi h f} \tag{4.1}$$

$\gamma(f)$ is known as the spectral density, and is, by construction, a periodic function of period 1. We shall use the following fundamental result:

$$\forall f, \quad \gamma(f) \geq 0 \tag{4.2}$$

Remember that if X_n is real, then $\gamma(f)$ is even, and limiting its representation to the interval $(0, 1/2)$ is sufficient.

The series

$$\rho(h) \;=\; \frac{R(h)}{R(0)} \tag{4.3}$$

is known as the (auto-)correlation function, and verifies $|\rho(h)| \leq 1$.

Empirical moments

Beginning with N successive observations, $X_1, \ldots, X_N$, of a WSS process, we are able to define the *empirical mean*:

$$\widehat{\mu}_N = \frac{1}{N} \sum_{n=1}^{N} X_n \tag{4.4}$$

the empirical covariance function:

$$\widehat{R}_N(h) = \begin{cases} \dfrac{1}{N} \displaystyle\sum_{n=1}^{N-|h|} (X_{n+|h|} - \widehat{\mu}_N)(X_n - \widehat{\mu}_N) & \text{if } \; |h| \leq N-1 \\ 0 \quad \text{if not} \end{cases} \tag{4.5}$$

and the empirical correlation function:

$$\widehat{\rho}_N(h) = \frac{\widehat{R}_N(h)}{\widehat{R}_N(0)} \tag{4.6}$$

Hence $|\widehat{\rho}_N(h)| \leq 1$. The following conditions may be demonstrated under very general conditions [4]:

- when N tends toward infinity, the series μ_N tends in probability toward μ,

[1]This condition is not particularly restrictive; for example, it is sufficient for $R(h)$ to tend toward 0 more quickly than $1/h$ when h tends toward infinity.

- the sequence $\widehat{R}_N(h)$ is a sequence of covariances (meaning that it is positive semi-definite) and, when N goes to infinity, it tends in probability to $R(h)$,

- if $\widehat{R}_N(0) \neq 0$, then, for any P, the Toëplitz covariance matrix constructed using $\widehat{R}_N(0), \ldots, \widehat{R}_N(P)$ is positive and invertible,

- for any value of P, it exists an AR-P process whose $(P+1)$ first values of the covariance function are $\widehat{R}_N(0), \ldots, \widehat{R}_N(P)$,

- using $\widehat{R}_N(h)$, the solution to the Yule-Walker equations (4.19) produces a polynomial with roots which are strictly inside the unit circle [4].

A sequence of independent, centered random variables of the same variance σ^2 is known as strong sense white noise. Weak sense white noise, on the other hand, is a sequence of uncorrelated, centered random variables of the same variance σ^2.

Consider the process X_n constructed by linear filtering of the sequence $\{Z_n\}$ following expression $X_n = \sum_h g_{n-h} Z_h$, where Z_n is a strong sense white noise and where $\sum_h |g_h| < +\infty$.

Let $\boldsymbol{\rho} = \begin{bmatrix}\rho(1) & \ldots & \rho(k)\end{bmatrix}^T$ and $\widehat{\boldsymbol{\rho}}_N = \begin{bmatrix}\widehat{\rho}_N(1) & \ldots & \widehat{\rho}_N(k)\end{bmatrix}^T$. In [4] it is shown that:

$$\sqrt{N}(\widehat{\boldsymbol{\rho}}_N - \boldsymbol{\rho}) \xrightarrow{d} \mathcal{N}(0, \boldsymbol{W}) \tag{4.7}$$

where the matrix $\boldsymbol{W}$ of dimension k has a generating element $W_{p,q}$:

$$\begin{aligned} W_{p,q} = \sum_{u=-\infty}^{\infty} & \{\rho(u+p) + \rho(u-p) - 2\rho(u)\rho(p)\} \\ & \times \{\rho(u+q) + \rho(u-q) - 2\rho(u)\rho(q)\} \end{aligned} \tag{4.8}$$

In the specific case where $X_n = Z_n$, expression (4.8) may be simplified and written:

$$\sqrt{N}\,\widehat{\boldsymbol{\rho}}_N \xrightarrow{d} \mathcal{N}(0, \boldsymbol{I}_k) \tag{4.9}$$

Hence, the Box-Pierce statistic

$$Q_{BP} = N \sum_{h=1}^{k} \widehat{\rho}_N^2(h)$$

follows, based on the hypothesis that the process is white, a χ_k^2 distribution with k degrees of freedom (exercise 4.1). The Ljung-Box statistic is sometimes preferred:

$$Q_{LB} = N(N+2) \sum_{h=1}^{k} \frac{1}{N-h} \widehat{\rho}_N^2(h)$$

Under the hypothesis that the process is white, this statistic also follows a χ_k^2 distribution with k degrees of freedom. These "portmanteau statistics" are widely used to test the whiteness hypothesis. The fact that the distribution of these two statistics is known, under the whiteness hypothesis, allows us to calculate a p-value (section 2.2.2) for the process whiteness test (exercise 4.2).

Example 4.1 (Whiteness test) Write a program which produces L draws of N samples of a strong sense white noise, which calculates the K first empirical correlations, which displays a histogram of values and which compares this histogram to the χ^2 distribution with k degrees of freedom (function `chi2pdf` in the `stats` toolbox).

> HINTS: Type the program shown below. The Box-Pierce statistic may also be replaced by the Ljung-Box statistic.

```
%===== BoxPierceTest.m
% Uses chi2pdf from the stats toolbox
clear all
L=10000; % number of samples
N=200;   % sample size
K=5;     % number of corr. coeff.
X=randn(N,L);
Xc = X - ones(N,1)*mean(X,1); % centered process
corrX = zeros(K,L);
for ir=1:L
    cov0 = Xc(:,ir)'*Xc(:,ir);
    for ik=1:K
        corrX(ik,ir) = (Xc(1:N-ik,ir)'*Xc(ik+1:N,ir)) / cov0;
    end
end
[BP,xbin] = hist(N*ones(1,K)*(corrX .^2),50);
distrBP = BP/(xbin(2)-xbin(1))/L;
bar(xbin,distrBP)
pdfchi2 = chi2pdf(linspace(0,max(xbin),20),K);
hold on, plot(linspace(0,max(xbin),20),pdfchi2,'r'), hold off
```

Example 4.2 (AR test on sunspots) Consider the file `sunspots.mat`. The sunspots are taken to be the result of an AR-2 process. Using the `lpc.m` function in MATLAB®, estimate the prediction residual. Write a program to test the hypothesis that this residual is white. Give the p-value.

> HINTS: in program `whitenesstest`, we have also drawn a confidence interval at 95% of the correlation components following (4.9). This test leads to the conclusion that the residual is white, and therefore that the AR modeling the sunspots is of order 2.

```
%===== whitenesstest.m
```

```
clear all
load sunspot.dat
x  = sunspot(:,2); N = length(x);
xc = x-mean(x);
[a,sigma] = lpc(xc,2);
residue = filter(a,1,xc);
covresidue = xcov(residue);
covresidue = covresidue(N:2*N-1);
corrresidue = covresidue/covresidue(1);
BP=N*sum(corrresidue(2:N) .^ 2);
pvalue = 1-chi2cdf(BP,N-1); % stats toolbox
stem((2:N),corrresidue(2:N))
hold on, plot([2,N],1.96*[-1 1;-1 1]/sqrt(N),':'), hold off
title(sprintf('p-value = %5.2f',pvalue),'fonts',14)
```

If `lpc.m` is not available, the function `xtoa` may be used (see [2, 3]):

```
function [a,sigma2]=xtoa(x,P)
%!========================================!
%! SYNOPSIS: [a,sigma2]=XTOA(x,P)          !
%! x      = signal Coefficients array      !
%! P      = model order                    !
%! a      = [1 a_1 a_2 ... a_P]            !
%! sigma2 = power of the input white noise !
%!========================================!
N=length(x); x=x(:); x=x-mean(x);
for k=1:P+1
    %===== biased estimate of R^*(k)
    rconj(k)=x(k:N)' * x(1:N-k+1) / N;
end
Rc=toeplitz(rconj); vaux=Rc \ eye(P+1,1);
a=vaux / vaux(1); sigma2=1 / vaux(1);
```

4.2 Linear prediction of WSS processes

The linear prediction of a random process consists of calculating an estimation $\widehat{X}_n$ of the value X_n of the process at instant n based on its immediate past of length N. The prediction error $X_N - \widehat{X}_n$ therefore represents anything which is new at instant n in relation to that which has already happened. Linear prediction is used in a wide variety of applications, such as compression, noise elimination, signal analysis and synthesis. The solution is given by the Yule-Walker equations.

4.2.1 Yule-Walker equations

Consider a WSS random process X_n with mean μ and covariance function $R(h) = \mathbb{E}\{(X_{n+h} - \mu)(X_n - \mu)^*\}$. Remember that $R(h) = R^*(-h)$. Specifically, if X_n has real values, $R(h)$ is real and $R(h) = R(-h)$. The general form of the linear prediction (affine) of X_n based on the N last previous values takes the form:

$$\widehat{X}_n = \gamma_{0,N} \times 1 + \gamma_{1,N} X_{n-1} + \cdots + \gamma_{N,N} X_{n-N} \tag{4.10}$$

The prediction problem is to determine the expression of the sequence γ_k from μ and $R(h)$, while minimizing the quadratic error. Expression (4.10) may also be written:

$$\widehat{X}_n = \alpha_{0,N} + \alpha_{1,N} X^c_{n-1} + \cdots + \alpha_{N,N} X^c_{n-N} \tag{4.11}$$

$$\text{where } X^c_n = X_n - \mu, \tag{4.12}$$

A priori, the coefficients $\alpha_{k,N}$ might be thought to depend on n, but this is not the case. Let:

$$\widehat{X}^c_n = \alpha_{1,N} X^c_{n-1} + \cdots + \alpha_{N,N} X^c_{n-N} \tag{4.13}$$

thus $\widehat{X}_n = \alpha_{0,N} + \widehat{X}^c_n$.

Theorem 1.2 states that the coefficients $\alpha_{0,N}, \ldots, \alpha_{N,N}$ that minimize the mean square error $\mathbb{E}\left\{|X_n - \widehat{X}_n|^2\right\}$ between the actual value X_n and the predicted value $\widehat{X}_n$ are such that the error or residue:

$$e_{n,N} = X_n - \widehat{X}_n = X_n - \alpha_{0,N} - \sum_{k=1}^{N} \alpha_{k,N} X^c_{n-k} \tag{4.14}$$

is orthogonal to 1 and to any X^c_{n-k} with $1 \leq k \leq N$. This orthogonality is written:

$$\mathbb{E}\left\{(X_n - \widehat{X}_n) \times 1\right\} = 0 \quad \Leftrightarrow \quad \alpha_{0,N} = \mu \tag{4.15}$$

Applying (4.15) to (4.13), we obtain $e_{n,N} = (X_n - \mu) - (\widehat{X}_n - \mu) = X^c_n - \widehat{X}^c_n$ and the orthogonality of $e_{n,N}$ with X^c_{n-k} is therefore written:

$$\mathbb{E}\left\{(X^c_n - \widehat{X}^c_n) X^{c*}_{n-k}\right\} = 0 \text{ for } 1 \leq k \leq N \tag{4.16}$$

Using the expectation's linearity, we get:

$$\mathbb{E}\left\{X^c_n X^{c*}_{n-k}\right\} - \mathbb{E}\left\{\widehat{X}^c_n X^{c*}_{n-k}\right\} = 0$$

By replacing $\widehat{X}^c_n$ by its expression (4.13), then by again using the expectation's linearity, we get:

$$\mathbb{E}\left\{X^c_n X^{c*}_{n-k}\right\} - \sum_{i=1}^{N} \alpha_{i,N} \mathbb{E}\left\{X^c_{n-i} X^{c*}_{n-k}\right\} = 0$$

Because the process is stationary, i.e. $\mathbb{E}\left\{X_n^c X_{n-k}^{c*}\right\} = R(k)$, we have:

$$R(k) - \sum_{i=1}^{N} \alpha_{i,N} R(k-i) = 0 \text{ for } 1 \leq k \leq N \tag{4.17}$$

The minimum mean square error is given by 1.29 which here has the expression:

$$\varepsilon_N^2 = \mathbb{E}\left\{|e_{n,N}|^2\right\} = \mathbb{E}\left\{(X_n^c - \widehat{X}_n^c)X_n^{c*}\right\} = R(0) - \sum_{i=1}^{N} \alpha_{i,N} R(-i) \tag{4.18}$$

A fundamental consequence of this is that, for a fixed N, neither the coefficients $\alpha_{k,N}$ nor the prediction error ε_N^2 are dependent on n.

Remark: the previous results show that no loss of generality results from supposing that $\mu = 0$. The prediction coefficients of $\widehat{X}_n^c$ are calculated using the covariance function $R(h)$ alone, then in order to obtain $\widehat{X}_n$ we use expressions (4.11) and (4.12), which use the mean μ.

Stacking the equation (4.18) and the N equations (4.17) in matrix form leads us to:

$$\begin{cases} R(0) - \sum_{k=1}^{N} \alpha_{k,N} R^*(k) = \varepsilon_N^2 \\ \Gamma_{N-1} \begin{bmatrix} \alpha_{1,N} \\ \vdots \\ \alpha_{N,N} \end{bmatrix} = \begin{bmatrix} R(1) \\ \vdots \\ R(N) \end{bmatrix} \end{cases} \tag{4.19}$$

where

$$\Gamma_{N-1} = \begin{bmatrix} R(0) & R^*(1) & \cdots & \cdots & R^*(N-1) \\ R(1) & R(0) & \ddots & \cdots & R^*(N-1) \\ \vdots & \ddots & \ddots & & \vdots \\ \vdots & & & & R^*(1) \\ R(N-1) & \cdots & \cdots & R(1) & R(0) \end{bmatrix} \tag{4.20}$$

These equations are called the *Yule-Walker equations*, or the *normal equations*. They allow us to calculate the prediction coefficients and the minimum prediction mean-square error using the covariance function of a WSS process. In practice, when μ and $R(h)$ are unknown, they are estimated using (4.4) and (4.5).

In the remainder of this section, all processes will be taken to be real, and the conjugation will therefore be omitted.

Example 4.3 (Prediction of an MA-1) Let $X_n = Z_n + b_1 Z_{n-1}$, where Z_n is a white process of variance σ^2. Let $R(h)$ be the autocovariance function of X_n. The process may be easily shown to be centered, and $R(0) = (1+b_1^2)\sigma^2$, $R(1) = R(-1) = b_1\sigma^2$ and $R(h) = 0$ for $|h| \geq 2$. Determine the prediction of X_n as a function of X_{n-1} and X_{n-2}.

HINTS: we obtain $\widehat{X}_n = \alpha_1 X_{n-1} + \alpha_2 X_{n-2}$ where, following (4.19), α_1 and α_2 are given by:

$$\begin{bmatrix} 1+b_1^2 & b_1 \\ b_1 & 1+b_1^2 \end{bmatrix} \begin{bmatrix} \alpha_1 \\ \alpha_2 \end{bmatrix} = \begin{bmatrix} b_1 \\ 0 \end{bmatrix}$$

and $\varepsilon^2 = (1 + b_1^2 - \alpha_1 b_1)\sigma^2$. Solving the equation, we obtain:

$$\begin{aligned} \begin{bmatrix} \alpha_1 \\ \alpha_2 \end{bmatrix} &= \frac{1}{(1+b_1^2)^2 - b_1^2} \begin{bmatrix} 1+b_1^2 & -b_1 \\ -b_1 & 1+b_1^2 \end{bmatrix} \begin{bmatrix} b_1 \\ 0 \end{bmatrix} \\ &= \frac{b_1}{(1+b_1^2)^2 - b_1^2} \begin{bmatrix} 1+b_1^2 \\ -b_1 \end{bmatrix} \end{aligned}$$

Theorem 4.1 (Predicting a causal AR-P process) *Let X_n be a causal AR-P process defined by equation:*

$$X_n + a_1 X_{n-1} + \cdots + a_P X_{n-P} = Z_n \tag{4.21}$$

with $A(z) = 1 + \sum_{k=1}^{m} a_k z^{-k} \neq 0$ for $|z| \geq 1$ and where Z_n is a WSS white process with the variance σ^2. Then, for any $N \geq P$, the prediction coefficients to the N-th order are given by:

$$\alpha_{k,N} = \begin{cases} -a_k & \textit{pour} \quad 1 \leq k \leq P \\ 0 & \textit{pour} \quad P < k \leq N \end{cases} \tag{4.22}$$

and the minimum prediction mean square error is equal to $\varepsilon_N^2 = \sigma^2$.

HINTS: X_n can be shown to be centered. Note [2] that, if $A(z) \neq 0$ for $|z| \geq 1$, then X_n is expressed causally as a function of Z_n, i.e. X_n only depends on the $Z_{n-\ell}$, with $\ell \geq 0$. As Z_n is white, Z_n and X_{n-h} are uncorrelated for any $h \geq 1$, i.e. $\mathbb{E}\{Z_n X_{n-h}\} = 0$ for $h \geq 1$. This is written $\mathbb{E}\left\{\left(X_n + \sum_{k=1}^{P} a_k X_{n-k}\right) X_{n-h}\right\} = 0$. This result implies that (i) the variable $(X_n + \sum_{k=1}^{P} a_k X_{n-k})$ is orthogonal to the sub-space $\mathcal{E}_{n-1,m}$ spanned by the variables $X_{n-1}, \ldots, X_{n-m}$ for any $m \geq 1$, and (ii) $\sum_{k=1}^{P} a_k X_{n-k}$ belongs to $\mathcal{E}_{n-1,m}$ for any $m \geq P$. Consequently, based on the projection theorem, $-\sum_{k=1}^{P} a_k X_{n-k}$ is the orthogonal projection X_n onto the sub-space $\mathcal{E}_{n-1,m}$, which demonstrates (4.22). In these conditions, the prediction error is Z_n and thus $\varepsilon_N^2 = \sigma^2$.

To conclude, the orthogonal projection of a causal AR-P causal onto its immediate past $N \geq P$ coincides with the orthogonal projection onto its past of length P, and the prediction coefficients are the coefficients of the recursive equation (4.21), changing the signs.

Note that the coefficients a_k of an AR model, equation (4.21), are linked in a linear manner to the covariances. This property is not valid for MA processes. The estimation of the coefficients of a model based on the estimated covariances is therefore simpler for an AR model than for an MA model.

Theorem 4.2 (Choleski) *Let X_n be a WSS process, with zero-mean, and covariance function $R(h)$. Then Γ_{N+1} defined by (4.20) can be written:*

$$\Gamma_{N+1} = A_{N+1}^{-1} D_{N+1} A_{N+1}^{-T} \tag{4.23}$$

where:

$$A_{N+1} = \begin{bmatrix} 1 & 0 & \cdots & \cdots & 0 \\ -\alpha_{1,1} & 1 & \ddots & & \vdots \\ \vdots & & \ddots & \ddots & \vdots \\ \vdots & & & \ddots & 0 \\ -\alpha_{N,N} & -\alpha_{N-1,N} & \cdots & -\alpha_{1,N} & 1 \end{bmatrix} \quad D_{N+1} = \begin{bmatrix} \varepsilon_0^2 & 0 & \cdots & 0 \\ 0 & \varepsilon_1^2 & \cdots & 0 \\ \vdots & & & \vdots \\ 0 & & \cdots & \varepsilon_N^2 \end{bmatrix}$$

where $\alpha_{i,j}$ is the i-th prediction coefficient of length j, ε_j^2 is the norm of the associated error and $\varepsilon_0^2 = R(0)$.

The following theorem will be admitted (4.3) [4].

Theorem 4.3 *If X_n is a WSS process such that ε_N does not tend toward 0 when N tends toward infinity, then for any N, the polynomial $A(z) = 1 - \sum_{k=1}^{N} \alpha_{k,N} z^{-k} \neq 0$ for $|z| \geq 1$.*

One direct consequence of theorem 4.3 is that any covariance sequence $R(h)$ may be associated, for any N, with a causal AR process of order N, of which the first $(N+1)$ coefficients are $R(0)$, …, $R(N)$ exactly.

The empirical covariance $\widehat{R}(h)$, given by expression (4.5), can easily be shown to be a sequence of covariances (see [2], property). Based on theorem 4.3 we see that, for any value of N, the coefficients solving the Yule-Walker equation associated with the sequence $\widehat{R}(0)$, …, $\widehat{R}(N)$ define a stable causal filter.

4.2.2 Levinson-Durbin algorithm

Let $\mathcal{E}_{n-1,N}$ be the sub-space spanned by the variables X_{n-1}, …, X_{n-N}. The Levinson-Durbin algorithm makes use of the specific geometric structure of second order stationary processes to establish a recursive formula, giving prediction coefficients for order $(N+1)$ based on the prediction coefficients obtained for order N. It is written:

Data: $R(h)$ for $h = 0$ to N
Result: $\alpha_{m,p}$, ε_p^2, k_p for $p = 1$ to N and $m = 1$ to p
Initialization:
$k_0 = 0$, $\alpha_{1,0} = 0$ and $\varepsilon_0^2 = R(0)$
for $p = 1$ *to* N **do**
  $k_p = \varepsilon_{p-1}^{-2}\left(R(p) - \sum_{s=1}^{p-1} \alpha_{s,p-1} R(p-s)\right)$;
  $\alpha_{p,p} = k_p$;
  for $m = 1$ *to* $p-1$ **do**
    $\alpha_{m,p} = \alpha_{m,p-1} - k_p \alpha_{p-m,p-1}$;
  end
  $\varepsilon_p^2 = \varepsilon_{p-1}^2(1 - k_p^2)$;
end

Algorithm 2: The Levinson algorithm

In practical terms, the Levinson algorithm is used to obtain an estimation of prediction and reflection coefficients using the estimation of covariance coefficients obtained from expression (4.5).

HINTS: consider a real, centered WSS process X_n with a covariance function $R(h)$. Let $\mathcal{E}_{n-1,N}$ be the space generated by $X_{n-1}, \ldots, X_{n-N}$. The forward orthogonal projection of X_n onto $\mathcal{E}_{n-1,N}$ is noted:

$$\widehat{X}_{n,N}^F = \sum_{k=1}^{N} \alpha_{k,N} X_{n-k} \tag{4.24}$$

and the `backward orthogonal projection` of X_{n-N-1} onto $\mathcal{E}_{n-1,N}$ is:

$$\widehat{X}_{n-N-1,N}^B = \sum_{k=1}^{N} \beta_{k,N} X_{n-k} \tag{4.25}$$

The two prediction errors, forward and backward, are respectively noted:

$$\epsilon_{n,N}^F = X_n - \widehat{X}_{n,N}^F \tag{4.26}$$

$$\epsilon_{n-N-1,N}^B = X_{n-N-1} - \widehat{X}_{n-N-1,N}^B \tag{4.27}$$

A geometric illustration is shown in Figure 4.1.

Applying the projection theorem in a similar way to that used in obtaining equation system (4.19), we obtain the following equation system for the

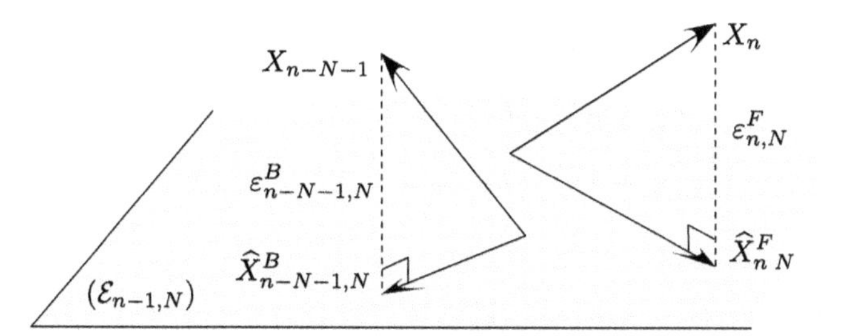

Figure 4.1 – *Forward/backward predictions: $\mathcal{E}_{n-1,N}$ is the linear space spanned by X_{n-1}, ..., X_{n-N}. The dotted arrows represent the forward prediction error $\epsilon^F_{n,N}$ and the backward prediction error $\epsilon^B_{n-N-1,N}$. The two errors are orthogonal to $\widehat{X}^F_{n,N}$ and to $\widehat{X}^B_{n-N-1,N}$.*

backward projection:

$$\left\{\begin{array}{l} R(0) - \sum_{k=1}^{N} \beta_{k,N} R(k) = \|\epsilon^B_{n-N-1,N}\|^2 \\ \Gamma_{N-1} \begin{bmatrix} \beta_{1,N} \\ \vdots \\ \beta_{N,N} \end{bmatrix} = \begin{bmatrix} R(1) \\ \vdots \\ R(N) \end{bmatrix} \end{array}\right. \tag{4.28}$$

Comparing equations (4.26) to (4.19), we deduce that $\alpha_{k,N} = \beta_{k,N}$ and that $\|\epsilon^F_{n,N}\|^2 = \|\epsilon^B_{n-N-1,N}\|^2$. Moreover, process $\epsilon^B_{n-N-1,N} = X_{n-N-1} - \widehat{X}^B_{n-N-1,N}$ appears as the filtering process for a WSS process, and is therefore itself WSS; hence, its norm is not dependent on n. This is written:

$$\forall n, n' \quad \varepsilon^2_N = \|\epsilon^F_{n,N}\|^2 = \|\epsilon^B_{n',N}\|^2$$

Replacing $\beta_{k,N}$ by $\alpha_{k,N}$ in(4.25), then using $k = N + 1 - h$, we obtain:

$$\widehat{X}^B_{n-N-1,N} = \sum_{k=1}^{N} \alpha_{k,N} X_{n-k} = \sum_{h=1}^{N} \alpha_{N+1-h,N} X_{n-N-1+h}$$

Assembling the backward and forward predictions, we have:

$$\left\{\begin{array}{lcl} \widehat{X}^F_{n,N} & = & \sum_{k=1}^{N} \alpha_{k,N} X_{n-k} \\ \widehat{X}^B_{n-N-1,N} & = & \sum_{h=1}^{N} \alpha_{N+1-h,N} X_{n-N-1+h} \end{array}\right. \tag{4.29}$$

We now wish to express the projection of X_n for order $N + 1$ using the projection of X_n for order N. Using equation (1.31) based on the direct sum

$\mathcal{E}_{n-1,N+1} = \mathcal{E}_{n-1,N} \oplus \epsilon^B_{n-N-1,N}$ (see Figure 4.1), we obtain:

$$\begin{aligned} \widehat{X}^F_{n,N+1} &= \widehat{X}^F_{n,N} + k_{N+1}\epsilon^B_{n-N-1,N} \\ &= \widehat{X}^F_{n,N} + k_{N+1}(X_{n-N-1} - \widehat{X}^B_{n-N-1,N}) \end{aligned} \tag{4.30}$$

with

$$k_{N+1} = \frac{(X_n, \epsilon^B_{n-N-1,N})}{\varepsilon^2_N} = \frac{(\epsilon^F_{n,N}, \epsilon^B_{n-N-1,N})}{\|\epsilon^F_{n,N}\|\|\epsilon^B_{n-N-1,N}\|} \tag{4.31}$$

where we make use of the fact that $X^F_{n,N} \perp \epsilon^B_{n-N-1,N}$. Following the Schwarz inequality, $|k_{N+1}| \leq 1$. Based on (4.31) and using $\epsilon^B_{n-N-1,N} = X_{n-N-1} - \widehat{X}^B_{n-N-1,N}$, we obtain:

$$k_{N+1} = \frac{R(N+1) - \sum_{k=1}^{N} \alpha_{k,N} R(n+1-k)}{\varepsilon^2_N} \tag{4.32}$$

Inserting the second equation of (4.30) into (4.29) we obtain:

$$\widehat{X}^F_{n,N+1} = \sum_{k=1}^{N} (\alpha_{k,N} - k_{N+1}\alpha_{N+1-k,N}) X_{n-k} + k_{p+1} X_{n-N-1}$$

which, by identification with the first equation of (4.29), gives us:

$$\left\{ \begin{array}{rcl} \alpha_{k,N+1} &=& \alpha_{k,N} - k_{N+1}\alpha_{N+1-k,N} \\ \alpha_{N+1,N+1} &=& k_{n+1} \end{array} \right. \tag{4.33}$$

Using (4.30), we obtain:

$$\begin{aligned} \epsilon^F_{n,N+1} &= X_n - \widehat{X}^F_{n,N+1} = (X_n - \widehat{X}^F_{n,N}) - k_{N+1}\epsilon^B_{n-N-1,N} \\ &= \epsilon^F_{n,N} - k_{N+1}\epsilon^B_{n-N-1,N} \end{aligned} \tag{4.34}$$

Taking the norm of the two members, we deduce that:

$$\varepsilon^2_{N+1} = \varepsilon^2_N + k^2_{N+1}\varepsilon^2_N - 2k_{N+1}(\epsilon^F_{n,N}, \epsilon^B_{n-N-1,N})$$

Using (4.31), we obtain:

$$\varepsilon^2_{N+1} = \varepsilon^2_N(1 - k^2_{N+1}) \tag{4.35}$$

Bringing together (4.32), (4.33) and (4.35), we arrive at the Levinson algorithm.

The sequence of coefficients k_N given by equation (4.32) is equal to the partial autocorrelation function defined by:

	correlation	partial correlation
Causal AR-P	evanescent	null from $(P+1)$
Invertible MA-Q	null from $Q+1$	evanescent
Causal and invertible ARMA-(P,Q)	evanescent	evanescent

Table 4.1 – Correlation vs. partial correlation

Definition 4.1 (Partial autocorrelation function) *Let X_n be a WSS random process with covariance function $R(h)$. The partial autocorrelation function is the sequence defined by:*

$$k_i = \begin{cases} \operatorname{Corr}(X_n, X_{n-1}) = \dfrac{(X_n, X_{n-1})}{\|X_n\| \, \|X_{n-1}\|} & \textit{for} \quad i = 1 \\ \operatorname{Corr}(\epsilon^F_{n,i-1}, \epsilon^B_{n-i,i-1}) = \\ \qquad \dfrac{(X_n - \widehat{X}^F_{n,i-1}, X_{n-i} - \widehat{X}^B_{n-i})}{\|X_n - \widehat{X}^F_{n,i-1}\| \, \|X_{n-i} - \widehat{X}^B_{n-i}\|} & \textit{for} \quad i \geq 2 \end{cases} \tag{4.36}$$

Due to stationarity, the sequence is not dependent on the choice of n.

In (4.36), the expression for $i = 1$ agrees with that for $i \geq 2$ in that $\epsilon^F_{n,0} = X_n$ and that $\epsilon^B_{n-1,0} = X_{n-1}$. Furthermore, in the expressions giving k_i, X_n and X_{n-i} are projected onto the same sub-space $\mathcal{E}_{n-1,i}$.

The result of this, equation (4.33), is notable and shows that the sequence of partial correlation coefficients is given by:

$$k_i = \alpha_{i,i} \tag{4.37}$$

where the $\alpha_{i,i}$ are calculated using the Levinson algorithm.

In the specific case of a causal AR-P process, we therefore have:

$$k_i = \begin{cases} \alpha_{i,i} & \text{for} \quad 1 \leq i \leq P \\ 0 & \text{for} \quad i > P \end{cases}$$

We can easily verify that the partial autocorrelation function of an MA$-Q$ does not go to 0, unlike that of an AR-P which is null above P. However, the absolute value is bounded by a decreasing exponential. Table 4.1 provides a summary of these different results.

Exercise 4.1 (Levinson algorithm) (see page 277) Write a function which estimates prediction coefficients using the recursion og the Levinson algorithm. Test this function.

4.2.3 Reflection coefficients and lattice filters

When linearly predicting X_n as a function of X_{n-1}, ..., X_{n-m+1} we write that the error prediction, equation (4.14), is orthogonal to the past[2], that is:

$$\varepsilon^F_{n,m-1} = X_n - \sum_{k=1}^{m-1} \alpha_{k,m-1} X_{n-k} \quad \perp \quad \{X_{n-1}, \dots, X_{n-m+1}\}$$

We are now going to perform what is called the *backward* "prediction" of X_{n-m} as a function of the $(m-1)$ values of the immediate future, that is X_{n-m+1}, ..., X_{n-1}. We have:

$$\varepsilon^B_{n-1,m-1} = X_{n-m} - \sum_{k=1}^{m-1} \beta_{k,m-1} X_{n-m+k} \quad \perp \quad \{X_{n-m+1}, \dots, X_{n-1}\}$$

$\varepsilon^B_{n-1,m-1}$ is a linear combination of X_{n-1}, X_{n-2}, ..., X_{n-m}, hence the index $(n-1)$ in its definition. Notice that, because of stationarity, $\varepsilon^B_{n-1,m-1} = X_{n+1-m} - \sum_{k=1}^{m-1} \beta_{k,m-1} X_{n+1-m+k}$ with a sequence of coefficients $\beta_{k,m-1}$ that do not depend on n.

By expressing the orthogonality, and by using the stationarity hypothesis, we can easily check that the same Yule-Walker equations are obtained for the *forward* and *backward* predictions and that:

$$\alpha_{k,m-1} = \beta_{k,m-1} \tag{4.38}$$

Notice that for $m = 0$, the *forward* and *backward* prediction errors are written:

$$\varepsilon^F_{n,0} = X_n \quad \text{and} \quad \varepsilon^B_{n,0} = X_n \tag{4.39}$$

respectively. By construction, $\varepsilon^F_{n,m-1}$ and $\varepsilon^B_{n-1,m-1}$ are orthogonal to the *same* subspace generated by X_{n-1}, X_{n-2}, ..., X_{n-m+1}, and therefore for any scalar λ:

$$\delta = \varepsilon^F_{n,m-1} + \lambda \varepsilon^B_{n-1,m-1} \quad \perp \quad \{X_{n-1}, X_{n-2}, \dots, X_{n-m+1}\}$$

Let us choose λ such that δ is also orthogonal to X_{n-m}, and refer to this particular value of λ as k_m. The vector $\delta_m = \varepsilon^F_{n,m-1} + k_m \varepsilon^B_{n-1,m-1}$ is therefore

[2] In this section, we assume the processes to be real and omit the star indicating conjugation.

orthogonal to the subspace generated by X_{n-1}, X_{n-2}, ..., X_{n-m}. Because δ_m is a linear combination of the type $X_n - u_n$ where u_n is a linear combination of X_{n-1}, X_{n-2}, ..., X_{n-m}, δ_m is, according to the projection theorem, the m-th step prediction error. Hence we can write it:

$$\varepsilon^F_{n,m} = \varepsilon^F_{n,m-1} + k_m \varepsilon^B_{n-1,m-1}$$

where k_m is defined by:

$$\varepsilon^F_{n,m-1} + k_m \varepsilon^B_{n-1,m-1} \perp X_{n-m}$$

which is expressed as $\mathbb{E}\left\{\varepsilon^F_{n,m-1} X_{n-m}\right\} + k_m \mathbb{E}\left\{\varepsilon^B_{n-1,m-1} X_{n-m}\right\} = 0$.

By replacing the prediction errors with their expressions as functions of X_n, we get:

$$k_m = -\frac{R(m) - \sum_{k=1}^{m-1} \alpha_{k,m-1} R(m-k)}{R(0) - \sum_{k=1}^{m-1} \alpha_{k,m-1} R(k)}$$

where $R(k)$ refers to the covariance function of the WSS process X_n. This proves expressions (4.36) and (4.40). Expression (4.41), which is recalled below, can be demonstrated in the same way:

$$\epsilon^B_{n,m} = \epsilon^B_{n-1,m-1} + k_m \epsilon^{F*}_{n,m-1}$$

In the following section, we shall present the direct and inverse filtering structures which connect the process to its two prediction errors. One of the advantages of this structure is that we have a simple stability criterion, based on the reflection coefficients.

The analysis filter: $X_n \mapsto \varepsilon^F_{n,m}$

The recurrence equation (4.34) was established during the demonstration of the Levinson algorithm:

$$\epsilon^F_{n,m+1} = \epsilon^F_{n,m} - k_{m+1} \epsilon^B_{n-m-1,m} \tag{4.40}$$

In the same way as we calculated expression (4.40), we may show that

$$\epsilon^B_{n-(m+1),m+1} = \epsilon^B_{n-m,m} - k_{m+1} \epsilon^F_{n-1,m} \tag{4.41}$$

Bringing together equations (4.40) and (4.41), we obtain a system of two filtering equations:

$$\left\{ \begin{array}{rcl} \epsilon^F_{n,m+1} & = & \epsilon^F_{n,m} - k_{m+1} \epsilon^B_{n-m-1,m} \\ \epsilon^B_{n-(m+1),m+1} & = & \epsilon^B_{n-m,m} - k_{m+1} \epsilon^F_{n-1,m} \end{array} \right. \tag{4.42}$$

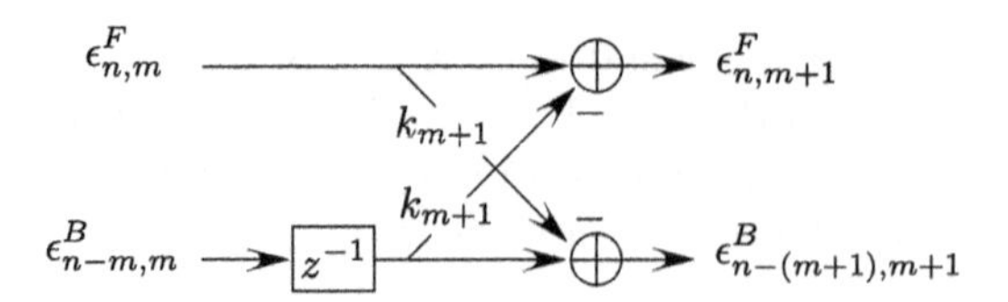

Figure 4.2 – *Modulus of the lattice analysis filter*

The input for this system is the pair $(\epsilon^F_{n,m}, \epsilon^B_{n-m,m})$, and the output is the pair $(\epsilon^F_{n,m+1}, \epsilon^B_{n-(m+1),m+1})$. This calculation modulus is illustrated in Figure 4.2.

If we cascade these blocks and use the initial condition $\varepsilon^F_{n,0} = \varepsilon^B_{n,0} = X_n$, we get the filtering diagram of Figure 4.3, called the *lattice filter*, which changes the signal X_n into the two prediction errors.

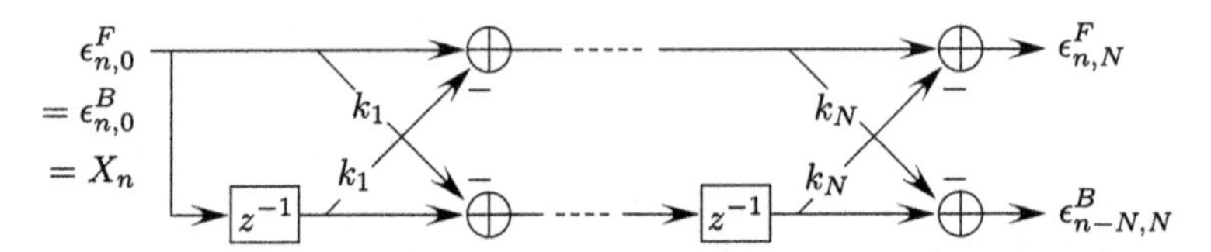

Figure 4.3 – *The lattice analysis filter*

Noting $e_{n,m} = \begin{bmatrix}\epsilon^F_{n,m} & \epsilon^B_{n-m,m}\end{bmatrix}^T$ and using the z-transformation, from system (4.42) we obtain:

$$\mathcal{E}_{m+1}(z) = \begin{bmatrix} 1 & -k_{m+1}z^{-1} \\ -k_{m+1}z^{-1} & 1 \end{bmatrix} \mathcal{E}_m(z) \tag{4.43}$$

The synthesis filter: $\epsilon^F_{n,m} \mapsto X_n$

The filter that changes $\epsilon^F_{n,m}$ into X_n is causal and stable. Starting off with equations (4.42), we can write:

$$\left\{ \begin{array}{rcl} \epsilon^F_{n,m} & = & \epsilon^F_{n,m+1} + k_{m+1}\epsilon^B_{(n-1)-m,m} \\ \epsilon^B_{n-m,m} & = & \epsilon^B_{n-(m+1),m+1} + k_{m+1}\epsilon^F_{n-1,m} \end{array} \right. \tag{4.44}$$

Relations (4.44) lead to a cell set-up with the input sequences $\epsilon^F_{n,m}$ and $\epsilon^B_{n,m-1}$ and the two output sequences $\epsilon^F_{n,m-1}$ and $\epsilon^B_{n,m}$. If we cascade these cells, we get the filtering diagram of Figure 4.4, called a *synthesis filter*, which changes the signal $\epsilon^F_{n,m}$ into the signal X_n.

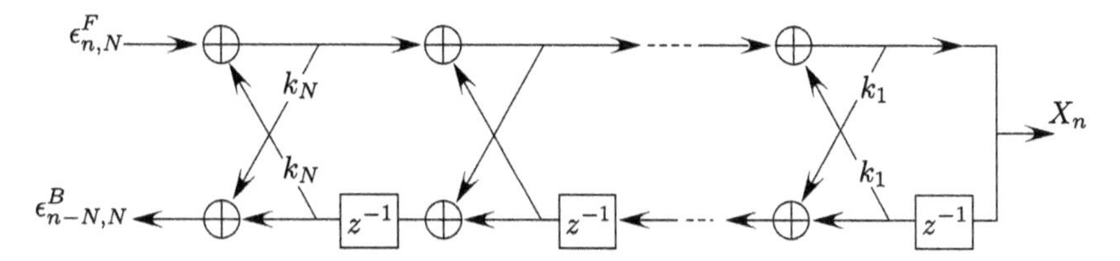

Figure 4.4 – *The lattice synthesis filter*

Stability

Let us begin by considering the analysis filter which connects process X_n with the forward and backward prediction errors. Iterating equation (4.43) and using the initial conditions $\varepsilon^F_{n,0} = \varepsilon^B_{n,0} = X_n$, we obtain two impulse response filters with input X_n. These filters are therefore stable.

However, the synthesis filter, which is the inverse of the analysis filter, is an infinite impulse response filter (IIR). Using expression (4.43), we may write that:

$$\mathcal{E}_m(z) = \frac{1}{1 - k^2_{m+1} z^{-2}} \begin{bmatrix} 1 & k_{m+1} z^{-1} \\ k_{m+1} z^{-1} & 1 \end{bmatrix} \mathcal{E}_{m+1}(z) \tag{4.45}$$

Iterating this expression N times, we obtain an IIR filter, where the moduli of the poles are the reflection coefficients. Consequently, this filter, the implementation of which is causal by construction, is stable if and only if $|k_i| < 1$. This condition is verified if we use the Levinson algorithm.

Exercise 4.2 (Lattice filtering) (see page 278)

1. Analysis: write the lattice filtering function that implements the formulas (4.42).

2. Synthesis: write the lattice filtering function that implements the formulas (4.44).

3. Write a program that checks the two previous functions, by generating an AR-P process using white noise and the `filter` function. Use the `atok` function to extract the reflection coefficients.

The two following functions `atok` and `ktoa` can be used to compute the coefficients a_i from the k_i and conversely.

```
function ki=atok(ai)
```

```
%!==============================================!
%! From the AR parameters (ai) to the reflection !
%! coefficients (ki)                             !
%! SYNOPSIS: ki = ATOK(ai)                       !
%!    ai = AR model (1 a1 ... aP)                !
%!    ki = reflection coefficients (k1 ... kP)   !
%!==============================================!
P=length(ai)-1; ki=zeros(P,1);
for ii=P:-1:1
    ki(ii)=ai(ii+1); bi=conj(ai(ii+1:-1:1));
    umodk2=1-ki(ii)*ki(ii)';
    ai=(ai-ki(ii)*bi)/umodk2;
    ai(ii+1)=[];
end
```

```
function ai = ktoa(ki)
%!============================================!
%! From the reflection coefficients (ki)      !
%! to the AR parameters (ai)                  !
%! SYNOPSIS: ai = KTOA(ki)                    !
%!    ki = reflection coefficients (k1 ... kP) !
%!    ai = AR-model parameters (1 a1 ... aP)   !
%!============================================!
P=length(ki); ai=[1;zeros(P,1)];
for ii=1:P
    ai(1:ii+1) = ai(1:ii+1) + ki(ii)*conj(ai(ii+1:-1:1));
end
```

4.3 Non-parametric spectral estimation of WSS processes

In this section, we consider the problem of the non-parametric estimation of the spectrum of a Wide-Sense Stationary (WSS) random process. The estimation is based on smoothing of the periodogram. The notion of stationarity will then be extended to the intercovariance function of two random processes, and a definition of the *cross-spectrum* will be given. This function is often used in signal processing to check for the presence of linear filtering between two observations.

The periodogram is derived from the Fourier transform of N successive samples. It provides an unbiased estimator of the spectral density but, unfortunately, its variance does not tend to 0 when the number of samples tends toward infinity. However a consistent estimator may be obtained by applying a smoothing approach.

Note: as both the time and frequency of signals will be manipulated, a lower-case letter will be used to denote the temporal representation, e.g. x_n, and an upper-case letter will be used to denote the frequency representation, e.g. X_k.

4.3.1 Correlogram

In order to estimate the spectral density the first simple idea is to mimic its expression from the covariance function. Indeed the spectral density, when it exists, is the discrete Fourier transform (DFT) of the covariance function. Therefore, for that estimation, we can take the discrete Fourier transform (DFT) of the empirical covariance $\widehat{R}(h)$ as defined in (4.5). This leads to the *spectral density estimate*:

$$\widehat{\gamma}_c(f) = \sum_{h=-(N-1)}^{N-1} \widehat{R}(h)e^{-2j\pi hf} \tag{4.46}$$

In the litterature $\widehat{\gamma}_c(f)$ is called the *correlogram*.

It is easy to show that, under very general conditions, the correlogram is very close to the periodogram defined by the expression (4.50). Therefore their performances can be derived simultaneously. It is shown in the next sections that the periodogram, and hence the correlogram, is a poor spectral density estimator. More specifically, they present variances which do not vanish to zero even when the number of samples goes to infinity. Without going in the details, it is clear that the covariance estimates for lags h close to N are based on a few number of samples and, thus, suffer of large errors the sum of which could not go to zero.

In spite of the bad performances of the correlogram, an efficient estimator can be derived from it. The main idea is to restrict the numbers of lags and apply a sequence of weights giving the Blackman-Tuckey estimator:

$$\widehat{\gamma}_{BT}(f) = \sum_{h=-(L-1)}^{L-1} w_h \widehat{R}(h)e^{-2j\pi hf} \tag{4.47}$$

The sequence of weights is referred to as the *time window*. This appproach reduces the variance but introduces a bias. We can expect that the variance will be of the order of L/N whereas the bias will be of $1/L$. As a rule of thumb, L is often chosen of the order of $N/10$. This discussion will be continued with more details, in particular addressing the window shape, in the section 4.3.3.

4.3.2 Periodogram

The discrete Fourier transformation (DFT) of N successive observations x_0, ..., x_{N-1} is expressed:

$$X_k = \frac{1}{\sqrt{N}} \sum_{n=0}^{N-1} x_n e^{-2j\pi kn/N}$$

where k ranges from 0 to $N-1$. This definition does not correspond to that which is generally used in signal processing, which does not include the term $1/\sqrt{N}$. However, it makes the transformation more symmetrical, e.g. the inverse contains the same factor $1/\sqrt{N}$.

Firstly, let $X_0 = N^{-1/2}\sum_n x_n$. Noting the useful identity

$$\frac{1}{N}\sum_{n=0}^{N-1} e^{-2j\pi kn/N} = \begin{cases} 1 & \text{if} \quad k=0 \\ 0 & \text{if} \quad k \neq 0 \end{cases}$$

we deduce that, for $k \neq 0$, the value X_k does not depend on the mean of the process (if we replace x_n by $x_n + \alpha$, only X_0 is modified).

If x_n is real then $X_k = X^*_{N-k}$. Moreover if N is even, hence X_0 and $X_{N/2}$ are real.

In what follows, we shall only consider the complex values associated with the indexes ranging[3] from 1 to $K = \lceil N/2 - 1 \rceil$.

We shall use the following fundamental result [4]:

Property 4.1 *Let x_n be a WSS random process and $\gamma(f)$ its spectral density. Thus, for the set of values of k ranging from 1 to $K = \lceil N/2 - 1 \rceil$, the K complex random variables X_k converge in distribution, when N tends to infinity, toward independent random variables with Gaussian, complex, circular, centered distributions, of respective variances $\gamma_k = \gamma(k/N)$. This is written:*

$$X_k \xrightarrow{d} \mathcal{N}_c(0, \gamma_k) \tag{4.48}$$

The probability density of X_k, deduced from expression (1.37), *is written:*

$$p_{X_k}(x_k) = \frac{1}{\pi\gamma_k} e^{-|x_k|^2/\gamma_k} \tag{4.49}$$

Without giving a detailed proof, note that for any random second order stationary process, the vectors formed from the exponentials $e^{2j\pi kn/N}$ are, asymptotically, the eigenvectors of the covariance matrices. In the Fourier base, the random variables X_k therefore have a tendency to decorrelate, becoming independent as they are asymptotically Gaussian.

Below, we shall consider that N is sufficiently large for the distribution of Z_k to be considered almost equal to its asymptotic distribution.

Definition 4.2 (Periodogram) *The following sequence is known as a periodogram:*

$$P_k = |X_k|^2 \tag{4.50}$$

In the next section we derive the probability density of P_k from that of X_k.

[3] The consistent estimation of X_0 and $X_{N/2}$ requires a smoothing process slightly different to that presented below.

Probability distribution of P_k

Let us show that, for any frequency index k between 1 and $N/2 - 1$, the variables P_k of the periodogram are random variables following an exponential distribution of parameter γ_k.

Based on the property 4.1, the joint probability density of the real part X_r and imaginary part X_i of X_k is given by:

$$p_{X_r,X_i}(x_r, x_i) = \frac{1}{\pi\gamma_k} e^{-(x_r^2+x_i^2)/\gamma_k}$$

where, for simplicity's sake, the index k has been omitted. Let $X_r = R\cos(\Theta)$ and $X_i = R\sin(\Theta)$, where $R \geq 0$ and $\Theta \in (0, 2\pi)$. The Jacobian of this transformation is R. The joint probability density of (R, Θ) is therefore written:

$$p_{R,\Theta}(r, \theta) = \frac{r}{\pi\gamma_k} e^{-r^2/\gamma_k} \mathbb{1}(r \geq 0)\mathbb{1}(\theta \in (0, 2\pi))$$

Integrating with respect to θ leads to the probability density of R:

$$p_R(r) = \frac{2r}{\gamma_k} e^{-r^2/\gamma_k} \mathbb{1}(r \geq 0)$$

From definition (4.50), we note that $P_k = R^2$, a transformation whose Jacobian is equal to $2R$. Hence, the probability density of P_k is given by:

$$p(v) = \frac{1}{\gamma_k} e^{-v/\gamma_k} \mathbb{1}(v \geq 0) \tag{4.51}$$

In conclusion, P_k follows an exponential distribution of parameter γ_k. Noting that an r.v. with an exponential distribution of parameter λ may be seen as λ times an r.v. with an exponential distribution of parameter 1, we may write:

$$P_k = \gamma_k E_k \tag{4.52}$$

where E_k is an r.v. with an exponential distribution of parameter 1. Expression (4.52) shows that in this situation, everything happens as if spectrum γ_k were affected by a multiplicative "noise" following an exponential distribution of parameter 1. Hence, using the logarithm (expression in dB), the estimation noise becomes additive.

As $\mathbb{E}\{E_k\} = 1$ and $\text{var}(E_k) = 1$, (4.52) shows that $\mathbb{E}\{P_k\} = \gamma_k$ and $\text{var}(P_k) = \gamma_k^2$. This means that the periodogram is an unbiased estimator of γ_k, but that, unfortunately, its variance does not tend toward 0 when N tends toward infinity. The periodogram is not, therefore, a consistent estimator of the spectral density.

We have indicated section 4.3.1 that the approach used in the Blackman-Tukey correlogram is equivalent to a convolution in the frequency domain. We are going to show that the smoothe periodogram forms a consistent estimator sequence of the spectral density.

Finally let us notice that an other, commonly used, approach consists of segmenting the data into several blocks with an overlap, typically of 50%, and calculating the mean of the periodograms obtained for each block. This is known as the Welch method: see [3]. This approach presents similar performances in terms of quadratic error than the smoothed periodogram.

4.3.3 Smoothed periodograms

The estimator of the smoothed periodogram is written:

$$\widehat{\gamma}(f) = \sum_{m=-M}^{M} W_m P_{k(f)+m} \tag{4.53}$$

where $k(f)$ is the integer which verifies $(Nf - 1/2) < k \leq (Nf + 1/2)$. For multiple frequencies of the form k/N, we have:

$$\widehat{\gamma}_k = \sum_{m=-M}^{M} W_m P_{k+m} \tag{4.54}$$

The sequence $\{W_m\}$ is known as the smoothing window. Its application is illustrated in Figure 4.5.

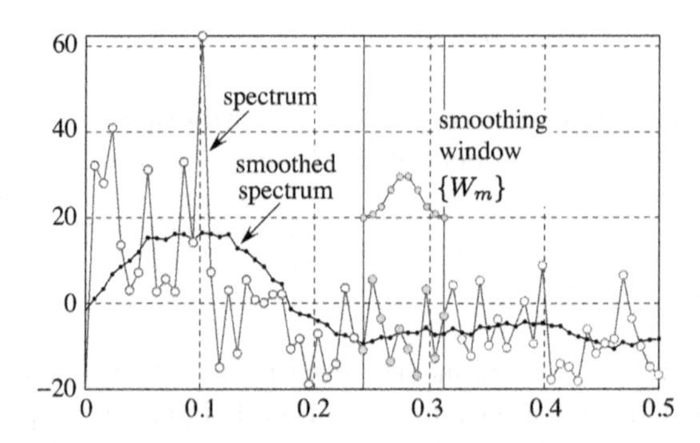

Figure 4.5 – *Application of a smoothing window to the spectrum*

Property 4.2 (Smoothing window) *The following hypotheses are applied in order to obtain a consistent estimator:*

1. $\sum_{m=-M}^{M} W_m = 1$

2. $W_{-m} = W_m$, *therefore* $\sum_{m=-M}^{M} mW_m = 0$ *(centered window),*

3. $\sum_{m=-M}^{M} W_m^2 \to 0$ *when* M *tends toward infinity and thus* W_m *tends toward* 0,

4. $M = N^{\beta}/2$ *with* $\beta \in (0,1)$*;* β *is typically between* 0.2 *and* 0.4 *inclusive. This guarantees that* M *and* $N/(2M+1)$ *both tend toward infinity when* N *tends toward infinity.*

The rectangular window $W_m = 1/(2M+1)$ verifies the first three properties.

Figure 4.6 shows the theoretical spectrum (dotted line) and the spectrum estimated using $N = 4096$ samples of an auto-regressive process defined by $x_n - 1.7x_{n-1} + 0.9x_{n-2} = w_n$, where w_n is a white noise. The smoothing window is a normalized Hamming window and $M = 20$. We see that at frequencies where the spectrum has a high second derivative, the differences are larger, in accordance with equation (4.60) demonstrated after.

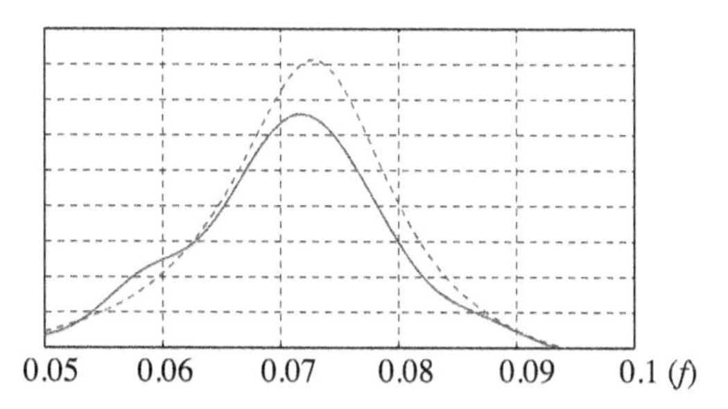

Figure 4.6 – *Spectral estimation. Full line: spectrum estimated using a smoothed periodogram. Dotted line: theoretical spectrum*

As we shall see in the following section, when N tends toward infinity, the bias and variance of $\widehat{\Gamma}_k$ tend toward 0. However, for a fixed value of N, as M increases, the bias increases whereas the variance decreases. The choice of a window and its size therefore involves a compromise between bias and variance. This compromise is discussed in the next section. We shall also see, on page 134, that, for sufficiently high values of N, the bias can be considered to be small in relation to the standard deviation, and it is possible to identify a confidence interval.

Bias/variance of the smoothed periodogram

Using expression (4.54) of the smoothed periodogram:

$$\widehat{\gamma}_k = \sum_{m=-M}^{M} W_m P_{k+m} \tag{4.55}$$

where, following (4.52), $P_{k-M}, \ldots, P_{k+M}$ are $(2M+1)$ independent, exponential random variables, of mean γ_{k+m} and respective variances γ_{k+m}^2. Hence we obtain for the expectation of $\widehat{\gamma}_k$

$$\mathbb{E}\{\widehat{\gamma}_k\} = \sum_{m=-M}^{M} W_m \gamma_{k+m} \tag{4.56}$$

and for its variance

$$\operatorname{var}(\widehat{\gamma}_k) = \sum_{m=-M}^{M} W_m^2 \gamma_{k+m}^2 \tag{4.57}$$

From (4.56) we see that the estimator includes a bias, given by:

$$\mathcal{B}_k = \gamma_k - \sum_{m=-M}^{M} W_m \gamma_{k+m} \tag{4.58}$$

Under very general conditions, the points in the periodogram in the window of length $(2M+1)$ become closer and closer, and, supposing (for example) that $\sum_h |h| \|R_h\| < +\infty$, we can show [4] that when N tends toward infinity, $\Gamma_{k+m} = \Gamma_k + O((2M+1)/N)$. The estimator is therefore asymptotically unbiased.

Choice of M

In practice, the choice of M is more complex than it first appears. The theoretical spectrum is unknown, and consequently formula (4.58), giving the bias, and formula (4.57), giving the variance, cannot be calculated. Hence, with no particular *a priori* assumptions, the choice of M is often simply related to the number N of observations. Thus we commonly use $M = N^{\beta}/2$ with $\beta \approx 0.3$.

One way of refining the choice of M is to assume that the spectral density has a certain regularity, for example that it is differentiable up to order 2. Then taking a Taylor expansion up to order 2 of $\gamma(f)$, we can write:

$$\gamma(f+\delta) \approx \gamma(f) + \delta\dot{\gamma}(f) + \frac{1}{2}\delta^2\ddot{\gamma}(f)$$

where $\dot{\gamma}(f)$ and $\ddot{\gamma}(f)$ respectively denote the first and second derivatives of the spectrum calculated at point f. For neighboring points separate from m/N, this expression leads us to take:

$$\gamma_{k+m} \approx \gamma_k + \frac{m}{N}\dot{\gamma}_k + \frac{1}{2}\frac{m^2}{N^2}\ddot{\gamma}_k \tag{4.59}$$

where $\dot{\gamma}_k = \dot{\gamma}(k/N)$ and $\ddot{\gamma}_k = \ddot{\gamma}(k/N)$.

According to (4.52), the random variable $P_{k+m} = \gamma_{k+m} E_m$ appears as the product of the deterministic value γ_{k+m} and a random variable E_m following an exponential distribution of parameter 1. Thus using (4.59) we get:

$$P_{k+m} \approx \left(\gamma_k + \frac{m}{N}\dot{\gamma}_k + \frac{m^2}{2N^2}\ddot{\gamma}_k \right) E_m$$

where E_m are independent random variables. Using (4.55), we obtain:

$$\widehat{\gamma}_k \approx \gamma_k \sum_{m=-M}^{M} W_m E_m + \dot{\gamma}_k \sum_{m=-M}^{M} W_m m E_m + \frac{\ddot{\gamma}_k}{2N^2} \sum_{m=-M}^{M} m^2 W_m E_m$$

From this, we deduce that

$$\mathbb{E}\left\{\widehat{\gamma}_k\right\} = \gamma_k + \frac{\ddot{\gamma}_k}{2N^2} \sum_{m=-M}^{M} m^2 W_m$$

using the facts that (i) $\sum_m W_m = 1$, (ii) $\sum_m m W_m = 0$ and (iii) $\mathbb{E}\left\{E_m\right\} = 1$. The bias of the estimator $\widehat{\Gamma}_k$ is therefore written:

$$\mathcal{B}_k = \frac{\ddot{\Gamma}_k}{2N^2} \sum_{m=-M}^{M} m^2 W_m$$

For a fixed value of N, we see that the bias increases with M.

The variance of $\widehat{\gamma}_k$ is given by:

$$\operatorname{var}\left(\widehat{\gamma}_k\right) = \mathbb{E}\left\{\left(\widehat{\gamma}_k - \mathbb{E}\left\{\widehat{\gamma}_k\right\}\right)^2\right\} = \gamma_k^2 \sum_{m=-M}^{M} W_m^2$$

using the fact that $\operatorname{var}(E_m) = 1$. We see that the quadratic error at frequency k is expressed:

$$\mathrm{MSE}_k = \gamma_k^2 \sum_{m=-M}^{M} W_m^2 + \frac{\ddot{\gamma}_k^2}{4N^4} \left(\sum_{m=-M}^{M} m^2 W_m \right)^2 \qquad (4.60)$$

For the rectangular window, $\sum_{m=-M}^{M} m^2 W_m = M(M+1)/3$ and thus $\mathcal{B}_k^2 \sim \ddot{\Gamma}_k^2 / N^{4\beta}$, whilst $\sum_{m=-M}^{M} W_m^2 = 1/(2M+1) = 1/N^{\beta}$. Both therefore tend toward zero as N tends toward infinity, but the squared bias decreases with a speed 4 times greater than that of the variance. Hence, for a sufficiently large value of N, confidence intervals are often determined based on the variance alone (see exercise 4.4).

Finally, a single numerical factor may be obtained for the bias, the variance and the mean square error by summation of the values of k. This gives us the "integrated" bias:

$$\mathcal{B} = \sum_{k=1}^{K} \mathcal{B}_k \qquad (4.61)$$

the "integrated" variance:

$$\mathcal{V} = \sum_{k=1}^{K} \text{var}\left(\widehat{\Gamma}_k\right) \tag{4.62}$$

and the "integrated" mean square error:

$$\text{MSE} = \sum_{k=1}^{K} \text{MSE}_k \tag{4.63}$$

The choice of M may be based on the minimization of the "integrated" mean square error MSE.

However, the second derivative of the spectrum is still unknown. In practice, this may be approximated by:

$$\ddot{\gamma}_k \quad = \quad \frac{\partial^2 \gamma_k}{\partial f^2} \approx N^2(\widehat{\gamma}_k - 2\widehat{\gamma}_{k-1} + \widehat{\gamma}_{k-2}) \tag{4.64}$$

using the fact that $\partial f \approx 1/N$.

Figure 4.7 shows the values of the integrated MSE as a function of M for 20 draws of a second order auto-regressive process. The number of samples is $N = 4096$. We see that the minimum mean square error occurs at around $M = 15$. Using the formula $2M + 1 = 31 \approx N^\beta$, this gives us $\beta \approx 0.4$.

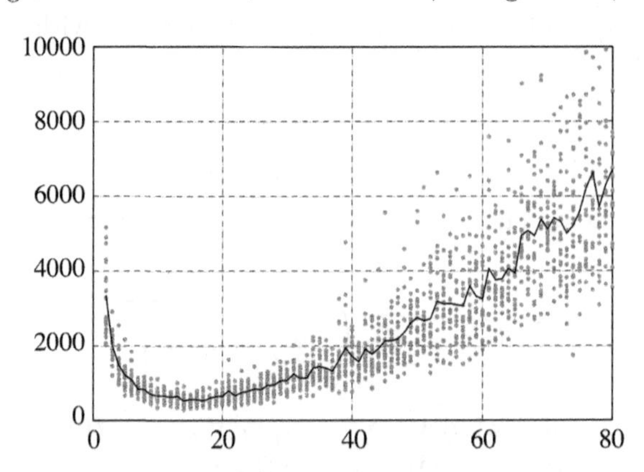

Figure 4.7 – *Integrated MSE (expression (4.63)) as a function of the size M of the smoothing window. The signal is a second order auto-regressive process. The number of samples is $N = 4096$. The points represent the values of the MSE over 20 draws. The full line shows the mean for these 20 draws*

Exercise 4.3 (Smoothed periodogram, bias/variance compromise)
(see page 279) In this exercise, we shall consider the effect of smoothing on the bias/variance compromise. To do this, consider the smoothed spectrum given by expression (4.53). Write a program:

- which produces N samples of an auto-regressive process defined by $x_n - 1.7x_{n-1} + 0.9x_{n-2} = w_n$, where w_n is a centered white noise of variance 1,
- which estimates the spectrum by a smoothed periodogram using a Hamming window of length $(2M+1)$,
- which calculates the mean square error between the smoothed periodogram and the theoretical spectrum, presumed to be known. What happens when the size $(2M+1)$ of the window is altered?

Choice of the window shape

Once the length of the frequency window has been determined, you have to select its shape. The selection is based on the fact that the window shape induces two adverse and opposite effects. One is due to the main lobe width, inducing a smoothing in the sharp region, and the other is the heights of the values out of the main lobe, that induces leakage. Leakage is defined as the effect of transferring energy from the bands where the energy is high to the bands where the energy is low. Therefore the choice of the windows depends on the spectrum we want to estimate, that is of course unknown. Practically the window shape is selected on a case-by-case approach using some *a priori* knowledge on the signal.

We can relate that to the time window used in the Blackman-Tukey correlogram (4.47). Indeed the convolution by a window in the frequency domain, as it is the case for the smoothed periodogram, is equivalent to a multiplication by a window in the time domain, as it is the case for the Blackman-Tukey correlogram. Under very general conditions the width in the time domain varies in opposite sense than the width in the frequency domain. That means that, using notations of expressions (4.47) and (4.53), we can write that $ML \sim N$. Therefore for the correlogram, L/N plays a similar role than $1/M$ for the periodogram. For example the correlogram variance is on order of L/N whereas the periodogram variance is on order of $1/M$. On the other hand the correlogram bias is on order of $1/L$ whereas the periodogram bias is on order of M/N.

The adverse and opposite effects of the windowing is easy to see in the Blackman-Tukey approach. Indeed the frequency window, associated to the time window, see expression (4.47), usually presents a main lobe and multiples secondary lobes. It is clear that, regarding the convolution, the main lobe width induces a smoothing for the sharp cut-off transitions whereas the secondary lobe heights induces a leakage. Unfortunately when we reduce the secondary lobe heights we increase the main lobe width. For more details on the use of windowing on spectral estimation see [3]. A comprehensive overview of the windowing effects is given in [29].

In summary, the choice of the value of M is related to the compromise bias/variance whereas the choice of the window shape to the compromise between sharp cut-off smoothing and leakage.

Approximate confidence interval

In this section, we derive a confidence interval for the smoothed periodogram. The calculation is made possible, requiring that the spectrum is almost constant along the frequency window of width $2M+1$. This assumption is approximately true if $(2M+1)/N$ is sufficiently small compared to the signal bandwidth. Based on item 4 of assumptions 4.2, $(2M+1)/N$ is of the order of $1/N^{1-\beta}$ and can be considered as negligible for N sufficiently large.

Under this assumption and using (4.55) and (4.52) we can write:

$$\widehat{\gamma}_k \approx \gamma_k \sum_{m=-M}^{M} W_m E_m \tag{4.65}$$

where $E_{-M}, \ldots, E_M$ is a sequence of $2M+1$ independent random variables, with the same exponential distribution of parameter 1. Consequently, $\widehat{\gamma}_k/\gamma_k$ is the weighted sum of $(2M+1)$ independent random variables, with the same exponential distribution of parameter 1. At first, as we have assumed, assumption 4.2, that $\sum_m W_m = 0$ and $\sum_{m=-M}^{M} W_m^2 \to 0$ when M tends to infinity, the bias is zero and its variance, equal to $\gamma_k^2 \sum_m W_m^2$, goes to zero when M tends to infinity. Therefore the sequence of the smoothed periodograms is asymptotically consistent.

On the other hand, $2\widehat{\gamma}_k/\gamma_k$ (multiplication by 2) appears as the sum of $2M+1$ r.v. with exponential distribution with parameter 2 which also coincides with the χ_2^2 distribution with 2 degrees of freedom. Let us denote:

$$L_W = \frac{1}{\sum_{m=-M}^{M} W_m^2} \tag{4.66}$$

and using the following approximation[4]:

$$\sum_{m=-M}^{M} W_m U_m \approx \chi_{2L_W}^2, \quad \text{where} \quad U_m \sim \chi_2^2 \text{ independent,} \tag{4.67}$$

we see that:

$$\frac{2L_W \widehat{\gamma}_k}{\gamma_k} \sim \chi_{2L_W}^2 \tag{4.68}$$

[4]We consider that the r.v. $\sum_{m=-M}^{M} W_m U_m$ approximately follows the distribution $c\chi_\nu^2$. To obtain c and ν, we identify the mean $c\nu = 2$ and the variance $2c^2\nu = 4/L_W$. We deduce that $c = 1/L_W$ and that $\nu = 2L_W$. The result is valid for the rectangular window.

This result allows us to build a confidence interval at $100\alpha\%$ of the form:

$$\frac{2L_W\widehat{\gamma}_k}{\chi^{2,[-1]}_{2L_W}(1/2+\alpha/2)} \le \gamma_k \le \frac{2L_W\widehat{\gamma}_k}{\chi^{2,[-1]}_{2L_W}(1/2-\alpha/2)} \tag{4.69}$$

Converting the expression into dB, we obtain the following confidence interval:

$$\begin{aligned}\widehat{\gamma}_k^{(dB)} &+ 10\log_{10} 2L_W - 10\log_{10}\chi^{2,[-1]}_{2L_W}(1/2+\alpha/2)\\ &\le \gamma_k^{(dB)} \le \\ \widehat{\gamma}_k^{(dB)} &+ 10\log_{10} 2L_W - 10\log_{10}\chi^{2,[-1]}_{2L_W}(1/2-\alpha/2)\end{aligned} \tag{4.70}$$

To conclude, presuming that the spectrum is, to all intents and purposes, constant over the whole frequency window of width $2M+1$, the estimator is unbiased as $\sum_m W_m = 1$, and its variance is equal to $\gamma_k^2 \sum_m W_m^2$. Note that this is based on the hypothesis that $\sum_{m=-M}^{M} W_m^2 \to 0$ when M tends toward infinity. The smoothed estimator therefore has a variance which tends toward 0 when M tends toward infinity.

Exercise 4.4 (Smoothed periodogram - confidence interval) (see page 281) Write a program:

- which produces N samples of an auto-regressive process defined by $x_n - 1.7x_{n-1} + 0.9x_{n-2} = w_n$, where w_n is a centered white noise of variance 1,
- which estimates the spectrum by smoothing the periodogram using a rectangular window of length $(2M+1)$ where $M = N^\beta/2$ with $\beta = 0.4$,
- which calculates a confidence interval at 95%,
- which compares these results with those obtained using the `pwelch` function in MATLAB®.

4.3.4 Coherence

Consider two random WSS processes x_n and y_n, with respective means μ_x and μ_y and respective spectra $S_{xx}(f)$ and $S_{yy}(f)$. x_n and y_n are said to be covariance-stationary if

$$\mathbb{E}\left\{(y_{n+h}-\mu_y)(x_n-\mu_x)^*\right\} = R_{yx}(h) \tag{4.71}$$

is solely dependent on h. We deduce that $R_{yx}(h) = R^*_{xy}(-h)$. The Fourier transformation of $R_{yx}(h)$ is known as the interspectrum, and is written:

$$\gamma_{yx}(f) = \sum_{h=-\infty}^{+\infty} R_{yx}(h)e^{-2j\pi nf} \tag{4.72}$$

This function is periodic of period 1 and verifies $\gamma_{xy}^*(f) = \gamma_{yx}(f)$, so $|\gamma_{xy}(f)| = |\gamma_{yx}(f)|$.

Unlike the spectral density, the interspectrum $\gamma_{yx}(f)$ is not necessarily positive or real. However, if processes x_n and y_n are real, then $\gamma_{yx}^*(f) = \gamma_{yx}(-f)$ and the modulus $|\gamma_{yx}(f)|$ of the interspectrum is an even function. This will be taken as given in the remainder of this section.

Estimation of the interspectrum

In a similar way to that used for spectral estimation based on a smoothed periodogram, we define the interperiodogram:

$$Q_k = Y_k X_k^* \tag{4.73}$$

A property analogous to 4.1 states that, when N tends toward infinity, the random vectors $\begin{bmatrix} X_k & Y_k \end{bmatrix}^T$, for values of k between 1 and $K = N/2 - 1$, are independent, Gaussian, circular complex and centered, of covariance:

$$\begin{bmatrix} \gamma_{xx}(k/N) & \gamma_{xy}(k/N) \\ \gamma_{yx}(k/N) & \gamma_{yy}(k/N) \end{bmatrix}$$

From this result, we deduce [4] that the estimator $\mathbb{E}\{Q_k\}$ is unbiased, but its variance does not tend toward 0 when N tends toward infinity. To obtain a consistent estimator, we apply a smoothing process of the form:

$$\widehat{\gamma}_{yx,k} = \sum_{m=-M}^{M} W_m Q_{k+m} \tag{4.74}$$

where the sequence W_m verifies the hypotheses 4.2. Under very general conditions, $\widehat{\gamma}_{yx,k}$ can be shown to converge toward $\gamma_{yx,k}$.

Coherence

For any frequency f such that $\gamma_{xx}(f)\gamma_{yy}(f) \neq 0$, the following function is known as the *magnitude squared coherence* (MSC) or simply coherence:

$$\text{MSC}(f) = \frac{|\gamma_{yx}(f)|^2}{\gamma_{xx}(f)\gamma_{yy}(f)} \tag{4.75}$$

Note that if x_n and y_n are real, then MSC(f) is an even function. In this case, we limit its representation to the interval $(0, 1/2)$.

Using the Schwarz inequality, it is easy to show that:

$$\text{MSC}(f) \leq 1 \tag{4.76}$$

This function has important practical applications for testing the presence of linearity. This is illustrated in the following example. Suppose that x_n and

y_n are, respectively, the input and output of a stable linear filter of complex gain $G(f)$. Hence:

$$\begin{aligned}\gamma_{yy}(f) &= |G(f)|^2\gamma_{xx}(f)\\ \gamma_{yx}(f) &= G(f)\gamma_{xx}(f)\end{aligned}$$

Thus, for any f, the coherence is equal to 1:

$$\mathrm{MSC}(f) = 1$$

This property is characteristic. We may show that:

$$\mathrm{MSC}(f) \leq 1$$

where the equality is verified if and only if x_n and y_n are the result of filtering of one of them by the other. In many practical applications, the function $\mathrm{MSC}(f)$ is used to see whether or not two given signals are related by linear filtering.

In order to estimate the coherence, we successively estimate spectra $\gamma_{xx}(f)$ and $\gamma_{yy}(f)$ along with the interspectrum $\gamma_{yx}(f)$. Using (4.53) and (4.74), we obtain an estimation of $\mathrm{MSC}_k = \mathrm{MSC}(k/N)$ of the form:

$$\widehat{\mathrm{MSC}}_k = \frac{|\widehat{\gamma}_{yx,k}|^2}{\widehat{\gamma}_{xx,k}\widehat{\gamma}_{yy,k}} \tag{4.77}$$

[4] shows that, when N tends toward infinity and for $\mathrm{MSC}_k \neq 0$:

$$\sqrt{\widehat{\mathrm{MSC}}_k} \xrightarrow{d} \mathcal{N}\left(\sqrt{\mathrm{MSC}_k}, \frac{(1-\mathrm{MSC}_k)^2}{2L_W}\right) \tag{4.78}$$

where L_W is given by (4.66). From this, we deduce a confidence interval at $100\alpha\%$ for the coherence, of the form:

$$\sqrt{\widehat{\mathrm{MSC}}_k} - c\frac{1-\widehat{\mathrm{MSC}}_k}{\sqrt{2L_W}} \leq \sqrt{\mathrm{MSC}_k} \leq \sqrt{\widehat{\mathrm{MSC}}_k} + c\frac{1-\widehat{\mathrm{MSC}}_k}{\sqrt{2L_W}} \tag{4.79}$$

where c verifies:

$$\int_{-\infty}^{c} \frac{1}{\sqrt{2\pi}} e^{-u^2/2} du = (1+\alpha)/2$$

Typically, for $\alpha = 0.95$, we have $c \approx 1.96$.

Exercise 4.5 (Magnitude square coherence) (see page 282) Consider the two processes defined by:

$$\begin{cases} x_{1,n} = g_{1,0}e_n + g_{1,1}e_{n-1} + g_{1,2}e_{n-2} + w_{1,n} \\ x_{2,n} = g_{2,0}e_n + g_{2,1}e_{n-1} + g_{2,2}e_{n-2} + w_{2,n} \end{cases}$$

where $g_{1,0} = g_{2,0} = 1$, $g_{1,1} = -1.2$, $g_{1,2} = 0.9$, $g_{2,1} = -0.4$ and $g_{2,2} = 0.3$; where e_n is a causal auto-regressive process of order 1, defined by the equation $e_n - 0.3e_{n-1} = z_n$, in which z_n is a centered white noise of variance 1; and where $w_{1,n}$ and $w_{2,n}$ are two centered white noises of respective variance $\sigma_1^2 = 4$ and $\sigma_2^2 = 1$. Processes e_n, $w_{1,n}$ and $w_{2,n}$ are taken to be uncorrelated.

1. Identify the expression of the coherence.
2. Using MATLAB® generate processes $x_{1,n}$ and $x_{2,n}$.
3. Using expression (4.77), estimate the coherence and a confidence interval at 90% following expression (4.79). Compare the result with the theoretical MSC.

Chapter 5

Inferences on HMM

In this chapter, we shall give a brief overview of Hidden Markov Models (HMM) [6]. These models are widely used in signal processing. They have a fundamental property which results in the existence of recursive algorithms, meaning that the number of operations and the size of the memory needed to calculate the required values do not increase with the number of samples. The best-known example of this is the Kalman filter [10].

Throughout this chapter, for simplicity, the notation $(n_1 : n_2)$ will be used to denote the sequence of integer values from n_1 to n_2 inclusive.

5.1 Hidden Markov Models (HMM)

A hidden Markov model (HMM) is a bivariate, discrete time process (X_n, Y_n), where X_n and Y_n are two real random vectors of finite dimension such that:

- X_n, $n \geq 1$, is a Markov process, i.e. for any function f the conditional expectation of $f(X_{n+1})$ given the σ-algebra generated by $\{X_s; s \leq n\}$ (the past until n) coincides with the conditional expectation of $f(X_{n+1})$ given the σ-algebra generated by $\{X_n\}$. If the conditional distributions have a density, that writes:

$$p_{X_{n+1}|X_{1:n}}(x_{n+1}; x_{1:n}) = p_{X_{n+1}|X_n}(x_{n+1}; x_n) \tag{5.1}$$

- Y_n, $n \geq 1$, is a process such that the conditional distribution of $Y_1, \ldots, Y_n$ given $X_1, \ldots, X_n$ is the product of the distributions of Y_k conditionally to X_k. If the conditional distributions have a density, we obtain:

$$p_{Y_{1:n}|X_{1:n}}(y_{1:n};x_{1:n}) = \prod_{k=1}^{n} p_{Y_k|X_k}(y_k;x_k) \tag{5.2}$$

– the initial r.v. X_1 has a known probability law. If this initial distribution has a probability density, it will be denoted $p_{X_1}(x_1)$.

The following expression of the joint distribution can be deduced from the previous assumptions:

$$\begin{aligned} p_{X_{1:n},Y_{1:n}}(x_{1:n},y_{1:n}) &= \\ & \prod_{k=1}^{n} p_{Y_k|X_k}(y_k;x_k) \prod_{k=2}^{n} p_{X_k|X_{k-1}}(x_k;x_{k-1}) p_{X_1}(x_1) \end{aligned} \tag{5.3}$$

Indeed using Bayes' rule and equation (5.2), we have:

$$\begin{aligned} p_{X_{1:n},Y_{1:n}}(x_{1:n},y_{1:n}) &= p_{Y_{1:n}|X_{1:n}}(y_{1:n};x_{1:n})\, p_{X_{1:n}}(x_{1:n}) \\ &= \prod_{k=1}^{n} p_{Y_k|X_k}(y_k;x_k)\, p_{X_{1:n}}(x_{1:n}) \end{aligned}$$

Using Bayes' rule once again and equation (5.1), we may write:

$$\begin{aligned} p_{X_{1:n}}(x_{1:n}) &= p_{X_n|X_{1:n-1}}(x_n;x_{1:n-1})\, p_{X_{1:n-1}}(x_{1:n-1}) \\ &= p_{X_n|X_{n-1}}(x_n;x_{n-1})\, p_{X_{1:n-1}}(x_{1:n-1}) \end{aligned}$$

Reiterating this writing process, we deduce that:

$$p_{X_{1:n}}(x_{1:n}) = \prod_{k=2}^{n} p_{X_k|X_{k-1}}(x_k;x_{k-1})\, p_{X_1}(x_1)$$

which completes the proof of expression (5.3).

Expression (5.3) may be represented by the Directed Acyclic Graph (DAG) shown in Figure 5.1, using the coding rule:

$$\mathbb{P}\{X_{1:n},Y_{1:n}\} = \prod_{i\in\mathcal{V}} \mathbb{P}\{i|\text{parents}(i)\}$$

where $\mathcal{V}$ denotes the set of all nodes in the graph. Note that the DAG of an HMM is a tree. In more general cases, we speak of a dynamic Bayesian network.

In practice, the variables $Y_{1:n}$ represent observations and variables $X_{1:n}$ represent "hidden" variables. Our objective is thus to make inferences concern-

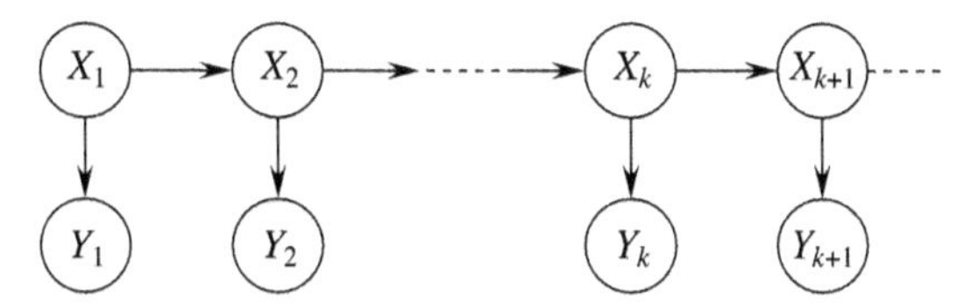

Figure 5.1 – *The Directed Acyclic Graph associated with an HMM takes the form of a tree*

ing the hidden variables based on the observations. In very general terms, we therefore need to calculate conditional distributions of the form $p_{X_{n_1:n_2}|Y_{m_1:m_2}}(x_{n_1:n_2}; y_{m_1:m_2})$. Thus, if we wish to "extract" all information concerning X_n based on the observations $Y_{1:n}$, we need to determine the distribution $p_{X_n|Y_{1:n}}(x_n; y_{1:n})$. Subsequently, any function of interest $f(X_n)$ may be calculated using the conditional expectation:

$$\mathbb{E}\{f(X_n)|Y_{1:n}\} = \int f(x)p_{X_n|Y_{1:n}}(x; Y_{1:n})dx$$

which is a "measurable" function of the observations.

A priori, the problem appears very simple. We only need to apply Bayes' rule and write that:

$$p_{X_{n_1:n_2}|Y_{m_1:m_2}}(x_{n_1:n_2}; y_{m_1:m_2}) = \frac{p_{X_{n_1:n_2},Y_{m_1:m_2}}(x_{n_1:n_2}, y_{m_1:m_2})}{p_{Y_{m_1:m_2}}(y_{m_1:m_2})}$$

Let us notice that the numerator and the denominator can be obtained by integrating the joint probability distribution a certain number of times. For example we have:

$$p_{X_n|Y_{1:n}}(x_n; y_{1:n}) = \frac{\int p_{X_{1:n},Y_{1:n}}(x_{1:n}, y_{1:n})dx_{1:n-1}}{\int p_{X_{1:n},Y_{1:n}}(x_{1:n}, y_{1:n})dx_{1:n}}$$

Unfortunately, with the exception of certain cases - such as the linear Gaussian case which leads to the Kalman filter, or the case where the variables $X_{1:n}$ have values within a discrete finite set - this expression is impossible to calculate. Other methods are therefore required for these cases. One notable solution [8] is based on the creation of samples, known as particles, which simulate the *a posteriori* distributions of X_n given the Y_n. This is known as *particle filtering*.

The factorized form of the expression (5.3) induces separation properties which are the basis of recursive algorithms. Let us consider a few examples. For any $k \leq n$, it is easy to show that:

$$p_{X_{1:n},Y_{1:n}}(x_{1:n}, y_{1:n}) = p_{X_{1:k},Y_{1:k}}(x_{1:k}, y_{1:k})p_{X_{k+1:n},Y_{k+1:n}|X_k}(x_{k+1:n}, y_{k+1:n}; x_k) \quad (5.4)$$

or that:

$$p_{X_{1:n},Y_{1:n}}(x_{1:n}, y_{1:n}) = p_{X_{1:k},Y_{1:k}}(x_{1:k}, y_{1:k}) \ldots$$
$$p_{X_{k+1}|X_k}(x_{k+1}; x_k) p_{X_{k+2:n},Y_{k+2:n}|X_{k+1}}(x_{k+2:n}, x_{k+2:n}; x_{k+1})$$

or that, for any $j \geq 1$:

$$p_{X_{1:n},Y_{1:n}}(x_{1:n}, y_{1:n})(x_{1:n}, y_{1:n}) = p(x_{1:k}, y_{1:k}) \ldots \tag{5.5}$$
$$p_{X_{k+1:k+j},Y_{k+1:k+j}|X_k}(x_{k+1:k+j}, y_{k+1:k+j}; x_k) \ldots$$
$$p_{X_{k+j+1:n},Y_{k+j+1:n}|X_{k+j}}(x_{k+j+1:n}, y_{k+j+1:n}; x_{k+j})$$

Taking $j = 1$ and integrating expression (5.5) over $x_{1:k-1}$ and over $x_{k+2:n}$, we obtain:

$$p_{X_{k:k+1},Y_{1:n}}(x_k, x_{k+1}, y_{1:n}) = p_{X_k|Y_{1:k}}(x_k; y_{1:k}) \tag{5.6}$$
$$p_{Y_{k+1}|X_{k+1}}(y_{k+1}; x_{k+1})\, p_{X_{k+1}|X_k}(x_{k+1}; x_k)\, p_{Y_{k+2:n}|X_{k+1}}(y_{k+2:n}; x_{k+1})$$

All of these results may be determined using graphic rules applied to the DAG associated to the considered Bayesian network. For more details, refer to [21].

5.2 Inferences on HMM

The list below covers a number of inference problems:

- *Learning*: the joint law (5.3) is taken to depend on a parameter θ, which may be constituted of parameters μ associated to the distributions of observations $p_{Y_k|X_k}(y_k; x_k, \mu)$, and/or parameters ϕ associated to the distributions of the hidden states $p_{X_{k+1}|X_k}(x_{k+1}; x_k, \phi)$. The aim is to estimate θ based on a series of observations $Y_{1:n}$. The EM algorithm presented in section (2.3.5) may be used.

- *Inferences* on X_{n+k} based on the observations $Y_{1:n}$. Here the joint law (5.3) is assumed to be known, and we wish to find the expression of $p_{X_{n+k}|Y_{1:n}}(x_{n+k}; y_{1:n})$:

  1. if $k = 0$, this is known as *filtering*;
  2. if $k > 0$, this is known as *prediction*;
  3. if $k < 0$, this is known as *smoothing*.

- *Estimation* of the sequence of hidden states $X_{1:n}$ based on observations $Y_{1:n}$ in the case where the values of X_n are found in $\mathcal{S} = \{1, \ldots, S\}$, which is finite and discrete. To do this, the *a posteriori* maximum can be used:

$$\widehat{X}_{1:n} = \arg \max_{s_{1:n} \in \mathcal{S}^n} \mathbb{P}\{X_{1:n} = s_{1:n} | Y_{1:n}\}$$

One rough approach consists of "testing" the S^n possible value combinations. However, a deeper examination allows us to obtain an algorithm with a complexity in terms of only $n \times S^2$. This is the Viterbi algorithm, presented in section 5.4.5.

5.3 Gaussian linear case: the Kalman filter

In general cases, filtering consists of calculating $p_{X_n|Y_{1:n}}(x_n|y_{1:n})$. A simple calculation shows that the HMM structure leads to a recursive algorithm, which calculates $p_{X_{n+1}|Y_{1:n+1}}(x_{n+1}|y_{1:n+1})$ in two steps based on $p_{X_n|Y_{1:n}}(x_n|y_{1:n})$:

1. a correction step, which calculates:

$$p_{X_{n+1}|Y_{1:n}}(x_{n+1}; y_{1:n}) = \int p_{X_n|Y_{1:n}}(x_n; y_{1:n}) p_{X_{n+1}|X_n}(x_{n+1}; x_n) dx_n$$

2. an update step, which calculates:

$$\begin{aligned} & p_{X_{n+1}|Y_{1:n+1}}(x_{n+1}; y_{1:n+1}) \\ = & \frac{p_{X_{n+1}|Y_{1:n}}(x_{n+1}; y_{1:n}) p_{Y_{n+1}|X_{n+1}}(y_{n+1}; x_{n+1})}{p_{Y_{n+1}|Y_{1:n}}(y_{n+1}; y_{1:n})} \end{aligned} \quad (5.7)$$

Note that the denominator $p_{Y_{n+1}|Y_{1:n}}(y_{n+1}; y_{1:n})$ is not dependent on x_{n+1}. Hence, using the fact that the integral of $p_{X_{n+1}|Y_{1:n+1}}(x_{n+1}; y_{1:n})$ with respect to x_{n+1} is equal to 1, we deduce that:

$$\begin{aligned} & p_{Y_{n+1}|Y_{1:n}}(y_{n+1}; y_{1:n}) \\ = & \int p_{X_{n+1}|Y_{1:n}}(x_{n+1}; y_{1:n}) p_{Y_{n+1}|X_{n+1}}(y_{n+1}; x_{n+1}) dx_{n+1} \end{aligned}$$

Therefore we only need to calculate the numerator of (5.7)

$$p_{X_{n+1}|Y_{1:n}}(x_{n+1}; y_{1:n}) p_{Y_{n+1}|X_{n+1}}(y_{n+1}; x_{n+1})$$

and then normalize by integration with respect to x_{n+1}. Hence we can replace, in the update step, expression (5.7) with:

$$p_{X_{n+1}|Y_{1:n+1}}(x_{n+1}; y_{1:n+1}) \quad \propto \quad p_{X_{n+1}|Y_{1:n}}(x_{n+1}; y_{1:n}) p_{Y_{n+1}|Y_n}(y_{n+1}; x_{n+1})$$

where the symbol $\propto$ means "proportional to".

Generally speaking, the correction and update expressions are intractable; however there are two important cases in which closed form expressions may be obtained. Firstly, the linear Gaussian case leads to the Kalman filter, which is discussed in this section. The second case arises when X_n takes its values in a finite state set, and will be discussed in section 5.4.

Let us consider the model defined for $n \geq 1$ by the two following equations:

$$\begin{cases} X_{n+1} = A_n X_n + B_n & \text{(evolution equation)} \\ Y_n = C_n X_n + U_n & \text{(observation equation)} \end{cases} \tag{5.8}$$

where $\{A_n\}$ and $\{C_n\}$ are two sequences of matrices with adequate dimensions. $\{B_n\}$ and $\{U_n\}$ are two centered Gaussian vector sequences, independent of each other, with the covariances R_n^B and R_n^U respectively. We also assume that the random vector X_1 is Gaussian with zero-mean and covariance Σ_1.

The first equation of (5.8) describes the evolution of the hidden state X_n.

The vector X_n is called the *state vector*, or just the *state*, and the first equation is called the *state equation* or *evolution equation.* The vector Y_n is the *measurement vector* and the second equation is called the *observation equation.*

Generally speaking, we wish to make inferences concerning states on the basis of observations. In this context, the evolution equation may be viewed as an *a priori* probability distribution of the states. For this reason, we can consider that we are in a Bayesian framework.

Let us return to equations (5.8). As an exercise, the following properties can be shown:

- based on the linear transformation property of Gaussian distribution, $(X_{1:n}, Y_{1:n})$ is jointly Gaussian;

- the sequence X_n is a Markov chain;

- the probability density of the conditional distribution of X_{n+1} given X_n has the expression:

$$p_{X_{n+1}|X_n}(x_{n+1}, x_n) \sim \mathcal{N}(A_n X_n; R_n^B)$$

- the probability distribution of $Y_{1:n}$ conditionally to $X_{1:n}$ verifies the property of independence expressed in equation (5.2);

- the probability density of the conditional distribution of Y_n given X_n has the expression:

$$p_{Y_n|X_n}(x_n, y_n) \sim \mathcal{N}(C_n X_n; R_n^U)$$

The bivariate process $(X_{1:n}, Y_{1:n})$ is therefore an HMM.

The fact that the probability distribution of X_n conditional on $Y_{1:n}$ is Gaussian constitutes a fundamental property. We therefore simply need to determine the expression of its mean and covariance. The Kalman algorithm provides a recursive means of calculating these two quantities. In this context, the following notation is generally used:

$$X_{n|k} = \mathbb{E}\{X_n|Y_{1:k}\} \tag{5.9}$$

$$P_{n|k} = \text{cov}(X_n|Y_{1:k}) \tag{5.10}$$

Note that, in accordance with property 1.10, the Gaussian character implies that the conditional expectation $X_{n|k}$ corresponds to the orthogonal projection of X_n onto the linear space spanned by Y_1, ..., Y_k. The Kalman filter can therefore be deduced on the basis of geometric arguments alone, associated with the projection theorem (see section 1.3). In this context, remember that $(X, Y) = \mathbb{E}\{XY\}$ denotes the scalar product of X and Y, and that $(X|Y_{1:n})$ is the orthogonal projection of X onto the linear space spanned by $Y_{1:n}$.

Before discussing the Kalman algorithm in greater detail, let us consider a classic example, concerning trajectography, leading to equations of the form set out in (5.8).

Example 5.1 Consider a vehicle moving along a straight line at the constant speed v. The position at the time $n+1$ is given by $d_{n+1} = d_n + vT$ where T denotes the sampling period. This motion equation can also be written:

$$\begin{cases} d_{n+1} = d_n + v_n T \\ v_{n+1} = v_n \end{cases}$$

with the initial conditions d_1 and v_1. Notice that the second equation of the system is reduced to $v_n = v_1$ if we assume that the vehicle has a constant speed. But if we have little faith in this hypothesis of a constant speed, the possible variability can be taken into account by assuming that:

$$\begin{cases} d_{n+1} = d_n + v_n T \\ v_{n+1} = v_n + b_n \end{cases}$$

where b_n is a random process acting as a *modeling noise*.

Hence the evolution equation has the expression, in matrix form:

$$\begin{bmatrix} d_{n+1} \\ v_{n+1} \end{bmatrix} = \begin{bmatrix} 1 & T \\ 0 & 1 \end{bmatrix} \begin{bmatrix} d_n \\ v_n \end{bmatrix} + \begin{bmatrix} 0 \\ b_n \end{bmatrix}$$

Let us now assume that the position d_n is observed at the output of a noised device delivering the value $y_n = d_n + u_n$, where u_n is a random process used as a model for the *measurement noise*. The set comprising the equation that describes the motion and the one that describes the observation leads to the following system of equations:

$$\begin{cases} \begin{bmatrix} d_{n+1} \\ v_{n+1} \end{bmatrix} = \begin{bmatrix} 1 & T \\ 0 & 1 \end{bmatrix} \begin{bmatrix} d_n \\ v_n \end{bmatrix} + \begin{bmatrix} 0 \\ b_n \end{bmatrix} \\ y_n = \begin{bmatrix} 1 & 0 \end{bmatrix} \begin{bmatrix} d_n \\ v_n \end{bmatrix} + u(n) \end{cases}$$

If we let $X_n = [d_n \quad v_n]^T$, and if we write the expression of the observation Y_n in vector form, we get:

$$\begin{cases} X_{n+1} &= AX_n + B_n \\ Y_n &= CX_n + U_n \end{cases}$$

which is similar to the expression (5.8).

Kalman filter algorithm

The Kalman algorithm offers a recursive solution to the HMM filtering problem in the linear Gaussian case, as defined in expressions (5.8). Its recursive characteristic should be understood in the sense that the number of operations and the amount of memory required by the algorithm do not increase as the number of observations increases. The Gaussian and linear characteristics of the model mean that the *a posteriori* law of X_n given $Y_{1:n}$ is Gaussian, and is therefore characterized completely by its mean $X_{n|n}$ and covariance $P_{n|n}$.

Exercise 5.1 gives a detailed proof of the Kalman algorithm in the scalar case. For the vectorial case, a fully similar proof leads to the algorithm (3).

Data: $Y_{n=1:N}$, $A_{n=1:N}$, $C_{n=1:N}$, $R^B_{n=1:N}$, $R^U_{n=1:N}$, Σ_1
Result: for $n = 1 : N$, $X_{n|n}$, $P_{n|n}$, ℓ
Initialization:
$X_{1|1} = 0$, $P_{1|1} = \Sigma_1$
$\Gamma_1 = C_1 \Sigma_1 C_1^T + R_1^U$
$\ell = -\frac{1}{2}(n_y \log(2\pi)) + Y_1^T \Gamma_1^{-1} Y_1 + \log\det\{\Gamma_1\})$
for $n = 2$ *to* N **do**

$$X_{n|n-1} = A_{n-1} X_{n-1|n-1} \quad \text{(prediction)} \tag{5.11}$$

$$P_{n|n-1} = A_{n-1} P_{n-1|n-1} A_{n-1}^T + R_{n-1}^B \tag{5.12}$$

$$\Gamma_n = C_n P_{n|n-1} C_n^T + R_n^U \tag{5.13}$$

$$G_n = P_{n|n-1} C_n^T \Gamma_n^{-1} \quad \text{(Kalman gain)} \tag{5.14}$$

$$i_n = Y_n - C_n X_{n|n-1} \quad \text{(innovation)} \tag{5.15}$$

$$X_{n|n} = X_{n|n-1} + G_n i_n \quad \text{(update)} \tag{5.16}$$

$$P_{n|n} = (I_{n_x} - G_n C_n) P_{n|n-1} \tag{5.17}$$

$$\ell = \ell - \frac{1}{2}(n_y \log(2\pi) + i_n^T \Gamma_n^{-1} i_n + \log\det\{\Gamma_n\}) \tag{5.18}$$

end

Algorithm 3: Kalman algorithm: n_x, n_y denote the dimensions of X_n and Y_n respectively. $X_{n|n}$: filter outputs, $P_{n|n}$: covariance of $X_{n|n}$, ℓ: log-likelihood of the observations.

The Kalman algorithm performs a computation of $X_{n|n}$ at the time n from the value $X_{n-1|n-1}$ in two successive steps: the *prediction* by equation (5.11) and the *update* by equation (5.16). And this calculation only requires the memorization of the finite dimension state.

The variable $i_n = Y_n - \mathbb{E}\{Y_n|Y_{1:n-1}\}$ is known as the innovation, and is the difference between that which is observed at instant n and that which can be "explained" by previous observations.

The variable noted $\ell = 2\log p_{Y_{1:n}}(Y_{1:n})$ in algorithm (3) is the log-likelihood of the observations, calculated using the series of innovations. In the case where $p_{Y_{1:n}}(Y_{1:n})$ depends on an unknown parameter θ, the Kalman algorithm can be used to carry out maximization with respect to θ (see exercise 5.4).

The expression of this algorithm calls for a few additive comments:

1. G_n is called the *Kalman gain*. It can be calculated beforehand, since it is determined by equations (5.12), (5.14) and (5.17), which do not depend on the observed data Y_n. However, except in the case of scalars (see exercise 5.1), the expression of G_n requires to solve a complex recursive equation referred as the *Riccati equation*. Without going into detail concerning the solution to this equation, note that MATLAB® includes a function `dare.m` (Control System toolbox) which calculates the limit of the gain, if such a limit exists, when n tends toward infinity.

2. Let $R_n^U = 0$. Using expressions (5.13) and (5.14), we obtain $G_n = P_{n|n-1}C_n^T(C_nP_{n|n-1}C_n^T)^{-1}$. From this, we see that $C_nG_n = I_{n_y}$. In the absence of observation noise, the Kalman gain is the right inverse of the observation matrix.

3. Now, let $R_n^B = 0$. The Kalman gain can be shown to tend toward 0. In this case, the evolution model is highly reliable, and the estimation $X_{n|n}$ is principally calculated using prediction $A_{n-1}X_{n-1|n-1}$.

4. The significance of equation (5.16) is obvious. According to the state equation of (5.8), the "best" value of X_n at time n is $A_{n-1}X_{n-1|n-1}$. Then this value is corrected by a quantity proportional to the difference between what we observe, saying Y_n and what we can expect from the previous observation, saying $C_nA_nX_{n-1|n-1}$. That may seen as an adaptive compromise between what we see, that is Y_n and what we know, that is the equation of evolution.

5. We wish to draw your attention to the fact that the algorithm requires us to know R_n^B and R_n^U. However, the theory can be generalized to include the situation where these quantities have to be estimated based on the observed data. In that case, the sequence of gains can no longer be calculated beforehand.

Exercise 5.1 (Kalman recursion in the scalar case) (see page 283) Consider the system described by the two equations:

$$\begin{cases} X_{n+1} &= a_n X_n + B_n \\ Y_n &= c_n X_n + U_n \end{cases} \tag{5.19}$$

where $\{a_n\}$ and $\{c_n\}$ are two sequences of scalars. $\{B_n\}$ and $\{U_n\}$ are two centered Gaussian sequences, independent of each other, with the variances $\sigma_B^2(n)$ and $\sigma_U^2(n)$ respectively. In particular, $\mathbb{E}\{B_n Y_k\} = 0$ and $\mathbb{E}\{U_n X_k\} = 0$.

We wish to determine the filtering distribution, i.e. the conditional distribution of X_n given $Y_{1:n}$. As we saw in section 1.4, this conditional distribution is Gaussian. Its mean corresponds to the orthogonal projection of X_n onto the sub-space spanned by $Y_{1:n}$, expression (1.45), and its covariance is given by (1.46). Using the notation from section 1.4, we can therefore write that $X_{n|n} = (X_n|Y_{1:n})$, that $X_{n+1|n} = (X_{n+1}|Y_{1:n})$ and that $\mathbb{E}\{Y_{n+1}|Y_{1:n}\} = (Y_{n+1}|Y_{1:n})$.

1. Show that $X_{n+1|n} = a_n X_{n|n}$.

2. Show that $(Y_{n+1}|Y_{1:n}) = c_{n+1} X_{n+1|n}$.

3. Use this result to deduce that G_n exists such that $X_{n+1|n+1} = X_{n+1|n} + G_{n+1} i_{n+1}$ where $i_{n+1} = Y_{n+1} - c_{n+1} X_{n+1|n}$.

4. Let $P_{n+1|n} = (X_{n+1} - X_{n+1|n}, X_{n+1} - X_{n+1|n})$. Show that:

$$\text{var}(i_{n+1}) = c_{n+1}^2 P_{n+1|n} + \sigma_U^2(n+1)$$

5. Show that the r.v. $i_{1:n}$ are independent. Use this to deduce the expression of $p_{Y_{1:n}}(y_{1:n})$ as a function of $i_{1:n}$ and $\text{var}(i_{1:n})$.

6. From this result, deduce that:

$$G_{n+1} = \frac{P_{n+1|n} c_{n+1}}{\sigma_U^2(n+1) + c_{n+1} P_{n+1|n} c_{n+1}}$$

7. Show that:

$$P_{n+1|n} = a_n^2 P_{n|n} + \sigma_B^2(n)$$

then that:

$$P_{n+1|n+1} = (1 - G_{n+1} c_{n+1}) P_{n+1|n}$$

Bringing together all of the previous results, verify that we obtain algorithm (3).

Exercise 5.2 (Denoising an AR-1 using Kalman) (see page 286) The discrete-time signal $Y_n = X_n + U_n$ is observed, where $n \geq 1$. The observation Y_n is obtained with the following program:

```
%===== genenoisyAR1.m
% Generation of the observed signal y
clear all
N=100; a=0.9; sigmab=2;
x=filter(1,[1 -a],sigmab*randn(N,1));
sigmau=3; y=x+sigmau*randn(N,1);
```

This program corresponds to the system described by:

$$\begin{cases} X_n = aX_{n-1} + B_n & \text{(state equation)} \\ Y_n = X_n + U_n & \text{(observation equation)} \end{cases}$$

B_n and U_n are assumed to be two white noises, uncorrelated with each other. Hence, X_n is an AR-1 process and Y_n a noisy AR-1. Let $\rho = \sigma_b^2/\sigma_u^2$. We want estimate X_n using the observed $Y_{1:n}$.

1. Determine, as a function of a and σ_b^2, the expression of the power $\mathbb{E}\left\{X_n^2\right\}$ corresponding to the stationary solution of the state equation. This value will serve as the initial expression for $P_{1|1}$.

2. Determine, as a function of a and ρ, the recursive equation of the Kalman gain, as well as the initial value G_1. Show that the Kalman gain tends toward a limit, and determine the expression of this limit.

3. Write a program that implements the Kalman filter for this model.

Exercise 5.3 (2D tracking) (see page 288) Consider a mobile element in the plane Ox_1x_2. Let $x_1(t)$ and $x_2(t)$ be continuous time functions representing the two components of its position, $\dot{x}_1(t)$ and $\dot{x}_2(t)$ their first derivatives (speeds) and $\ddot{x}_1(t)$ and $\ddot{x}_2(t)$ their second derivatives (accelerations).

If the acceleration is null, i.e. $\ddot{x}_1(t) = 0$ and $\ddot{x}_2(t) = 0$, the trajectory of the mobile element is a straight line. When the acceleration is non-null, this may mean that the vehicle has increased or decreased its speed in a straight line, and/or that the vehicle has deviated from the straight line.

To take account of possible acceleration, $\ddot{x}_1(t)$ and $\ddot{x}_2(t)$ are modeled by two Gaussian, centered white noises of the same variance σ^2. This allows us to track a mobile element with a trajectory which does not follow a straight line.

Let T be the sampling period. For $i = 1$ and $i = 2$, let $X_{i,n} = x_i(nT)$ and $\dot{X}_{i,n} = \dot{x}_i(nT)$, and let $X_n = \begin{bmatrix} X_{1,n} & X_{2,n} & \dot{X}_{1,n} & \dot{X}_{2,n} \end{bmatrix}^T$.

We presume that only the position is observed, and that this observation is subject to additional noise. The position is therefore written $Y_n = CX_n + U_n$ with:

$$C = \begin{bmatrix} 1 & 0 & 0 & 0 \\ 0 & 1 & 0 & 0 \end{bmatrix}$$

and U_n is the observation noise.

1. Use a first-order Taylor approximation to show that:

$$X_{n+1} = AX_n + B_n \text{ with } A = \begin{bmatrix} 1 & 0 & T & 0 \\ 0 & 1 & 0 & T \\ 0 & 0 & 1 & 0 \\ 0 & 0 & 0 & 1 \end{bmatrix}$$

   and where B_n is a noise. Give the covariance matrix as a function of T and σ^2.

2. Suppose that $T = 1/10$ s and that the speed is of the order of 30 m/s (approximately 90 km/h). Explain the connection between σ and a possible acceleration over a duration T. The speed may be considered to vary by a quantity proportional to v_0.

3. Write a function implementing the Kalman algorithm in the case where the parameters of the model are not dependent on time.

   Write a program:

   – which creates noised observations Y_n from a trajectory of any form, for example that obtained using:

```
%===== trajectorygeneration.m
% Trajectory generation
M = 100; M0 = 2; N = M*M0; L0 = N*v0*T;
Xtrue = resample(L0*randn(M0,2),M,1);
```

   – which uses the Kalman algorithm to estimate $X_{n|n}$ and $P_{n|n}$,

   – which displays the confidence ellipse of the estimated position. Remember [2] that the confidence region at $100\alpha\%$ of a random Gaussian vector of dimension 2, mean μ and covariance matrix C is an ellipse with equation $(x-\mu)^T C^{-1}(x-\mu) = -2\log(1-\alpha)$.

Exercise 5.4 (Calibration of an AR-1) (see page 291) In the case of model (5.8), calibration consists of estimating the matrices based on a series of observations, for example using a maximum likelihood-type estimator.

Consider the AR-1 from exercise 5.2. We wish to estimate the three parameters a, σ_b and σ_u, maximizing the likelihood of the observations $Y_{1:N}$. We shall use the `KalmanFilter.m` function coded in exercise 5.3.

Write a program which estimates these parameters by an exhaustive search. As an exercise, you may also wish to use the `fminsearch` function in MATLAB® (Optimization Toolbox) or the EM algorithm [26].

Exercise 5.5 (Calculating the likelihood of an ARMA) (see page 291)
Consider the HMM model defined by:

$$\begin{cases} X_{n+1} &= AX_n + RZ_{n+1} \\ Y_n &= CX_n \end{cases} \tag{5.20}$$

where Z_n is a sequence of independent random variables, with a Gaussian distribution, of mean 0 and variance σ^2, and X_n is a vector of length r, where

$$A = \begin{bmatrix} -a_1 & 1 & 0 & \dots & 0 \\ -a_2 & 0 & 1 & \dots & 0 \\ \vdots & & & & \\ -a_{r-1} & 0 & 0 & \dots & 1 \\ -a_r & 0 & 0 & \dots & 0 \end{bmatrix}, \quad R = \begin{bmatrix} 1 \\ b_1 \\ b_2 \\ \vdots \\ b_{r-1} \end{bmatrix}, \quad C = \begin{bmatrix} 1 & 0 & 0 & \dots & 0 \end{bmatrix}$$

1. Show that Y_n verifies the recursive equation:

$$Y_n + \sum_{m=1}^{r} a_m Y_{n-m} = Z_n + \sum_{k=1}^{r-1} b_k Z_{n-k}$$

Consequently, if $a_j = 0$ for $j > p$ and $b_j = 0$ for $j > q$, and taking $r = \max(p, q+1)$, Y_n verifies:

$$Y_n + \sum_{m=1}^{p} a_m Y_{n-m} = Z_n + \sum_{k=1}^{q} b_k Z_{n-k} \tag{5.21}$$

Thus, if $a(z) = 1 + \sum_{m=1}^{p} a_m z^{-m} \neq 0$ for $|z| \geq 1$, Y_n is an ARMA-(p,q) [2] which is expressed causally as a function of Z_n.

2. Use the Kalman algorithm to calculate, recursively, the log-likelihood $\ell_n = \log p_{Y_{1:n}}(y_{1:n}; \theta)$ of an ARMA associated with parameter $\theta = \{a_{1:p}, b_{1:q}, \sigma^2\}$.

3. Let $\theta = \{a_{1:p}, b_{1:q}, \sigma^2\}$ and $\widetilde{\theta} = \{a_{1:p}, b_{1:q}, 1\}$. Determine the relationship giving the likelihood ℓ associated with θ as a function of the likelihood $\widetilde{\ell}$ associated with $\widetilde{\theta}$.

4. Write a program which calculates the likelihood of an ARMA-(p,q) using the Kalman algorithm.

5. Write a function which uses $a_1, \dots, a_p, b_1, \dots, b_q$ and σ^2 to calculate the first n covariance coefficients of an ARMA. Find the likelihood. Compare this result to those obtained using the program of the previous question.

5.4 Discrete finite Markov case

In cases where the states X_n of an HMM take their value in a finite set of values, the inference on X_n has a closed form expression. This expression is based on two smoothing algorithms, as defined in section 5.2, known as the Baum-Welch or forward-backward algorithms.

Consider an HMM with hidden states X_n, where n ranges from 1 to N, which have values in a finite set $S = \{1, 2, \ldots, S\}$. The distribution of the observations Y_n given $X_n = i$ is taken to have a probability density denoted $g_n(y_n|i)$. Let us denote $p_n(i|j) = \mathbb{P}\{X_n = i|X_{n-1} = j\}$ and $\omega(i) = \mathbb{P}\{X_1 = i\}$.

Exercise 5.6 (Discrete HMM generation) (see page 295) Consider an HMM with $S = 4$ hidden states, with an initial distribution $\omega = \begin{bmatrix}1/2 & 1/4 & 1/8 & 1/8\end{bmatrix}$ and the following transition probability matrix[1]:

$$P = \begin{bmatrix} 0.4 & 0.1 & 0.3 & 0.2 \\ 0.1 & 0.4 & 0.3 & 0.2 \\ 0.3 & 0.1 & 0.4 & 0.2 \\ 0.1 & 0.3 & 0.1 & 0.5 \end{bmatrix} \quad \text{where} \quad P_{j,i} = p(i|j)$$

Let $F_s(x)$ be the cumulative function associated with the conditional distribution of X_n given $X_{n-1} = s$. More precisely:

$$F_s(x) = \sum_{i=1}^{S} \mathbb{1}(x \le i) p(i|s) = \sum_{i=1}^{S} \mathbb{1}(x \le i) P_{s,i}$$

1. Show that:

$$X_{n+1} = \sum_{j=1}^{S} j \times \mathbb{1}(U_n \in [F_{X_n}(j-1), F_{X_n}(j)])$$

   where U_n is a sequence of independent r.v.s with a uniform distribution over $(0, 1)$.

2. Consider that the distribution of the observations conditionally to the states is Gaussian, with respective means μ_i and covariance C_i with $i = 1$ to S. Using the results of exercise 3.2, write a function which generates a sequence of data following the proposed HMM.

Now let us determine two recursive formulas, known as the forward-backward formulas. We shall then consider the way in which these formulas are used in 1 and 2 instant smoothing algorithms, and in the algorithm used to estimate $X_{1:N}$ on the basis of $Y_{1:N}$.

[1]The number in line j denotes the initial state, and the number in column i gives the final state. Thus, $\sum_i P_{j,i} = 1$, i.e. the sum of the elements in a line is equal to 1.

5.4.1 Forward-backward formulas

Let:

$$\alpha_n(i) = \mathbb{P}\{X_n = i | Y_{1:n}\} \quad (5.22)$$

$$\beta_n(i) = p_{Y_{n+1:N}|X_n=i}(y_{n+1:N}|i)\frac{L(y_{1:n})}{L(y_{1:N})} \quad (5.23)$$

where

$$L(y_{1:n}) = p_{Y_{1:n}}(y_{1:n}) \quad (5.24)$$

is the likelihood associated with the observations $Y_{1:n}$ when the distribution of $Y_{1:n}$ has a density. In the case of discrete r.v.s, $p_{Y_{1:n}}(y_{1:n})$ should be replaced by $\mathbb{P}\{Y_{1:n} = y_{1:n}\}$.

Forward recursion

We wish to determine the recursion giving $\alpha_n(i)$ as a function of $\alpha_{n-1}(i)$. We can write:

$$\mathbb{P}\{X_n = i | Y_{1:n}\} = \sum_{j=1}^{S} \mathbb{P}\{X_n = i, X_{n-1} = j | Y_{1:n}\} \quad (5.25)$$

Based on expression (5.3), we have:

$$\begin{aligned}\mathbb{P}\{X_{1:n} = x_{1:n} | Y_{1:n}\} L(y_{1:n}) &= \dots \\ &\mathbb{P}\{X_{1:n-1} = x_{1:n-1} | Y_{1:n-1}\} L(y_{1:n-1}) \times \dots \\ &p_n(x_n|x_{n-1}) \times g_n(y_n|x_n)\end{aligned}$$

Summing on x_k with $k = 1$ to $n-2$, we obtain:

$$\begin{aligned}\mathbb{P}\{X_{n-1} = x_{n-1}, X_n = x_n | Y_{1:n}\} &= \frac{L(y_{1:n-1})}{L(y_{1:n})} \times \dots \\ \mathbb{P}\{X_{n-1} = x_{n-1} | Y_{1:n-1}\} &\times p_n(x_n|x_{n-1}) \times g_n(y_n|x_n)\end{aligned}$$

Applying this expression to (5.25), we obtain:

$$\alpha_n(i) = g_n(y_n|i) \times \frac{L(y_{1:n-1})}{L(y_{1:n})} \times \sum_{j=1}^{S} \alpha_{n-1}(j) p_n(i|j) \quad (5.26)$$

Let us notice that the initial value writes $\alpha_1(i) = g_1(Y_1|i)\,\omega(i)/L(y_1)$.

REMARK: the sum of the $\alpha_n(i)$ for i ranging from 1 to S is equal to 1. For this reason, we simply need to calculate the following terms:

$$\widetilde{\alpha}_n(i) = g_n(y_n|i) \times \sum_{j=1}^{S} \alpha_{n-1}(j) p_n(i|j) \quad (5.27)$$

and then sum these terms over i in order to obtain the normalization constant. This constant is expressed:

$$c_n = \frac{L(y_{1:n})}{L(y_{1:n-1})} = \sum_{i=1}^{S} \widetilde{\alpha}_n(i) \tag{5.28}$$

This result allows us to find an important formula giving the log-likelihood of the N observations:

$$\ell_N = \log L(y_{1:N}) = \sum_{n=1}^{N} \log c_n \tag{5.29}$$

In summary, the forward algorithm is written:

Data: $n = 1 : N$, $y_n \in \mathbb{R}^d$,
$i, j = 1 : S$ ω_i, $g_n(y|i)$, $p_n(i|j)$
Result: $\alpha_n(i)$, c_n, ℓ_n
Initialization:
for $i = 1$ *to* S **do**
| $\widetilde{\alpha}_1(i) = g_1(y_1|i)\omega(i)$;
end
$c_1 = \sum_{i=1}^{S} \widetilde{\alpha}_1(i)$; $\ell_1 = \log c_1$;
for $i = 1$ *to* S **do**
| $\alpha_1(i) = \widetilde{\alpha}_1(i)/c_1$;
end
for $n = 2$ *to* N **do**
| **for** $i = 1$ *to* S **do**
| | $\widetilde{\alpha}_n(i) = g_n(y_n|i) \sum_{j=1}^{S} \alpha_{n-1}(j) p_n(i|j)$;
| **end**
| $c_n = \sum_{i=1}^{S} \widetilde{\alpha}_n(i)$;
| **for** $i = 1$ *to* S **do**
| | $\alpha_n(i) = \widetilde{\alpha}_n(i)/c_n$;
| **end**
| $\ell_n = \ell_{n-1} + \log c_n$;
end

Algorithm 4: Forward recursion

Backward recursion

We now wish to determine the recursion giving $\beta_n(i)$ as a function $\beta_{n+1}(i)$. As an exercise, following an approach very similar to that used to obtain (5.26), we can show that:

$$\beta_n(i) = \frac{L(y_{1:n})}{L(y_{1:n+1})} \times \sum_{j=1}^{S} \beta_{n+1}(j) p_{n+1}(j|i) g_{n+1}(y_{n+1}|j) \tag{5.30}$$

with the final value $\beta_N(i) = 1$ for any i. The backward algorithm is written:

Data: $n = 1 : N$, $i, j = 1 : S$,
$Y_n \in \mathbb{R}^d$, $g_n(y|i)$, $p_n(i|j)$,
c_n (from algorithm 4)
Result: $\beta_n(i)$
Initialization:
for $i = 1$ *to* S **do**
  $\beta_N(i) = 1$;
end
for $n = N - 1$ *to* 1 **do**
  for $i = 1$ *to* S **do**
    $\beta_n(i) = \frac{1}{c_{n+1}} \sum_{j=1}^{S} \beta_{n+1}(j) p_{n+1}(j|i) g_{n+1}(y_{n+1}|j)$;
  end
end

Algorithm 5: Backward recursion

5.4.2 Smoothing with one instant

In this section, for the sake of simplicity, expression $\mathbb{P}\{X_{1:N} = x_{1:N}|Y_{1:N}\} \times L(y_{1:n})$ will be noted $\mathbb{P}\{X_{1:N} = x_{1:N}, Y_{1:N} = y_{1:N}\}$, which is more concise, but only correct in cases where X_n and Y_n are random variables with discrete values. Remember that $L(y_{1:n})$ is obtained using expression (5.24).

We wish to calculate, recursively:

$$\gamma_n(i) = \mathbb{P}\{X_n = i|Y_{1:N}\}$$

Using expression (5.4), we can write that:

$$\begin{aligned} \mathbb{P}\{X_{1:N} = x_{1:N}, Y_{1:N} = y_{1:N}\} &= \mathbb{P}\{X_{1:n} = x_{1:n}, Y_{1:n} = y_{1:n}\} \dots \\ &\times \mathbb{P}\{X_{n+1:N} = x_{n+1:N}, Y_{n+1:N} = y_{n+1:N}\} / \mathbb{P}\{X_n = x_n\} \end{aligned} \tag{5.31}$$

Summing over $x_{1:n-1}$ and $x_{n+1:N}$ and replacing x_n by i, we obtain:

$$\begin{aligned} \mathbb{P}\{X_n = i, Y_{1:N} = y_{1:N}\} &= \mathbb{P}\{X_n = i, Y_{1:n} = y_{1:n}\} \dots \\ &\times \mathbb{P}\{Y_{n+1:N} = y_{n+1:N}|X_n = i\} \end{aligned}$$

Using expressions (5.22), we obtain:

$$\mathbb{P}\{X_n = i, Y_{1:N} = y_{1:N}\} = \alpha_n(i) \times \beta_n(i) \times L(y_{1:N})$$

Therefore:

$$\gamma_n(i) = \alpha_n(i) \times \beta_n(i) \tag{5.32}$$

5.4.3 Smoothing with two instants

We wish to calculate, recursively:

$$\xi_n(i,j) = \mathbb{P}\{X_{n+1} = i, X_n = j | Y_{1:N}\}$$

Starting with expression (5.6), which may be rewritten:

$$\begin{aligned} p_{X_{n:n+1},Y_{1:N}}(x_n, x_{n+1}, y_{1:N}) &= p_{X_n|Y_{1:n}}(x_n, y_{1:n}) \\ &\underbrace{p_{Y_{n+1}|X_{n+1}}(x_{n+1}, y_{n+1})}_{g_{n+1}(y_{n+1}|x_{n+1})} \underbrace{p_{X_{n+1}|X_n}(x_{n+1}, x_n)}_{p_{n+1}(x_{n+1},x_n)} \underbrace{p_{Y_{n+2:N}|X_{n+1}}(x_{n+1}, y_{n+2:N})}_{\propto \beta_{n+1}(x_{n+1})} \end{aligned}$$

we have:

$$\xi_n(i,j) = \alpha_n(j)\beta_{n+1}(i)p_{n+1}(i|j)g_{n+1}(y_{n+1}|i) \tag{5.33}$$

Bringing together the forward and backward algorithms along with expressions (5.32) and (5.33), we obtain a means of calculating smoothing formulas for cases with one and two instants.

5.4.4 HMM learning using the EM algorithm

As we shall see, the one and two instant smoothing formulas are required in order to calculate the auxiliary function of the EM algorithm associated with the estimation of $\theta = (\omega_i, p(i|j), \rho_i)$, where ρ_i is a parameter of the observation distribution $g(y|i) = g(y;\rho_i)$. In this case, we presume that these distributions are not dependent on n.

Let us prove that the auxiliary function of the EM algorithm, as defined by equation (2.97), associated with the distribution of the discrete HMM has the following expression:

$$\begin{aligned} Q(\theta, \theta') &= \sum_{n=1}^{N}\sum_{i=1}^{S} \log g(y_n;\rho_i)\gamma'_n(i) + \sum_{n=1}^{N-1}\sum_{i=1}^{S}\sum_{j=1}^{S} \log p(i|j)\xi'_n(i,j) \\ &\quad + \sum_{i=1}^{S} \log(\omega_i)\gamma'_1(i) \end{aligned} \tag{5.34}$$

where the prime indicates that the calculated quantity is associated with the value θ' of the parameter.

Indeed the joint probability law of $(X_{1:N}, Y_{1:N})$ is written:

$$
\begin{aligned}
p_{X_{1:N},Y_{1:N}}(x_{1:N}, y_{1:N}) &= \prod_{n=1}^{N} g(y_n|X_n = x_n) \times \\
& \prod_{n=1}^{N-1} \mathbb{P}\{X_{n+1} = x_{n+1}|X_n = x_n\} \mathbb{P}\{X_1 = x_1\}
\end{aligned}
$$

Taking its logarithm we have:

$$
\begin{aligned}
\log p_{X_{1:N},Y_{1:N}}(x_{1:N}, y_{1:N}) &= \sum_{n=1}^{N} \log g(y_n|X_n = x_n) \\
&+ \sum_{n=1}^{N-1} \log \mathbb{P}\{X_{n+1} = x_{n+1}|X_n = x_n\} + \log \mathbb{P}\{X_1 = x_1\}
\end{aligned}
$$

Using the identity $f(x_k) = \sum_{i=1}^{S} f(i)\mathbb{1}(x_k = i)$, we obtain:

$$
\begin{aligned}
\log p_{X_{1:N},Y_{1:N}}(x_{1:N}, y_{1:N}) &= \sum_{j=1}^{S}\sum_{n=1}^{N} \log g(y_n|j)\mathbb{1}(X_n = i) \\
+\sum_{i=1}^{S}\sum_{j=1}^{S}\sum_{n=1}^{N-1} \log p(i|j)\mathbb{1}(X_{n+1} = i, X_n = j) &+ \sum_{j=1}^{S} \log \omega_j \mathbb{1}(X_1 = j)
\end{aligned}
$$

Taking the conditional expectation with respect to $Y_{1:N}$ under parameter θ' and using the fact that $\mathbb{E}\{\mathbb{1}(X = i)\} = \mathbb{P}\{X = i\}$, we have:

$$
\begin{aligned}
\mathbb{E}_{\theta'}\{\mathbb{1}(X_n = i)|Y_{1:N}\} &= \mathbb{P}_{\theta'}\{X_n = i|Y_{1:N}\} = \gamma'_n(i) \\
\mathbb{E}_{\theta'}\{\mathbb{1}(X_{n+1} = i, X_n = j)|Y_{1:N}\} &= \mathbb{P}_{\theta'}\{X_{n+1} = i, X_n = j|Y_{1:N}\} = \xi'_n(i,j) \\
\mathbb{E}_{\theta'}\{\mathbb{1}(X_1 = i)|Y_{1:N}\} &= \mathbb{P}_{\theta'}\{X_1 = i|Y_{1:N}\} = \gamma'_1(i)
\end{aligned}
$$

This demonstrates (5.34).

Re-estimation formulas

The maximization of $Q(\theta, \theta')$ with respect to θ is carried out as follows. Canceling the first derivative with respect to ω_i, under the constraint that $\sum_i \omega_i = 1$, we obtain:

$$
\omega_i = \frac{\gamma'_1(i)}{\sum_{j=1}^{S} \gamma'_1(j)} \tag{5.35}
$$

Canceling the first derivative of Q with respect to $p(i|j)$, under the constraint that $\sum_i p(i|j) = 1$ for any j, we have:

$$p(i|j) = \frac{\sum_{n=1}^{N-1} \xi'_n(i,j)}{\sum_{i=1}^{S} \sum_{n=1}^{N-1} \xi'_n(i,j)} \tag{5.36}$$

In addition, we assume that the distribution $g(y|i)$ is Gaussian, with mean μ_i and covariance C_i. Canceling the first derivative of Q with respect to μ_i, we have:

$$\mu_i = \frac{\sum_{n=1}^{N} Y_n \gamma'_n(i)}{\sum_{n=1}^{N} \gamma'_n(i)} \tag{5.37}$$

Similarly, canceling the first derivative of Q with respect to C_i, we have:

$$C_i = \frac{\sum_{n=1}^{N} \gamma'_n(i)(Y_n - \mu_i)(Y_n - \mu_i)^T}{\sum_{n=1}^{N} \gamma'_n(i)} \tag{5.38}$$

Exercise 5.7 (EM algorithm for HMM) (see page 296) Consider an HMM with $S = 4$ discrete states, including the initial state $\omega = \begin{bmatrix}1/2 & 1/4 & 1/8 & 1/8\end{bmatrix}$, with the following matrix of transition probabilities:

$$P = \begin{bmatrix} 0.4 & 0.1 & 0.3 & 0.2 \\ 0.1 & 0.4 & 0.3 & 0.2 \\ 0.3 & 0.1 & 0.4 & 0.2 \\ 0.1 & 0.3 & 0.1 & 0.5 \end{bmatrix} \quad \text{where} \quad P_{j,i} = p(i|j)$$

and where the densities $g(y;\mu_i, C_i)$ are Gaussian, with respective means μ_i and respective covariances C_i. Let us remark that μ_i and C_i are assumed to be independent of n. Let $\theta = (\mu_i, C_i, \omega_i, p(i|j))$ and let $Y_{1:N}$ be a sequence of N observations.

1. Use MATLAB® to write a function which implements algorithms (4) and (5). The inputs consist of the observations, along with ω, $p(i|j)$, μ_i and C_i. The function will perform the sequences α and β and the likelihood.

2. Write a function to estimate θ using the EM algorithm.

3. Test this algorithm using the generator obtained in exercise 5.6.

5.4.5 The Viterbi algorithm

Consider an HMM of which the states have values in a finite set $\mathcal{S} = \{1, \dots, S\}$ of S values. The transition distributions $p_n(i|j) = \mathbb{P}\{X_n = i|X_{n-1} = j\}$ are assumed to be known, as are the probability densities Y_n conditionally to $X_n = i$, denoted $g_n(y|i)$.

We observe y_1, ..., y_N and we wish to determine the sequence x_1, ..., x_N which maximizes $\mathbb{P}\{X_{1:N} = x_{1:N}|y_{1:N}\}$. Maximizing $\mathbb{P}\{X_{1:N} = x_{1:N}|y_{1:N}\}$ with respect to $x_{1:N}$ is equivalent to maximizing the joint distribution of $(X_{1:N}, Y_{1:N})$.

The "brute force" approach involves calculating the joint law for the S^N possible configurations of the sequences $x_{1:N}$. As we shall see, the Viterbi algorithm reduces the number of calculations requiring only NS^2 steps, which is considerably lower than S^N.

We assume that, at time step $n-1$, we have S optimal sequences of length $n-1$ ending with the S possible values of x_{n-1}. In what follows, the sequence finishing with the value $j \in \{1, \ldots, S\}$ at instant $(n-1)$ will be referred to as the j-th path of length $(n-1)$. Let $\text{met}_{n-1}(j)$ be the associated joint probability, known as the path metric. Using (5.3), taking the logarithm and noting $d_n(i|j) = \log \mathbb{P}\{X_{1:n-2} = x_{1:n-2}, X_{n-1} = j, X_n = i, y_{1:n}\}$, we have:

$$d_n(i|j) = \text{met}_{n-1}(j) + \log g_n(y_n|i) + \log p_n(i|j)$$

$d_n(i|j)$ is known as the branch metric (going from j to i at step n). There are S possible ways of extending the j-th path of length $n-1$. However, as we wish to find the maximum metric, only the ascendant giving the maximum metric should be retained. Consequently, the i-th path of length n has the following metric:

$$\text{met}_n(i) = \max_j d_n(i|j)$$

The ascendant giving the maximum value is written:

$$\text{asc}_n(i) = \arg\max_j d_n(i|j)$$

This ascendant should be retained in order to calculate the optimal sequence for step N.

Figure 5.2 shows a calculation diagram for $S = 3$ and $N = 6$, with an oriented graph of $SN = 18$ nodes. This type of representation is known as a lattice. At step n, we calculate the $S^2 = 9$ branch metrics going from $(n-1)$ to n. From these 9 possible extensions, we only retain the $S = 3$ optimal metrics reaching the S nodes of step n, along with their ascendants.

In conclusion, for each stage, we determine the S possible paths of length n with their associated metrics. The calculation is continued up to step N. The optimal sequence is obtained at the end of the process via backtracking.

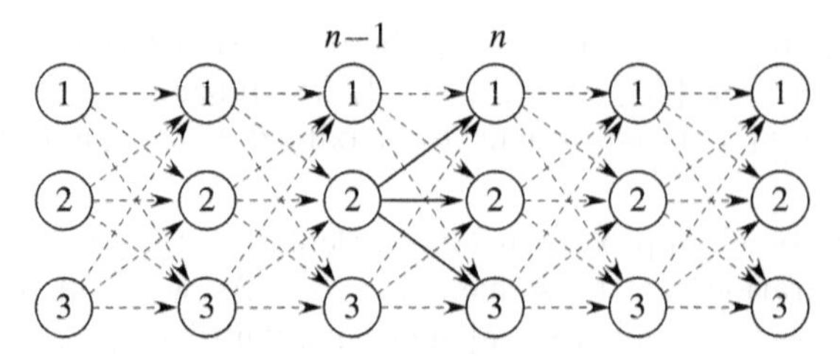

Figure 5.2 – *Lattice associated with $S = 3$ states for a sequence of length $N = 6$*

Data: $g(Y(n), i)$, $p(n, i, j)$,
for n from 1 to N, for i and j from 1 to S
Result: $X(n)$ for n from 1 to N
Initialization: $\text{met}(0, j) = 0$ for j from 1 to S:
for $n = 1$ *to* N **do**
  for $i = 1$ *to* S **do**
    for $j = 1$ *to* S **do**
      $d(i, j) =$
      $\text{met}(n-1, j) + \log g(n, Y(n), i) + \log p(n, i, j)$;
    end
  end
  for $i = 1$ *to* S **do**
    $\text{met}(n, i) = \max_j d(i, j)$;
    $\text{asc}(n, i) = \arg\max_j d(i, j)$;
  end
end
`/* backtracking */`
$X(N) = \arg\max_i \text{met}(N, i)$;
for $n = N - 1$ *to* 1 **by step** -1 **do**
  $X(n) = \text{asc}(n+1, X(n+1))$;
end

Algorithm 6: Viterbi algorithm

The Viterbi algorithm is a dynamic programming algorithm in a lattice. In theory, the optimal sequence is deduced based on the totality of the N observations. In practice, if the observations are time indexed, as in digital communications (see section 6.3.2), we use stopping criteria in order to take intermediate decisions before the final observations have been received.

Example 5.2 Consider an HMM with observations of which the log-likelihoods for each hidden state are elements of the matrix `logG`, the transition matrix for which is given by `A`. Use this data to write a code, in MATLAB®, which extracts the sequence of states using algorithm (6). Note that the choice of likelihoods clearly favors the sequence 1, 2, 3, 4, 5.

The data is produced by the following program:

```
%===== testViterbi.m
clear all
S = 5;
%===== because the diagonal is largely over
% the non diagonal elements,
% the sequence 1 to 5 is clearly promoted
logY = 200*diag(rand(S,1)+1);
logG = -10*randn(3*S,S) + repmat(logY,3,1);
N = size(logG,1);
A = rand(S); logA = log(A);
X = fviterbi(logG,logA);X
```

Use the function:

```
function hatX = fviterbi(logG,logA)
%!====================================================!
%! Viterbi algorithm                                  !
%! SYNOPSIS: hatX = FVITERBI(logG,logA)               !
%! Inputs                                             !
%!     logA(i,j) = log p(i|j)                         !
%!     logG(n,i) = array (N x S) = [ log g(y_n|i) ]   !
%! Outputs                                            !
%!     hatX = "optimal sequence"                      !
%!     pmax = max. of probability                     !
%!====================================================!
N = size(logG,1); S = size(logG,2);
met = zeros(N+1,S); asc = zeros(N,S);
hatX = zeros(N,1); d = zeros(S);
for in = 1:N
    for js=1:S
        for is=1:S
            d(is,js) = met(in,js)+...
                logG(in,is)+logA(is,js);
        end
    end
    for is=1:S
        [met(in+1,is),asc(in,is)] = max(d(is,:));
    end
end
[pmax, hatX(N)] = max(met(N+1,:));
for in=N-1:-1:1
    hatX(in) = asc(in+1,hatX(in+1));
end
```

Chapter 6

Selected Topics

6.1 High resolution methods

In this section, we are going to present methods that belong to the category of what are called subspace methods, and that will provide us with a way of estimating the frequencies of a harmonic mixture by finding the P minima of a single-variable function.

6.1.1 Estimating the fundamental of periodic signals: MUSIC

The algorithm presented hereafter, the acronym of which is *MUSIC*, for *MUltiple SIgnal Characterization*, works better than the DTFT when the frequency differences are much smaller than the inverse of the number of observed points (resolution limit of Fourier). Furthermore, it can be applied to a broader observation model than that of sines corrupted by noise.

Based on the example of a sum of P real sines corrupted by white noise, we are going to end up with equation (6.3) responsible, because of how it is written, for the notable properties of the covariance matrix.

Sum of P real sines corrupted by white noise

Consider the real observation $x(n)$, $n \in \{0, \ldots, N-1\}$, of the type:

$$x(n) = \sum_{k=1}^{P} a_k \cos(2\pi f_k n) + \sum_{k=1}^{P} b_k \sin(2\pi f_k n) + b(n) \tag{6.1}$$

1. $\{f_k\}$ is a sequence of P frequencies, all of them different from one another, belonging to the interval $(0, 1/2)$,

2. $\{a_k\}$ and $\{b_k\}$ are two sequences of P real values,

3. $b(n)$ is a centered, white noise, with the unknown variance σ^2.

Once the sequence f_k has been estimated, the sequences a_k and b_k can be estimated using the least squares method [2, 3].

Let us now develop $s(n+\ell)$. We successively get:

$$\begin{aligned} s(n+\ell) &= \sum_{k=1}^{P} a_k \cos(2\pi n f_k)\cos(2\pi \ell f_k) - \sum_{k=1}^{P} a_k \sin(2\pi n f_k)\sin(2\pi \ell f_k) \\ &\quad + \sum_{k=1}^{P} b_k \cos(2\pi n f_k)\sin(2\pi \ell f_k) + \sum_{k=1}^{P} b_k \sin(2\pi n f_k)\cos(2\pi \ell f_k) \end{aligned}$$

Let $\boldsymbol{\theta} = [\theta_1 \quad \ldots \quad \theta_P]^T$ where $\theta_k = 2\pi f_k$ and let:

$$\boldsymbol{u}_\ell(\boldsymbol{\theta}) = \begin{bmatrix} \cos(2\pi\ell f_1) \\ \vdots \\ \cos(2\pi\ell f_P) \\ \sin(2\pi\ell f_1) \\ \vdots \\ \sin(2\pi\ell f_P) \end{bmatrix} \text{ and } \boldsymbol{s}(n) = \begin{bmatrix} a_1\cos(2\pi n f_1) + b_1 \sin(2\pi n f_1) \\ \vdots \\ a_P\cos(2\pi n f_P) + b_P \sin(2\pi n f_P) \\ -a_1\sin(2\pi n f_1) + b_1 \cos(2\pi n f_1) \\ \vdots \\ -a_P\sin(2\pi n f_P) + b_P \cos(2\pi n f_P) \end{bmatrix}$$

With these notations, we have:

$$s(n+\ell) = \boldsymbol{u}_\ell^T(\boldsymbol{\theta})\boldsymbol{s}(n) \tag{6.2}$$

If we stack M values $s(n+\ell)$ for $\ell = 0, \ldots, M-1$, we can write:

$$\begin{bmatrix} s(n) \\ \vdots \\ s(n+M-1) \end{bmatrix} = \begin{bmatrix} \boldsymbol{u}_0^T(\boldsymbol{\theta}) \\ \vdots \\ \boldsymbol{u}_{M-1}^T(\boldsymbol{\theta}) \end{bmatrix} \boldsymbol{s}(n) = \boldsymbol{A}(\boldsymbol{\theta})\boldsymbol{s}(n)$$

Finally, if we add noise, we end up with the observation model:

$$\boldsymbol{x}(n) = \boldsymbol{A}(\boldsymbol{\theta})\boldsymbol{s}(n) + \boldsymbol{b}(n) \tag{6.3}$$

where we have assumed:

$$\boldsymbol{x}(n) = \begin{bmatrix} x(n) & x(n+1) & \ldots & x(n+M-1) \end{bmatrix}^T$$

$$\boldsymbol{b}(n) = \begin{bmatrix} b(n) & b(n+1) & \ldots & b(n+M-1) \end{bmatrix}^T$$

and where $\boldsymbol{A}(\boldsymbol{\theta})$ is an $(M \times 2P)$ matrix with the expression:

$$\boldsymbol{A}(\boldsymbol{\theta}) = \begin{bmatrix} \boldsymbol{C}(\boldsymbol{\theta}) & \boldsymbol{S}(\boldsymbol{\theta}) \end{bmatrix} \tag{6.4}$$

$$\text{with} \quad \boldsymbol{C}(\boldsymbol{\theta}) = \begin{bmatrix} 1 & \cdots & 1 \\ \vdots & & \vdots \\ \cos(\ell\theta_1) & \cdots & \cos(\ell\theta_P) \\ \vdots & & \vdots \\ \cos((M-1)\theta_1) & \cdots & \cos((M-1)\theta_P) \end{bmatrix}$$

$$\text{and} \quad \boldsymbol{S}(\boldsymbol{\theta}) = \begin{bmatrix} 0 & \cdots & 0 \\ \vdots & & \vdots \\ \sin(\ell\theta_1) & \cdots & \sin(\ell\theta_P) \\ \vdots & & \vdots \\ \sin((M-1)\theta_1) & \cdots & \sin((M-1)\theta_P) \end{bmatrix}$$

Notice that, in expression (6.3), the observation is of the type "signal plus noise", and that the signal part, that is $\boldsymbol{A}(\boldsymbol{\theta})\boldsymbol{s}(n)$, is the product of two terms, one of them, $\boldsymbol{A}(\boldsymbol{\theta})$, depending only on the paramater $\boldsymbol{\theta}$ we wish to estimate, and the other, $\boldsymbol{s}(n)$, depending only on n.

Based on a sequence of N observations $x(0)$, ..., $x(N-1)$, consider the $(M \times M)$ matrix defined by:

$$\widehat{\mathbf{R}}_N = \frac{1}{N-M+1} \sum_{n=0}^{N-M} \boldsymbol{x}(n)\boldsymbol{x}^H(n) \tag{6.5}$$

If we change over to the mathematical expectation on the two sides of (6.5), then use (6.3) and the white noise hypothesis, we get:

$$\mathbb{E}\left\{\widehat{\mathbf{R}}_N\right\} = \frac{1}{N-M+1} \sum_{n=0}^{N-M} \mathbb{E}\left\{\boldsymbol{x}(n)\boldsymbol{x}^H(n)\right\} = \boldsymbol{A}(\boldsymbol{\theta})\boldsymbol{R}_s\boldsymbol{A}^H(\boldsymbol{\theta}) + \sigma^2\boldsymbol{I}_M$$

where $\boldsymbol{I}_M$ is the $M \times M$ identity matrix and where:

$$\boldsymbol{R}_s = \frac{1}{N-M+1} \sum_{n=0}^{N-M} \boldsymbol{s}(n)\boldsymbol{s}^H(n)$$

is a $(2P \times 2P)$ matrix.

We will assume rather than prove that if N is much greater than M, $\widehat{\mathbf{R}}_N$ is a good estimate of the $M \times M$ matrix defined by:

$$\boldsymbol{R} = \boldsymbol{A}(\boldsymbol{\theta})\boldsymbol{R}_s\boldsymbol{A}^H(\boldsymbol{\theta}) + \sigma^2\boldsymbol{I}_M \tag{6.6}$$

The MUSIC algorithm uses the fact that the eigendecomposition of the matrix $\boldsymbol{R}_0 = \boldsymbol{A}(\boldsymbol{\theta})\boldsymbol{R}_s\boldsymbol{A}^H(\boldsymbol{\theta})$ can be obtained directly from that of $\boldsymbol{R}$, and hence from that of its estimate $\widehat{\mathbf{R}}_N$. Before we present the MUSIC algorithm, we are going to determine an important property inferred from the general form of the observation model provided by expression (6.3).

General Form of the Observation Model

Consider the size M complex observation model:

$$\boldsymbol{x}(n) = \underbrace{\boldsymbol{A}(\boldsymbol{\theta})\,\boldsymbol{s}(n)}_{\text{signal}} + \boldsymbol{b}(n) \tag{6.7}$$

where the $M \times P$ matrix $\boldsymbol{A}(\boldsymbol{\theta})$ with $M > P$ is of the type:

$$\boldsymbol{A}(\boldsymbol{\theta}) = \begin{bmatrix}\boldsymbol{a}(\theta_1) & \dots & \boldsymbol{a}(\theta_P)\end{bmatrix} \tag{6.8}$$

where θ_k belongs to a scalar domain Θ. Notice that the same function $\boldsymbol{a}(\theta)$ is used to define the P columns of the matrix $\boldsymbol{A}(\boldsymbol{\theta})$. This model includes of course the case of a sum of P real sines. This is done simply by decomposing each sine with the frequency f_k as the sum of two complex exponentials with the frequencies $\pm f_k$. In that case, the matrix $\boldsymbol{A}(\boldsymbol{\theta})$ is given by expression (6.4).

We now come back to the general model (6.7). The P column vectors of the matrix $\boldsymbol{A}(\boldsymbol{\theta})$ generate in $\mathbb{C}^M$ a P' dimension subspace, with $P' \leq P$, called the *signal subspace*. Its complementary, dimension $(M - P')$ subspace is called the *noise subspace*.

Let $\boldsymbol{s}(n)$ be a centered, length P, complex vector, and let $\boldsymbol{R}_s = \mathbb{E}\{\boldsymbol{s}(n)\boldsymbol{s}^H(n)\}$ be its covariance matrix. Let $\boldsymbol{b}(n)$ be a centered, length M, complex noise such that:

$$\mathbb{E}\left\{\boldsymbol{b}(n)\boldsymbol{b}^H(n)\right\} = \sigma^2\mathbf{I}_M$$

$\boldsymbol{s}(n)$ and $\boldsymbol{b}(n)$ are assumed to be uncorrelated. This means that $\boldsymbol{x}(n)$ is centered and that its covariance matrix has the expression:

$$\boldsymbol{R} = \mathbb{E}\left\{\boldsymbol{x}(n)\boldsymbol{x}^H(n)\right\} = \boldsymbol{A}(\boldsymbol{\theta})\boldsymbol{R}_s\boldsymbol{A}^H(\boldsymbol{\theta}) + \sigma^2\boldsymbol{I}_M = \boldsymbol{R}_0 + \sigma^2\boldsymbol{I}_M \tag{6.9}$$

where we have assumed $\boldsymbol{R}_0 = \boldsymbol{A}(\boldsymbol{\theta})\boldsymbol{R}_s\boldsymbol{A}^H(\boldsymbol{\theta})$. The form of expression (6.9) leads to the following properties:

$\boldsymbol{R}_0$ is a positive matrix with a rank $\leq P$

This is because $\boldsymbol{R}_s$ is a positive matrix, the rank of which is less than or equal to P. Therefore $\boldsymbol{A}(\boldsymbol{\theta})\boldsymbol{R}_s\boldsymbol{A}(\boldsymbol{\theta})^H$ is itself positive with a rank smaller than or equal to P, since it is generated by the P column vectors of $\boldsymbol{A}$.

If the ranks of $\boldsymbol{A}(\boldsymbol{\theta})$ and $\boldsymbol{R}_s$ are equal to P, in other words if $\boldsymbol{A}(\boldsymbol{\theta})$ and $\boldsymbol{R}_s$ are full rank matrices, then $\boldsymbol{R}_0$ is a full rank matrix.

$\boldsymbol{R}$ and $\boldsymbol{R}_0$ have the same eigenvectors

This is because $\boldsymbol{R}_0$ has $P' \leq P$ strictly positive eigenvalues and $(M - P')$ null eigenvalues. This result is a direct consequence of the previous result. The set of these eigenvectors is an orthonormal basis of $\mathbb{C}^M$.

Let $\boldsymbol{v}_1, \ldots, \boldsymbol{v}_{P'}$ be the eigenvectors associated with the strictly positive eigenvalues of $\boldsymbol{R}_0$. We have $\boldsymbol{R}_0\boldsymbol{v}_i = \lambda_i\boldsymbol{v}_i$. If we multiply equation (6.9) on the right by $\boldsymbol{v}_i$, we get $\boldsymbol{R}\boldsymbol{v}_i = (\sigma^2+\lambda_i)\boldsymbol{v}_i$. Therefore $\mu_i = \sigma^2+\lambda_i > \sigma^2$ is an eigenvalue of $\boldsymbol{R}$ associated with the eigenvector $\boldsymbol{v}_i$.

Let $\boldsymbol{g}_1, \ldots, \boldsymbol{g}_K$, with $K = M - P'$, be the eigenvectors associated with the null eigenvalues of $\boldsymbol{R}_0$. We have $\boldsymbol{R}_0\boldsymbol{g}_i = 0$. If we multiply equation (6.9) on the right by $\boldsymbol{g}_i$, we get $\boldsymbol{R}\boldsymbol{g}_i = \sigma^2\boldsymbol{g}_i$. Therefore σ^2 is an eigenvalue of $\boldsymbol{R}$ with multiplicity K, associated with the eigenvectors $\boldsymbol{g}_1, \ldots, \boldsymbol{g}_K$.

Remember that $\boldsymbol{v}_j^H\boldsymbol{g}_k = 0$ for any pair (j, k), which is a consequence of the orthogonality of the eigendecomposition of a positive matrix.

COMMENT: we know that the periodogram of a sum of sines and of a noise tends to σ^2 for the frequency values that are different from the frequencies of the sine components. This result is similar to the previous one stating that the eigenvalues of the noise subspace are all equal to σ^2. The Fourier transform performs some kind of orthogonal decomposition that approximately separates the space in a signal subspace and a noise subspace. In the noise subspace, each component then has the same power σ^2.

As a conclusion, we have the following theorem:

Theorem 6.1 *Let $\boldsymbol{x}(n)$ be the size M complex observation model:*

$$\boldsymbol{x}(n) = \boldsymbol{A}(\boldsymbol{\theta})\boldsymbol{s}(n) + \boldsymbol{b}(n) \tag{6.10}$$

where the $(M \times P)$ matrix $\boldsymbol{A}(\boldsymbol{\theta})$ with $M > P$ is of the type:

$$\boldsymbol{A}(\boldsymbol{\theta}) = \begin{bmatrix}\boldsymbol{a}(\theta_1) & \ldots & \boldsymbol{a}(\theta_P)\end{bmatrix}$$

$\boldsymbol{s}(n)$ is a centered process with the covariance matrix $\boldsymbol{R}_s = \mathbb{E}\left\{\boldsymbol{s}(n)\boldsymbol{s}^H(n)\right\}$, $\boldsymbol{b}(n)$ is a centered white noise with the covariance matrix $\mathbb{E}\left\{\boldsymbol{b}(n)\boldsymbol{b}^H(n)\right\} = \sigma^2\mathbf{I}_M$. $\boldsymbol{s}(n)$ and $\boldsymbol{b}(n)$ are assumed to be uncorrelated. If $\boldsymbol{R}_0 = \boldsymbol{A}(\boldsymbol{\theta})\boldsymbol{R}_s\boldsymbol{A}^H(\boldsymbol{\theta})$ then $\boldsymbol{x}(n)$ is centered and its covariance matrix is such that:

$$\boldsymbol{R} = \boldsymbol{R}_0 + \sigma^2\boldsymbol{I} = \boldsymbol{V}\Lambda\boldsymbol{V}^H + \sigma^2\boldsymbol{G}\boldsymbol{G}^H \tag{6.11}$$

where the matrix $\boldsymbol{V}$ is comprised of $P' \leq P$ orthonormal eigenvectors of $\boldsymbol{R}_0$ associated with strictly positive eigenvalues, where $\Lambda = \mathrm{diag}(\lambda_1, \ldots, \lambda_{P'})$ is the diagonal matrix with these eigenvalues on its diagonal, and where the matrix $\boldsymbol{G}$ is comprised of the $M - P'$ unit eigenvectors of $\boldsymbol{R}_0$ associated with null eigenvalues. We have $\boldsymbol{V}^H\boldsymbol{G} = 0$. Notice that $\boldsymbol{G}^H\boldsymbol{G} = \boldsymbol{I}$ and therefore that $\boldsymbol{G}\boldsymbol{G}^H$ is the orthogonal projector onto the noise subspace.

Theorem 6.1 implies that:

Property 6.1 *The subspace generated by the columns of $\boldsymbol{V}$ coincides with the subspace generated by the columns of $\boldsymbol{A}(\boldsymbol{\theta})\boldsymbol{R}_s$, and both are contained in the space generated by the columns of $\boldsymbol{A}(\boldsymbol{\theta})$. This means that if P' refers to the rank of $\boldsymbol{R}_s$ where $P' \leq P$, then there is a full rank $(P \times P')$ matrix $\boldsymbol{T}$ such that $\boldsymbol{V} = \boldsymbol{A}(\boldsymbol{\theta})\boldsymbol{T}$.*

Estimation Based on the Noise Subspace

Consider the general complex situation presented in theorem 6.1. Starting off with the orthogonality property of $\boldsymbol{A}(\boldsymbol{\theta})$ with the noise subspace, we are going to construct an estimation of $\theta_1, \ldots, \theta_P$ based on the P minima of a *single-variable* function. In the case where $\boldsymbol{a}(\theta)$ is of the complex exponential type, this function is in the form of a trigonometric polynomial.

If we multiply equation (6.11) on the right by $\boldsymbol{G}$, we get $\boldsymbol{R}_0\boldsymbol{G} = 0$ and therefore:

$$\boldsymbol{A}(\boldsymbol{\theta})\boldsymbol{R}_s\boldsymbol{A}^H(\boldsymbol{\theta})\boldsymbol{G} = 0 \tag{6.12}$$

The MUSIC estimator searches for the value of $\boldsymbol{\theta}$ such that $\boldsymbol{A}^H(\boldsymbol{\theta})\boldsymbol{G} = 0$, which means that (6.12) is true. The converse is true if the $(M \times P)$ matrix $\boldsymbol{A}(\boldsymbol{\theta})\boldsymbol{R}_s$ is a full rank matrix.

Remember that for any matrix $\boldsymbol{M}$, we have the equivalence:

$$\boldsymbol{M} = 0 \Longleftrightarrow \text{trace}(\boldsymbol{M}\boldsymbol{M}^H) = 0 \tag{6.13}$$

We simply have to notice that $\text{trace}(\boldsymbol{M}\boldsymbol{M}^H) = \sum_i \sum_k |m_{ik}|^2$, where m_{ik} refers to the generic element of $\boldsymbol{M}$.

Using (6.13) then leads us to the following expression for the MUSIC estimator of the $\boldsymbol{\theta}$ parameter:

Property 6.2 (MUSIC estimator) *The MUSIC estimator of the $\boldsymbol{\theta}$ parameter associated with the model* (6.7) *is:*

$$\widehat{\boldsymbol{\theta}}_{\text{MUSIC}} = \arg\min_{\boldsymbol{\theta} \in \Theta^P} \text{trace}\left(\boldsymbol{A}(\boldsymbol{\theta})\boldsymbol{A}^H(\boldsymbol{\theta})\boldsymbol{G}\boldsymbol{G}^H\right) \tag{6.14}$$

where $\boldsymbol{G}\boldsymbol{G}^H$ is the orthogonal projector onto the noise subspace obtained from the decomposition of the covariance matrix estimate.

If we replace (6.8) in expression (6.14), we get:

$$\begin{aligned}
\text{trace}\left(\boldsymbol{A}(\boldsymbol{\theta})\boldsymbol{A}^H(\boldsymbol{\theta})\boldsymbol{G}\boldsymbol{G}^H\right) &= \sum_{k=1}^{P} \text{trace}(\boldsymbol{a}(\theta_k)\boldsymbol{a}^H(\theta_k)\boldsymbol{G}\boldsymbol{G}^H) \\
&= \sum_{k=1}^{P} \boldsymbol{a}^H(\theta_k)\boldsymbol{G}\boldsymbol{G}^H\boldsymbol{a}(\theta_k)
\end{aligned}$$

Because $\boldsymbol{G}\boldsymbol{G}^H$ is a positive matrix, the minimization is equivalent to finding the P arguments of the P minima that are closest to 0 of the *single*-variable function:

$$J(\theta) = \boldsymbol{a}(\theta)^H \boldsymbol{G}\boldsymbol{G}^H \boldsymbol{a}(\theta) \tag{6.15}$$

Numerical Computation of the P minima

The simplest and broadest method for finding the minima of $J(\theta)$ consists of calculating $J(\theta)$ on a set of values θ sampled on a fine grid. The minima are identified by typing:

```
idxMin=find(diff(sign(diff(J)))==2);
```

This method will be used in the FFT-MUSIC algorithm.

We are now going to see methods for which $\boldsymbol{a}(\theta)$ is comprised of complex exponentials, which means we can use the FFT.

Case where the components are complex exponentials

Consider the particular case where the function can be written:

$$\boldsymbol{a}(\theta) = \begin{bmatrix} 1 & e^{j\theta} & \dots & e^{j(M-1)\theta} \end{bmatrix}^T$$

This includes the case defined by expression (6.4). All we have to do is group together the two columns $\boldsymbol{c}(\theta_k)$ and $\boldsymbol{s}(\theta_k)$ obtained from the sequences $\cos(\ell\theta_k)$ and $\sin(\ell\theta_k)$, then to set $\boldsymbol{a}(\theta_k) = \boldsymbol{c}(\theta_k) + j\boldsymbol{s}(\theta_k)$.

According to (6.15), an estimation of θ_1, ..., θ_P is obtained by determining the P minima of the single-variable function:

$$Q(e^{j\theta}) = \mathbf{a}^H(\theta)\boldsymbol{G}\boldsymbol{G}^H \mathbf{a}(\theta)$$

Notice that $Q(e^{j\theta})$ can also be interpreted as the value, calculated on the unit circle, of the polynomial:

$$Q(z) = \boldsymbol{a}_z^H(1/z^*)\boldsymbol{G}\boldsymbol{G}^H \boldsymbol{a}_z(z) \tag{6.16}$$

where $\boldsymbol{a}_z(z) = \begin{bmatrix} 1 & z & \dots & z^{M-1} \end{bmatrix}^T$ and where $z = e^{j\theta}$. With this notation, we then have $\boldsymbol{a}_z^H(1/z^*) = \begin{bmatrix} 1 & z^{-1} & \dots & z^{-(M-1)} \end{bmatrix}$ and:

$$Q(z) = z^{-(M-1)} \begin{bmatrix} z^{M-1} & \dots & z & 1 \end{bmatrix} \boldsymbol{G}\boldsymbol{G}^H \begin{bmatrix} 1 \\ z \\ \vdots \\ z^{M-1} \end{bmatrix} = z^{-(M-1)}\widetilde{Q}(z) \tag{6.17}$$

$\widetilde{Q}(z)$ is a $2(M-1)$ degree polynomial in z the roots of which come in pairs, since, by construction, if z_0 is a root, then $1/z_0^*$ is a root.

Calculation of the coefficients of the polynomial $\widetilde{Q}(z)$

Let $\boldsymbol{P} = \boldsymbol{G}\boldsymbol{G}^H$ be the matrix with its generating element given by:

$$p_{ij} = \sum_{k=1}^{K} g_{ik} g_{jk}^* \text{ for } i, j = 1, \ldots, M$$

where g_{ik} is the i-th component of the k-th column vector of $\boldsymbol{G} = [\boldsymbol{g}_1 \cdots \boldsymbol{g}_k \cdots \boldsymbol{g}_K]$. The coefficients q_d of the polynomial $\widetilde{Q}(z) = \sum_{d=0}^{2M-2} q_d z^d$ are obtained from the relation (6.17), which is written:

$$\begin{cases} q_d = \sum_{k=0}^{d} p_{M-d+k,k+1} & \text{for} \quad d = 0, \ldots, (M-1) \\ q_d = q_{2M-d-2}^* & \text{with} \quad d = M, \ldots, (2M-2) \end{cases}$$

As you can see, the coefficients q_d are calculated as the sum of the diagonal terms of the matrix $\boldsymbol{G}\boldsymbol{G}^H$ according to the diagram in Figure 6.1.

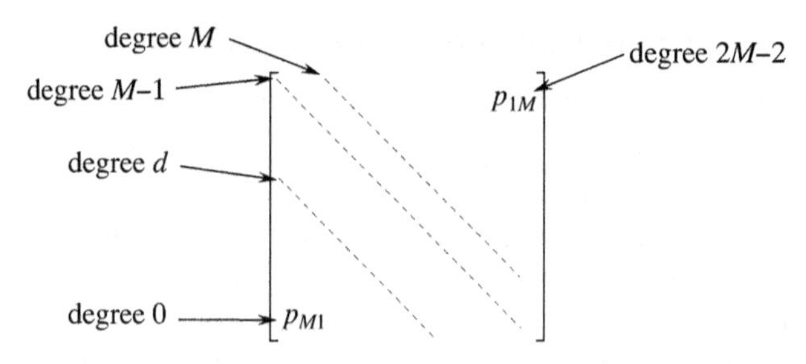

Figure 6.1 – *Calculation of the coefficients of $\widetilde{Q}(z)$ based on $\boldsymbol{P} = \boldsymbol{G}\boldsymbol{G}^H$*

Implementation of the MUSIC Algorithm

The following sums up the MUSIC algorithm in the case of an observation that is the sum of P complex exponential components. In the case where the signal $x(n)$ is the sum of P real sines, all we have to do is apply the algorithm by considering $x(n)$ as a linear combination of complex exponentials.

1. Choose $M > P$.

2. Calculate:

$$\widehat{R}_N = \frac{1}{N-M+1} \sum_{n=0}^{N-M} \boldsymbol{x}(n) \boldsymbol{x}^H(n)$$

with $\boldsymbol{x}(n) = \begin{bmatrix} x(n) & \ldots & x(n+M-1) \end{bmatrix}^H$

3. Calculate the eigendecomposition of $\widehat{R}_N$. Use it to find the $(M \times (M-P))$ matrix $\boldsymbol{G}$, constructed from the $(M-P)$ eigenvectors associated with the $(M-P)$ smallest eigenvalues. Calculate the $(M \times M)$ matrix $\boldsymbol{G}\boldsymbol{G}^H$.

4. Use the previous result to find the coefficients of the polynomial (equation (6.17)):

$$\widetilde{Q}(z) = \begin{bmatrix} z^{M-1} & \dots & z & 1 \end{bmatrix} \boldsymbol{G}\boldsymbol{G}^H \begin{bmatrix} 1 & z & \dots & z^{M-1} \end{bmatrix}^T$$

Once the polynomial $\widetilde{Q}(z)$ is obtained, the estimation of the P values $f_k = \theta_k/2\pi$ can then be achieved, among other possibilities:

1. by calculating the $2(M-1)$ roots of $\widetilde{Q}(z)$, then keeping the P stable roots that are closest to the unit circle. This is called the *root-music* method;

2. or by finding the P minima of $\widetilde{Q}(e^{j\theta})$. This is achieved simply by calculating, with the help of the `fft` function, $\widetilde{Q}(e^{j\theta})$ for $\theta = 2\pi k/L$, where $k \in \{0, \dots, L-1\}$. This is called the *fft-music* method.

The two methods lead basically to the same values if the roots of the polynomial $\widetilde{Q}(z)$ are very close to the unit circle. Remember that they would be on the unit circle if the signal were a perfect mixture of exponentials without noise. The FFT method then has a small advantage, because root finding algorithms often require more computation time. However, in the presence of significant noise, it would seem the root finding method is better suited.

Performances depend on the choice of M. What should be remembered is that M has to tend to infinity when N tends to infinity, but not as fast as N, which is the case for example for $M = N^\gamma$, with $\gamma < 1$, such as $\gamma = 4/5$, or $\gamma = 2/3$.

ROOT-MUSIC The `music(xm,p)` function given below implements the MUSIC algorithm for the estimation of the frequencies of a sum of P complex exponentials by using a root finding method for the polynomial $\widetilde{Q}(z)$. It first calculates $\widehat{R}_n$ based on the expression (6.5) where we chose $M = N^{4/5}$ (try also for example $M = N^{2/3}$). Then it calculates the vectors of the noise subspace with the use of the MATLAB® function `svd`[1] and uses the result to find the coefficients of the polynomial $\widetilde{Q}(z)$ (given by (6.17)). Finally, the function returns the roots of $\widetilde{Q}(z)$ using the MATLAB® function `roots`. Based on the values of these roots, it can then estimate the frequencies f_1, ..., f_P using the `angle` function.

[1]You can also use the MATLAB® function `eig`, since the matrix we are dealing with is square, positive, and therefore its singular value decomposition (`svd` function) and its eigendecomposition (`eigen values`) coincide. However, the `eig` function does not sort the eigenvalues, and must be followed by the MATLAB® function `sort`.

```
function [z]=music(xm,p)
%!==================================================!
%! MUSIC: estimation of the roots of a polynomial !
%! for p complex exponentials in a white noise    !
%! SYNOPSIS: [z]=MUSIC(xm,p)                      !
%!     xm = complex observations                  !
%!     p  = number of complex exponentials        !
%!     z  = estimation of the p complex roots     !
%!          corresponding to the p frequencies    !
%!==================================================!
xm=xm(:); xm=xm-mean(xm); N=length(xm);
%===== M must be > p
M=fix(N^(4/5)); R2x=zeros(M,M);
for ii=1:N-M+1
    idb=ii; ifn=idb+M-1;
    C2x=xm(idb:ifn)*xm(idb:ifn)';
    R2x=R2x+C2x;
end
R2x=R2x/(N-M+1); [u1 d1 v1]=svd(R2x);
GG=v1(:,p+1:M); PP=GG*GG'; QQ=zeros(2*M-1,1);
%===== constructing Q(z)
for d=0:M-1
    for k=0:d
        QQ(d+1) = QQ(d+1) + PP(M-d+k,k+1);
    end
end
QQ(M+1:2*M-1)=conj(QQ(M-1:-1:1));
zz=roots(QQ(2*M-1:-1:1));
%===== the roots with moduli > 1
%      and with non-zero imaginary parts
v=find(abs(zz)<=1 & imag(zz)~=0);
z=sort(zz(v)); mr=length(z); z=z(mr-p+1:mr);
```

This function can be used for signals containing P real sines, simply by considering $2P$ complex exponentials. The function returns a sequence of complex conjugate values (see example 6.1).

FFT-MUSIC The `musicFFT(xm,p)` function implements the MUSIC algorithm for the estimation of the frequencies of a sum of P complex exponentials by searching for the minima of the polynomial on the unit circle. It first calculates $\widehat{R}_n$ based on the expression (6.5) where we chose $M = N^{4/5}$. Then it uses the result to find the coefficients of the polynomial $\widetilde{Q}(z)$ (given by (6.17)). Finally, it returns the P values of z on the unit circle that minimize $\widetilde{Q}(z)$. This is achieved by the FFT computation of `Lfft` values of the polynomial $\widetilde{Q}(z)$. Then the P minima are found first by determining the sign of the derivative between two consecutive values with the use of the command `dQQfft=filter([-1 1],1,QQfft)` (you can also use the `diff` function which returns one less value). When the sign of the derivative goes from -1 to $+1$, it means we just went by a minimum. To run this test, the `filter` function is used once more by executing `dsdQQfft=filter([1 -1],1,sdQQfft)` and comparing the result with the value $(+1) - (-1) = 2$.

The choice of `Lfft` modifies the accuracy of the calculation of the obtained minima. We chose `Lfft=16*1024` in the function function `musicFFT.m`. Restricting the search to the minima by calculating more values of $\widetilde{Q}(z)$ could save you some time, but only in the neighborhoods of the obtained values.

```
function [zout,QQfft]=musicFFT(xm,p)
%!=======================================================!
%! MUSIC: estimation of the frequencies in a mixture !
%! of p complex exponentials in a white noise        !
%! SYNOPSIS: [zout,QQfft]=MUSICFFT(xm,p)             !
%!      xm   = complex observation sequences         !
%!      p    = number of complex exponentials        !
%!      zout = estimation of the p complex roots     !
%!              on the unit circle                   !
%!=======================================================!
xm=xm(:); xm=xm-mean(xm); N=length(xm);
%===== M must be > p
M=fix(N^(4/5)); R2x=zeros(M,M);
for ii=1:N-M+1
    idb=ii; ifn=idb+M-1;
    C2x=xm(idb:ifn)*xm(idb:ifn)'; R2x=R2x+C2x;
end
R2x=R2x/(N-M+1); [u1 d1 v1]=svd(R2x);
GG=v1(:,p+1:M); PP=GG*GG'; QQ=zeros(2*M-1,1);
%===== constructing Q(z)
for d=0:M-1
    for k=0:d
        QQ(d+1) = QQ(d+1) + PP(M-d+k,k+1);
    end
end
QQ(M+1:2*M-1)=conj(QQ(M-1:-1:1)); QQ=QQ(2*M-1:-1:1);
Lfft=16*1024; QQfft=abs(fft(QQ,Lfft));
%===== computation of QQfft(n)-QQfft(n-1)
dQQfft=filter([1 -1],1,QQfft);
%===== if QQfft increases sdQQfft=+1
sdQQfft=sign(dQQfft); dsdQQfft=filter([1 -1],1,sdQQfft);
pos=find(dsdQQfft==2);  [QQfftpos indQQ]=sort(QQfft(pos));
pos=pos(indQQ); pos=pos(1:p);
zout=exp(2*j*pi*(pos-1)/Lfft);
```

Example 6.1 (Implementation and simulations)

1. Write a program that generates a size $N = 60$ sample of the $P = 2$ sines defined by the values:

f_k	0.2	0.21
a_k	1	0.2
b_k	0	0

with a noise added to it such that the SNR is equal to 20 dB.

2. We wish to evaluate the performances depending on the signal-to-noise ratio. This is achieved by setting $N = 60$, $f_1 = 0.2$, $f_2 = 0.21$ and the amplitude ratio to 0.2. The signal-to-noise ratio varies between 10 and 20 dB. The square deviation between the estimator and the real value is a useful performance indicator. In practice, the analytical expression of the result is impossible to obtain, which is why simulations are done by performing a large number of trials. This is called a *Monte-Carlo* simulation.

   Write a program that conducts such a simulation based on 300 trials.

HINTS:

1. Type:

```
%===== applmusic.m
clear
nfft=1024; freq=(0:nfft-1)/nfft;
SNRdB=20;             % SNR
N=80; tps=(0:N-1); % N = number of samples
fq=[0.2 0.21]; p=length(fq); alpha=[1 0.2];
%===== signal
sig=alpha * cos(2*pi*fq'*tps);
vseff=std(sig);
siggmab=sqrt(10^(-SNRdB/10))*vseff/sqrt(2);
b=siggmab*randn(1,N); x=sig+b; % noised signal
%=====
[racm,QQfft]=musicFFT(x,2*p);  % MUSIC
fqm=angle(racm)/(2*pi);
fqaux=sort(fqm);                 % sort the frequencies
fqmo=fqaux(p+1:2*p); regang=2*pi*fqmo*tps;
RR=[cos(regang)' sin(regang)']; ab=RR \ x';
alphamo=sqrt(ab(1:p) .^2 + ab(p+1:2*p) .^2);
%===== displaying the DTFT
subplot(211); plot(freq,20*log10(abs(fft(x,nfft))/N))
set(gca,'xlim',[0 0.5]); zoom xon; grid
LQ=length(QQfft);
subplot(212); plot([0:LQ-1]/LQ,-20*log10(abs(QQfft)));
set(gca,'xlim',[0 1/2]); grid
%===== displaying the results
disp(sprintf('SNR: \t %5.4g dB',SNRdB));
```

```
disp(sprintf('Number of samples: \t %3i',N));
disp(sprintf('Number of sines: \t %3i',p));
disp('True values :')
disp(sprintf('Freq.=%5.4g\t ampl.=%5.4g\t \n',...
              [fq;alpha]))
disp('Estimated values :')
disp(sprintf('\t freq.=%5.4g\t ampl.=%5.4g\t \n',...
            [fqmo';alphamo']))
```

The parameters cannot be properly estimated with the DTFT since $f_2 - f_1 = 0.01 < 1/N \simeq 0.017$. As you can see, in Figure 6.2, the DTFT only shows one maximum instead of two around the frequencies f_1 and f_2.

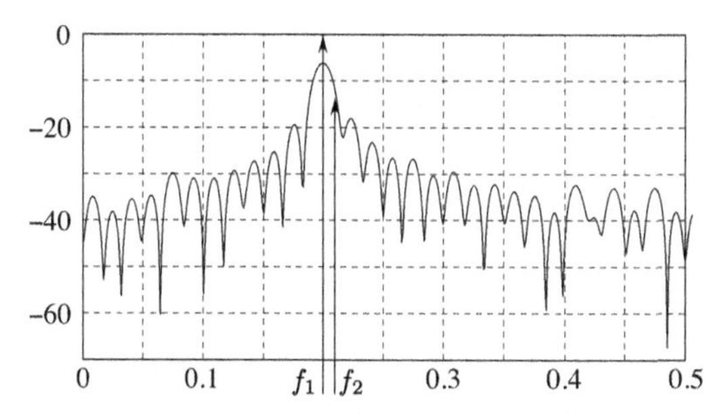

Figure 6.2 – *DTFT of the signal* $x(n) = \cos(2\pi f_1 n) + 0.2\cos(2\pi f_2 n) + b(n)$ *with* $f_1 = 0.2$, $f_2 = 0.21$, $N = 60$. *The signal-to-noise ratio is equal to* 20 *dB*

You can also test the `musicFFT.m` function, simply by typing `racm=musicFFT(x,2*p);` instead of `racm=music(x,2*p);` before executing the program again.

2. `simumusic.m` implements a simulation to evaluate the performances for the measurement of f_1 and f_2 for several values of the signal-to-noise ratio. These results, obtained from 300 trials, are shown in Figure 6.3.

   The graph on the left shows the mean square deviation of the estimation of f_1, and the one on the right shows the mean square deviation of the estimation of f_2, for several values of the signal-to-noise ratio. Notice that in both cases, the square deviation "decreases" as the signal-to-noise ratio increases. Furthermore, performances are better for f_1 than they are for f_2, because the amplitude of the sine with the frequency f_2 is much smaller than the one associated with f_1.

```
%===== simumusic.m
```

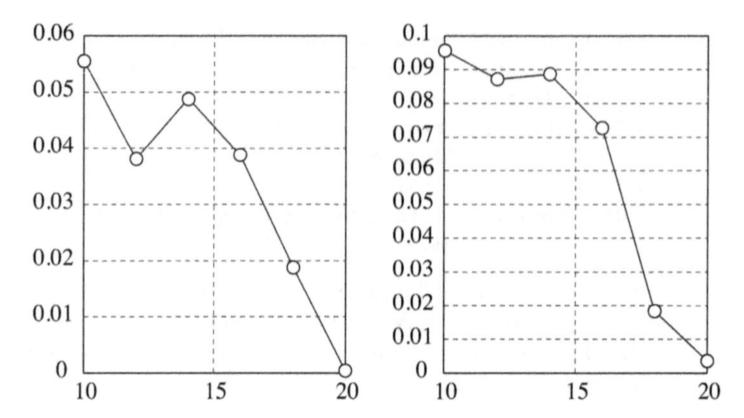

Figure 6.3 – *Performances of the* MUSIC *algorithm. Signal* $x(n) = \cos(2\pi f_1 n) + 0.02\cos(2\pi f_2 n) + b(n)$ *with* $f_1 = 0.2$, $f_2 = 0.21$, $N = 60$. *Square root of the square deviation of the estimation, for each of the two frequencies, evaluated for a length* 300 *simulation, for several values of the signal-to-noise ratio in dB*

```
clear;
N=80; tps=(0:N-1);       % N = number of samples
fq=[0.2 0.21]; alpha=[1 0.2]; p=length(fq);
%===== signal
sig=alpha * cos(2*pi*fq'*tps);
vseff=std(sig)/sqrt(2);
%===== SNR
SNR=(10:2:20); lSNR=length(SNR);
eqmfq=zeros(lSNR,p);
L=30;                    % Number of trials
for ii=1:lSNR
    SNRdB=SNR(ii);
    siggmab=sqrt(10^(-SNRdB/10))*vseff;
    fqmo=zeros(p,L); alphamo=zeros(p,L);
    for jj=1:L
        b=siggmab*randn(1,N);
        %==== noised signal
        x=sig+b; racm=music(x,2*p);
        fqm=angle(racm)/(2*pi); fqaux=sort(fqm);
        fqmo(:,jj)=fqaux(p+1:2*p);
    end
    dfqmo=fqmo-fq'*ones(1,L);
    eqmfq(ii,:)=std(dfqmo');
end
%===== displaying results
subplot(121); plot(SNR,eqmfq(:,1),'o'); grid
hold on; plot(SNR,eqmfq(:,1)); hold off
```

```
subplot(122); plot(SNR,eqmfq(:,2),'o'); grid
hold on; plot(SNR,eqmfq(:,2)); hold off
```

You can also conduct the simulation for the function `musicFFT.m`.

Remember that, in practice, the periodogram is better suited than the MUSIC algorithm as soon as the product RT_s is greater than 3, whatever the value of the signal-to-noise ratio. However, if the product $RT_s < 3$ and if the signal-to-noise ratio is greater than 30 dB, then the MUSIC algorithm should be used rather than the periodogram.

6.1.2 Introduction to array processing: MUSIC, ESPRIT

Figure 6.4 shows a linear antenna comprising $M = 3$ sensors assumed to be identical and separated by the distance d. A source located in the direction θ sends a wave with the carrier frequency F_0. This source is assumed to be far enough for the wave's phase lines to be considered parallel lines. The received signal is sampled at the frequency F_s and $f_0 = F_0/F_s$.

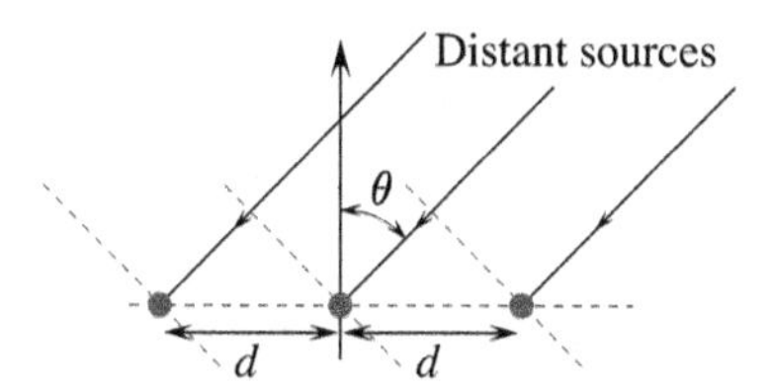

Figure 6.4 – *Linear antenna with three sensors and $P = 1$ distant source assumed to be far away. The dashed lines represent the equiphase lines*

Under these conditions, if $s(n) = e(n)e^{2j\pi f_0 n}$ represents the complex (sampled) signal received by sensor 1, then the signal received by sensor 2, located at a distance d from sensor 1, has the expression $x_2(n) = e(n-\tau)e^{2j\pi f_0(n-\tau)}$, where the delay $\tau = d\sin\theta/c$ and where $\theta \in (-\pi/2, \pi/2)$. c refers to the propagation speed. In radiocommunications, $c = 3 \times 10^8$ m/s.

Narrow Band Hypothesis

Let us assume that $b_s \ll f_0$, where b_s refers to the frequency band of the signal $e(n)$. Then the signal $s(n)$ is said to be a narrow band signal around f_0. In that case, if we use $f_0 = c/\lambda_0$ where λ_0 refers to the wavelength and $\tau = d\sin\theta/c$, then $b_s\tau \ll f_0 d/c = d/\lambda_0$.

When $c = 3 \times 10^8$, the value of $f_0 d/c$ can be small, meaning that the condition $b_s\tau \ll 1$ is met, and hence that $e(n)$ varies little during the time τ.

We can then infer that $e(n-\tau) \approx e(n)$. As a conclusion, when the signals are narrow band and when the propagation speed is very high:

$$x_1(n) = s(n) \quad \text{and} \quad x_2(n) = s(n)e^{2j\pi f_0 \tau} \tag{6.18}$$

This result is no longer true for acoustic propagation (SONAR type propagation), for which the value of c is too small, meaning that the value of $f_0 d/c$ is always high. In that case, it is not always possible to have $b_s\tau \ll 1$. From now on, we will assume that the conditions are those of the first case, meaning that the approximation $e(n-\tau) \approx e(n)$ is valid.

Expressions (6.18) are true for two sensors, and can easily be extended to a set of M sensors, which leads to:

$$\boldsymbol{x}(n) = \begin{bmatrix} x_1(n) & \dots & x_M(n) \end{bmatrix}^T = \boldsymbol{a}(\alpha)s(n)$$

where

$$\boldsymbol{a}(\alpha) = \begin{bmatrix} 1 & e^{2j\pi\alpha} & \dots & e^{2j\pi(M-1)\alpha} \end{bmatrix}^T \tag{6.19}$$

with $\alpha = f_0\tau = df_0 \sin\theta/c$. The function $\boldsymbol{a}(\alpha)$ is called the array manifold. From now on, we will only be considering the case of a linear and uniform antenna for which $\boldsymbol{a}(\alpha)$ is of the type (6.19).

If we use the expression $\lambda_0 = c/f_0$ for the wavelength, we get:

$$\alpha = \frac{d}{\lambda_0}\sin(\theta) \tag{6.20}$$

To achieve an univocal identification of α, α must belong to the interval $(-1/2, +1/2)$, and therefore we must have:

$$\frac{d}{\lambda_0} \leq \frac{1}{2}$$

The distance between two sensors must be smaller than half the wavelength.

If we now assume that there are P sources, located in P directions $\theta_1, \dots, \theta_P$ of the plane, and that the observation is the sum of these P contributions and of a noise, then we can write:

$$\underset{(M\times 1)}{\boldsymbol{x}(n)} = \underset{(M\times P)}{\boldsymbol{A}(\boldsymbol{\alpha})} \times \underset{(P\times 1)}{\boldsymbol{s}(n)} + \underset{(M\times 1)}{\boldsymbol{b}(n)} \tag{6.21}$$

where $\boldsymbol{s}(n) = \begin{bmatrix} s_1(n) & \dots & s_P(n) \end{bmatrix}^T$ and where

$$\boldsymbol{A}(\boldsymbol{\alpha}) = \begin{bmatrix} \boldsymbol{a}(\alpha_1) & \boldsymbol{a}(\alpha_2) & \dots & \boldsymbol{a}(\alpha_P) \end{bmatrix}$$

$\boldsymbol{b}(n)$ is an $(M \times 1)$ noise vector. The noise is assumed to be centered and white. We then have $\mathbb{E}\{\boldsymbol{b}(n)\} = \boldsymbol{0}$ and

$$\mathbb{E}\left\{\boldsymbol{b}(n)\boldsymbol{b}^H(k)\right\} = \delta_{n,k}\sigma^2\boldsymbol{I}_M$$

Notice that the signal model provided by equation (6.21) is identical to the one given by equation (6.7).

We will not extensively discuss the problem of determining the P arrival directions based on the observation of N shots $\{\boldsymbol{x}(1), \ldots, \boldsymbol{x}(N)\}$. We will merely give a few answers.

Classic Beamforming

We saw that the wave originating from source p reaches the M sensors with delays proportional to α_p, hence the idea of multiplicating $\boldsymbol{x}(n)$ on the left by the weighting sequence $\boldsymbol{w}^H = \begin{bmatrix} 1 & e^{2j\pi\alpha} & \ldots & e^{2j\pi(M-1)\alpha} \end{bmatrix}$, then to find the maximum. In the absence of noise, we have:

$$S_{\text{B}}(\alpha) = \frac{1}{N}\sum_{n=1}^{N} |\boldsymbol{a}^H(\alpha)\boldsymbol{x}(n)|^2 = \frac{1}{N}\sum_{n=1}^{N}\sum_{k=1}^{P} |\boldsymbol{a}^H(\alpha)\boldsymbol{a}(\alpha_k)s_k(n)|^2$$

where the function:

$$\boldsymbol{a}^H(\alpha)\boldsymbol{a}(\alpha_p) = \frac{\sin(M(\alpha - \alpha_p))}{\sin(\alpha - \alpha_p)} e^{j(M-1)(\alpha-\alpha_p)}$$

has maxima in $\alpha \approx \alpha_p$ with $p \in \{1, \ldots, P\}$. Everything happens as if the antenna's gain was focused in the direction α_p. This is called *beamforming.* Therefore, to estimate the P arrival directions, we have to find the P maxima of the function:

$$S_{\text{B}}(\alpha) = \boldsymbol{a}^H(\alpha)\widehat{\boldsymbol{R}}_N\boldsymbol{a}(\alpha) \tag{6.22}$$

where $\widehat{\boldsymbol{R}}_N = N^{-1}\sum_{n=1}^{N} \boldsymbol{x}(n)\boldsymbol{x}^H(n)$. The separations between the maxima increase, or in other words the resolution is enhanced when $\min_{i\neq j} |\alpha_i - \alpha_j| \gg 1/M$. This condition can also be written:

$$\min_{i\neq j} \frac{Md}{\lambda_0} 2|\sin((\theta_i - \theta_j)/2)\cos((\theta_i + \theta_j)/2)| \gg 1$$

If we need to distinguish arrival angles separated by e, for small values of e, we can write that:

$$M \times e \times \frac{d}{\lambda_0} \gg 1 \text{ with } \frac{d}{\lambda_0} = \frac{1}{2} \tag{6.23}$$

For $e = 1/10$ radian and for $d/\lambda_0 = 1/2$, we have to set $M \gg 20$.

In the case where condition (6.23) is not well satisfied, the performances of this method become mediocre. Different methods can be used, such as the MUSIC algorithm, seen in section 6.1.1.

The Capon Method

In order to enhance the resolution obtained by the classic approach, Capon suggests in [5] searching for the weighting vector $\boldsymbol{w}_k$ which, on the one hand, minimizes:

$$\mathbb{E}\left\{|\boldsymbol{w}_k^H \boldsymbol{x}(n)|^2\right\} = \boldsymbol{w}_k^H \boldsymbol{R} \boldsymbol{w}_k$$

where $\boldsymbol{R} = \mathbb{E}\left\{\boldsymbol{x}(n)\boldsymbol{x}^H(n)\right\}$ and, on the other, satisfies the constraint $\boldsymbol{w}_k^H \boldsymbol{a}(\alpha_k) = 1$. The idea is to perform a weighted sum of the observations in order to cancel every component, except for the one coming from the direction α_k for which the gain is equal to 1. $\boldsymbol{w}_k$ therefore performs a spatial filtering that focalizes the antenna on the source located in the direction α_k.

The solution is obtained using the Lagrange multiplier method, which consists of solving the following equivalent problem:

$$\left\{\begin{array}{l} \min_{\boldsymbol{w}_k} \left[\boldsymbol{w}_k^H \boldsymbol{R} \boldsymbol{w}_k - \lambda(\boldsymbol{w}_k^H \boldsymbol{a}(\alpha_k) - 1)\right] \\ \boldsymbol{w}_k^H \boldsymbol{a}(\alpha_k) - 1 = 0 \end{array}\right. \tag{6.24}$$

By setting to zero the derivative with respect to $\boldsymbol{w}_k$ of the first expression, we get $\boldsymbol{R}_x \boldsymbol{w}_k - \lambda \boldsymbol{a}(\alpha_k) = 0$, which leads to:

$$\boldsymbol{w}_k = \lambda \boldsymbol{R}_x^{-1} \boldsymbol{a}(\alpha_k)$$

By expressing the fact that $\boldsymbol{w}_k^H \boldsymbol{a}(\alpha_k) = 1$, we get $\lambda = (\boldsymbol{a}^H(\alpha_k)\boldsymbol{R}^{-1}\boldsymbol{a}(\alpha_k))^{-1}$. If we replace it in $\boldsymbol{w}_k$, we get:

$$\boldsymbol{w}_k = \frac{1}{\boldsymbol{a}^H(\alpha_k)\boldsymbol{R}^{-1}\boldsymbol{a}(\alpha_k)} \boldsymbol{R}^{-1}\boldsymbol{a}(\alpha_k)$$

Replacing this in the expression of the criterion leads us to:

$$\boldsymbol{w}_k^H \boldsymbol{R} \boldsymbol{w}_k = \frac{\boldsymbol{a}^H(\alpha_k)\boldsymbol{R}^{-1}\boldsymbol{R}\boldsymbol{R}^{-1}\boldsymbol{a}(\alpha_k)}{(\boldsymbol{a}^H(\alpha_k)\boldsymbol{R}^{-1}\boldsymbol{a}(\alpha_k))^2} = \frac{1}{\boldsymbol{a}^H(\alpha_k)\boldsymbol{R}^{-1}\boldsymbol{a}(\alpha_k)}$$

In an estimation problem, the matrix $\boldsymbol{R}$ is replaced by:

$$\widehat{\boldsymbol{R}}_N = \frac{1}{N}\sum_{n=1}^{N} \boldsymbol{x}(n)\boldsymbol{x}^H(n)$$

The *Capon method* consists of determining the P maxima of the function:

$$S_{\text{Capon}}(\alpha) = \frac{1}{\boldsymbol{a}^H(\alpha)\boldsymbol{R}^{-1}\boldsymbol{a}(\alpha)} \tag{6.25}$$

This expression should be compared with (6.22).

MUSIC

This approach was presented in detail in section 6.1.1. Just remember that, starting with equation (6.21), we end up with the following expression of the covariance matrix:

$$\boldsymbol{R} = \boldsymbol{R}_0 + \sigma^2 \boldsymbol{I}$$

where $\boldsymbol{R}_0 = \boldsymbol{A}\boldsymbol{R}_s\boldsymbol{A}^H$ is assumed to be a full rank matrix. We showed with equation (6.11) that:

$$\boldsymbol{R} = \boldsymbol{V}\Lambda\boldsymbol{V}^H + \sigma^2 \boldsymbol{G}\boldsymbol{G}^H$$

where $\boldsymbol{G}\boldsymbol{G}^H$ refers to the orthogonal projector onto the noise subspace. The MUSIC algorithm then consists of finding the P maxima of the function:

$$S_{\text{MUSIC}}(\alpha) = \frac{1}{\boldsymbol{a}^H(\alpha)\boldsymbol{G}\boldsymbol{G}^H\boldsymbol{a}(\alpha)} \tag{6.26}$$

ESPRIT

The ESPRIT algorithm, short for *Estimation of Signal Parameters via Rotational Invariant Techniques*, was first suggested in [24]. It uses the fact that the antenna can be decomposed in two identical sub-antennas. It also assumes that $\boldsymbol{R}_s$ is a full rank matrix, and therefore that there is (see property 6.1) a full rank $(P \times P)$ matrix $\boldsymbol{T}$ such that:

$$\boldsymbol{V} = \boldsymbol{A}(\boldsymbol{\alpha})\boldsymbol{T}$$

Let $\boldsymbol{A}_1 = [\boldsymbol{I}_{M-1} \;\; 0]\boldsymbol{A}$ and $\boldsymbol{A}_2 = [0 \;\; \boldsymbol{I}_{M-1}]\boldsymbol{A}$. In other words, $\boldsymbol{A}_1$ represents the first $(M-1)$ lines of $\boldsymbol{A}$ and $\boldsymbol{A}_2$ the last $(M-1)$. The expression of $\boldsymbol{A}(\alpha)$ implies that $\boldsymbol{A}_2 = \boldsymbol{A}_1\Omega$ where $\Omega = \text{diag}\left(e^{2j\pi\alpha_1}, \ldots, e^{2j\pi\alpha_P}\right)$.

Let $\boldsymbol{V}_1 = [\boldsymbol{I}_{M-1} \;\; 0]\boldsymbol{V}$ and $\boldsymbol{V}_2 = [0 \;\; \boldsymbol{I}_{M-1}]\boldsymbol{V}$. Because of the expression $\boldsymbol{V} = \boldsymbol{A}\boldsymbol{T}$ $\boldsymbol{V}_1 = \boldsymbol{A}_1\boldsymbol{T}$ and $\boldsymbol{V}_2 = \boldsymbol{A}_2\boldsymbol{T}$, and therefore:

$$\boldsymbol{V}_2 = \boldsymbol{V}_1\boldsymbol{T}^{-1}\Omega\boldsymbol{T}$$

The pseudo-inverse of $\boldsymbol{V}_1$ is denoted by $\boldsymbol{V}_1^{\#} = (\boldsymbol{V}_1^H\boldsymbol{V}_1)^{-1}\boldsymbol{V}_1^H$. By definition $\boldsymbol{V}_1^{\#}\boldsymbol{V}_1 = \boldsymbol{I}$. This means that we can write:

$$\boldsymbol{T}\boldsymbol{V}_1^{\#}\boldsymbol{V}_2 = \Omega\boldsymbol{T}$$

Notice that $\boldsymbol{V}_1^{\#}\boldsymbol{V}_2$ is a $(P \times P)$ matrix.

We are now going to show that, if $\boldsymbol{T}\boldsymbol{A} = \boldsymbol{B}\boldsymbol{T}$, then $\boldsymbol{A}$ and $\boldsymbol{B}$ have the same eigenvalues. This is because if λ is an eigenvalue of $\boldsymbol{A}$ associated with the eigenvector $\boldsymbol{u}$, meaning that $\boldsymbol{A}\boldsymbol{u} = \lambda\boldsymbol{u}$, then we have $\boldsymbol{T}\boldsymbol{A}\boldsymbol{u} = \lambda\boldsymbol{T}\boldsymbol{u}$ on one hand, and $\boldsymbol{T}\boldsymbol{A}\boldsymbol{u} = \boldsymbol{B}\boldsymbol{T}\boldsymbol{u}$ on the other, and therefore $\boldsymbol{B} \times \boldsymbol{T}\boldsymbol{u} = \lambda \times \boldsymbol{T}\boldsymbol{u}$ which leads to the fact that λ is an eigenvalue of $\boldsymbol{B}$ associated with the eigenvector $\boldsymbol{T}\boldsymbol{u}$. Because of this property, and because Ω is a diagonal matrix, we have the following result:

Property 6.3 (ESPRIT estimation) *The ESPRIT estimator of the parameter* $\boldsymbol{\alpha}$ *associated with the model* (6.7) *is* $\widehat{\boldsymbol{\alpha}}_{\text{ESPRIT}} = [\alpha_1, \ldots, \alpha_P]$ *where the* α_1, ..., α_P *are the* P *eigenvalues of the matrix* $\boldsymbol{V}_1^{\#}\boldsymbol{V}_2$ *where* $\boldsymbol{V}_1 = [\boldsymbol{I}_{M-1}\ 0]\boldsymbol{V}$, $\boldsymbol{V}_2 = [0\ \boldsymbol{I}_{M-1}]\boldsymbol{V}$ *and where* $\boldsymbol{V}$ *is defined by equation* (6.11).

The ESPRIT algorithm uses this property: the covariance matrix is estimated based on a sequence of observations $\boldsymbol{x}(n)$, and its eigendecomposition leads to the matrix $\boldsymbol{V}$ associated with the signal subspace of the P highest eigenvalues. We then have to determine the eigenvalues of $\boldsymbol{V}_1^{\#}\boldsymbol{V}_2$. One of the strong points of ESPRIT compared to MUSIC is that it does not require the search for maxima.

Example 6.2 (Comparison of MUSIC and ESPRIT)

1. Write a MATLAB® function that implements the ESPRIT algorithm.
2. Write a function that implements the MUSIC algorithm for antenna processing:
   - either with the FFT, by computing the function $S_{\text{MUSIC}}(\alpha)$ given by expression (6.26) then by determining its P maxima,
   - either by finding the P roots closest to the unit circle with a modulus smaller than 1 of the polynomial $Q(z)$ given by expression (6.17).
3. Write a program that simulates the three sources coming in from the three directions $\theta_1 = -30°$, $\theta_2 = 15°$ and $\theta_3 = 20°$, on an antenna comprising $M = 8$ sensors. The distance between two sensors will be set equal to half the wavelength.

   Evaluate for 100 trials the square deviations for the estimations as functions of the signal-to-noise ratio for $T = 20$ snapshots.

   By referring to condition (6.23), notice that the difference $e = \theta_3 - \theta_1$ is smaller than 1/10th of a radian and that therefore $Med/\lambda_0 = 0.4$.

HINTS:

1. This function implements the ESPRIT algorithm. Type:

```
function z=esprit_doa(xm,P)
%!==============================================!
%! ESPRIT for DOA                                !
%! SYNOPSIS: z=ESPRIT_DOA(xm,P)                  !
%! xm : (M x N) observations of                  !
%!        N snapshots on M sensors (complex data) !
%!  P: number of sources (P<M)                   !
%!  z: eigenvalues of pinv(V1)*V2                !
%!==============================================!
```

```
[M, N]=size(xm);
xm=xm-(xm*ones(N,N)/N); Rx=zeros(M,M);
for ii=1:N, Rx=Rx+xm(:,ii)*xm(:,ii)'; end
Rx=Rx/N;
[UU dd VV]=svd(Rx);
S1=UU(1:M-1,1:P); S2=UU(2:M,1:P);
PHI=S1 \ S2; vp=eig(PHI); z=-angle(vp)';
```

2. This function implements the MUSIC algorithm. Type:

```
function [z]=music_doa(xm,P,method)
%!=======================================================!
%! FFT-MUSIC and ROOTS-MUSIC for DOA                      !
%! SYNOPSIS: [z]=MUSIC_DOA(xm,P,method)                   !
%!    xm     = (M x N) observations of                    !
%!             N Snapshots on M Sensors (complex data)    !
%!    P      = number of Sources (P<M)                    !
%!    method = 'FFT' ou 'ROOTS'                           !
%!    z      = angles of the roots                        !
%!=======================================================!
[M, N]=size(xm);
xm=xm-(xm*ones(N,N)/N); Rx=zeros(M,M);
for ii=1:N, Rx=Rx+xm(:,ii)*xm(:,ii)'; end
Rx=Rx/N; [UU dd VV]=svd(Rx); GG=VV(:,P+1:M);
if strncmp(method,'ROOTS',3)
    %===== FFT-ROOTS
    PP=GG*GG';
    QQ=zeros(2*M-1,1);
    for d=0:M-1
        for k=0:d, QQ(d+1)=QQ(d+1)+PP(M-d+k,k+1); end
    end
    QQ(M+1:2*M-1)=conj(QQ(M-1:-1:1));
    zz=roots(QQ(2*M-1:-1:1));
    %===== keep the closest complex roots to
    %      the unit circle
    v=find(abs(zz)<=1 & imag(zz)~=0);
    rac=sort(zz(v)); mr=length(rac);
     rac=rac(mr-P+1:mr);
    z=-angle(rac)';
elseif method=='FFT'
    %===== FFT-MUSIC
    Lfft=1024; GGf=abs(fft(GG,Lfft)) .^2;
    weights=ones(M-P,1); Smusic=1 ./ (GGf*weights);
    diffSmusic=diff(Smusic);
    S0=max(Smusic);
    %===== look for P maxima
    uM=zeros(1,P);
    for pp=1:P
        [Smax, indmax]=max(Smusic);
        uM(pp)=(indmax-1)/Lfft;
        if uM(pp)>1/2, uM(pp)=-1+uM(pp); end
```

```
        %===== suppress the found maximum
        kkmin=max([indmax-2 1]);
        while diffSmusic(kkmin)>0&kkmin>1,
            kkmin=kkmin-1;
        end
        kkmax=min([indmax Lfft-1]);
        while diffSmusic(kkmax)<0&kkmax<Lfft-1,
            kkmax=kkmax+1;
        end
        Smusic(kkmin:kkmax)=zeros(kkmax-kkmin+1,1);
    end
    z=-uM*2*pi;
else
    error('Method must be: ROOTS or FFT')
end
```

3. Type the following program:

```
%===== testEspritMusic.m
% RMS error for DOA Estimation as a function of the SNR
% (ESPRIT and MUSIC methods)
clear
%===== distance between sensors / wavelength
d_on_lambda0 = 1/2;
M = 8;                          % number of sensors
%===== trying two sequences of vtheta
vtheta = sort([-30 10 20]); % true DOA
%vtheta = sort([-30 15 20]); % true DOA
P = length(vtheta);             % number of sources
T = 20;                         % number of snapshots
%===== SNR list in dB
LISTEsnr=[10:2:18]; ln=length(LISTEsnr);
nbruns=100; %===== number of runs (> 2)
eqmE=zeros(ln,P); eqmM=zeros(ln,P);
for bb=1:ln
    snr=LISTEsnr(bb);
    for rr=1:nbruns
        x=Fgene(d_on_lambda0,M,vtheta,snr,T);
        zzE=esprit_doa(x,P);
        sinthetaE=zzE/(2*pi*d_on_lambda0);
        ind=find(sinthetaE>1);
        sinthetaE(ind)=ones(length(ind),1);
        ind=find(sinthetaE<-1);
        sinthetaE(ind)=-ones(length(ind),1);
        deltathetaE=...
            sort(asin(sinthetaE)*180/pi)-vtheta;
        eqmE(bb,:)=...
            eqmE(bb,:)+deltathetaE.*deltathetaE;
        %===== trying the ROOTS or FFT methods
        zzM=music_doa(x,P,'FFT');
        sinthetaM=zzM/(2*pi*d_on_lambda0);
```

```
            ind=find(sinthetaM>1);
            sinthetaM(ind)=ones(length(ind),1);
            ind=find(sinthetaM<-1);
            sinthetaM(ind)=-ones(length(ind),1);
            deltathetaM=...
                sort(asin(sinthetaM)*180/pi)-vtheta;
            eqmM(bb,:)=...
                eqmM(bb,:)+deltathetaM.*deltathetaM;
        end
        eqmE(bb,:)=eqmE(bb,:)/nbruns;
        eqmM(bb,:)=eqmM(bb,:)/nbruns;
end
for pp=1:P
    subplot(P,1,pp); subplot(3,1,pp);
    plot(LISTEsnr,eqmE(:,pp),'b');
    subplot(3,1,pp); hold on;
    plot(LISTEsnr,eqmM(:,pp),'r'); hold off
end
```

which uses the signal generating function:

```
function x=Fgene(d_on_lambda0,M,vtheta,snr,N)
%!=====================================================!
%! Generates N snapshots on M sensors for DOA vtheta   !
%! SYNOPSIS: x=FGENE(d_on_lambda0,M,vtheta,snr,N,OPT) !
%!    d_on_lambda0 = sensors distance                  !
%!                   and wavelength ratio              !
%!    M      = number of Sensors                       !
%!    vtheta = DOA of sources (P=length(vtheta)<M)     !
%!    snr    = signal to noise ratio in dB             !
%!    N      = number of snapshots                     !
%!=====================================================!
P=length(vtheta);
lz=-2j*pi*(0:M-1)'*sin(vtheta*pi/180)*d_on_lambda0 ;
Az=exp(lz);
%===== P random sources
so=randn(P,N); Pso=sum(trace(so*so'))/N;
so=so/sqrt(Pso); sr=zeros(M,N);
for tt=1:N, sr(:,tt)=Az*so(:,tt); end
nvb=sqrt(1/2)*10^(-snr/20);
x=sr+nvb*(randn(M,N)+j*randn(M,N));
```

Exercise 6.1 (MUSIC 2D) (see page 300) We consider an antenna comprising M sensors assumed to be identical. The vectors $\boldsymbol{r}_m = [x_m\ y_m\ z_m]^T$, with $m \in \{1, \ldots, M\}$, describe the 3D location of the m-th sensor. $K < M$ sources are assumed to be narrow band with a wavelength λ_0. When the plane wave hypothesis is valid (the sources are far from the antenna), the response of the

m-th sensor to the k-th source is given by:

$$a_m(\zeta_k, \varphi_k) = \exp\left(-j\, \boldsymbol{r}_m^T\, \boldsymbol{\beta}(\zeta_k, \varphi_k)\right)$$

where the wave-vector:

$$\boldsymbol{\beta}_k(\zeta_k, \varphi_k) = \frac{2\pi}{\lambda_0}\begin{bmatrix}\sin(\zeta_k)\cos(\varphi_k) & \sin(\zeta_k)\sin(\varphi_k) & \cos(\zeta_k)\end{bmatrix}$$

and where ζ_k is the elevation and φ_k the azimuth (see Figure 6.5) of the direction of propagation of the k-th source.

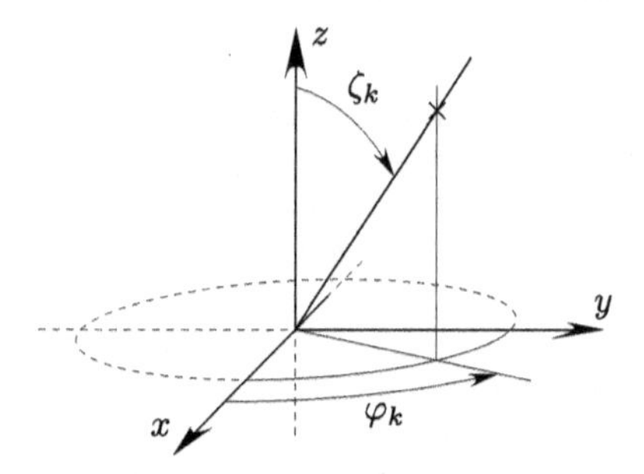

Figure 6.5 – *Angular references*

Under the narrow band assumption, the observed signal may be written:

$$\boldsymbol{x}(n) = \begin{bmatrix}\boldsymbol{a}(\zeta_1, \varphi_1) & \ldots & \boldsymbol{a}(\zeta_K, \varphi_K)\end{bmatrix}\boldsymbol{s}(n) + \boldsymbol{b}(n)$$

1. Determine the expression of the MUSIC function (see formula 6.26).
2. Write a program:
   - that simulates $N = 100$ samples of $\boldsymbol{x}(n)$ for $K = 3$ with $\zeta = [30° \; 40° \; 50°]$, $\varphi = [60° \; 50° \; 20°]$ and with an antenna comprising $M = 25$ sensors located on a grid in the plane xOy;
   - that displays the MUSIC function in 2D.

6.2 Digital Communications

6.2.1 Introduction

Digital communications offer many challenges to people working in signal processing. The new services provided for cellular communications or the internet pose problems which have to do partly with digital signal processing.

Simply put, "making a digital communication" consists of transmitting a continuous-time signal constructed from a message comprised of a sequence

$\{d_k\}$ of bits. To conduct this transmission operation, a device called a *modulator* is used for emitting. When the signal is received, the opposite operation is conducted by a device called a *demodulator*. Therefore, in two-way communications, a modulator and a demodulator are necessary at both ends of the communications line. The word *modem* comes from the contraction of these two words.

According to the characteristic features of the channel, low-pass or high-pass, the modulation is implemented in *baseband*, or on a *carrier frequency*.

Baseband modulation

In *baseband* modulations, the modulation operation consists of producing a signal $x_e(t)$ (Figure 6.6) defined by:

$$x_e(t) = \sum_{k=-\infty}^{+\infty} a_k h_e(t - kT) \tag{6.27}$$

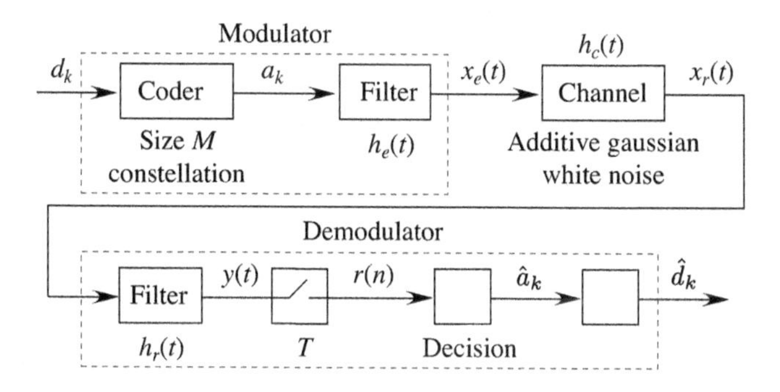

Figure 6.6 – *Baseband modulation and demodulation*

In expression (6.27):

- a_k is a sequence of symbols with possible values in a finite set of M values, called a *constellation*. The sequence a_k is constructed from the sequence of the bits d_k by an coding algorithm that associates a sequence of bits with each symbol of the constellation. Most of the time, the constellation is comprised of $M = 2^N$ real values, and hence, to each possible value in the constellation corresponds a code word with the length:

$$N = \log_2(M) \text{ bits} \tag{6.28}$$

- T represents the time interval between the transmission of two consecutive symbols. The *symbol rate*, or *modulation rate*, expressed in *Bauds*, is defined by:

$$R = \frac{1}{T} \tag{6.29}$$

– The choice of the impulse function $h_e(t)$ depends on the characteristic features of the transmission channel.

"On the other end of the line", the operations that have to be performed to reconstruct the original sequence consist of a filtering, followed by a sampling and a test designed to retrieve the symbols a_k as best as possible. These symbols are then decoded to obtain the sequence of bits d_k. The exercises in this section shed light on the methods used all along the transmission channel.

Carrier frequency modulation

In *carrier frequency* modulations F_0 (Figure 6.7), the transmitted signal has the expression:

$$\begin{aligned} x_e(t) &= \text{Re}\left(\alpha(t)e^{2j\pi F_0 t}\right) \\ &= \text{Re}(\alpha(t))\cos(2\pi F_0 t) - \text{Im}(\alpha(t))\sin(2\pi F_0 t) \end{aligned} \tag{6.30}$$

where

$$\alpha(t) = \sum_{k=-\infty}^{+\infty} a_k h_e(t - kT) \tag{6.31}$$

The symbols a_k are complex. The complex signal $\alpha(t)$ is called the *complex envelope of the real signal $x_e(t)$ with respect to the frequency F_0*. The real and imaginary parts of $\alpha(t)$ are called the *phase component* and the *quadrature component* of $x_e(t)$ respectively.

The complex envelope can be quite useful both in theoretical calculations and in simulation programs. This is due to the fact that the error probabilities, in the presence of additive white noise, do not depend on the choice of the carrier frequency F_0. Therefore, it is useless in a simulation to generate the modulated signal.

The diagram in Figure 6.7 shows an implementation of the modulator for which $h_e(t)$ was assumed to be real. As for the demodulator, the received signal is first processed so as to extract the real and imaginary parts from the complex envelope. Each of the two signals is then filtered and sampled. The two results make up the real and imaginary parts of a sequence of complex observations used by the decision-making system to determine the transmitted sequence of symbols, then the transmitted sequence of bits.

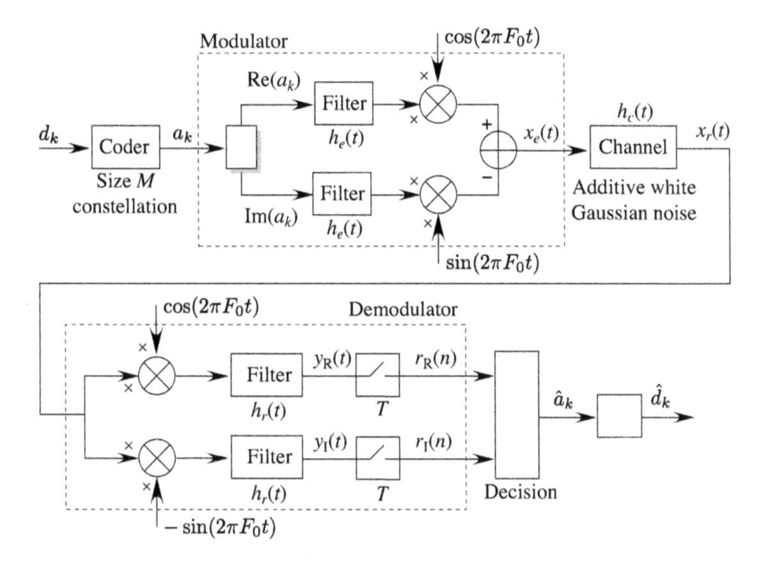

Figure 6.7 – *Carrier frequency modulation and demodulation*

Relation between symbol rate and bit rate

As a result of the operation putting together the bits in groups of N, the time T between the transmission of the two consecutive symbols is equal to N times the time interval T_b between two consecutive bits. If the binary rate is denoted by $D = 1/T_b$, and if we use expressions (6.28) and (6.29), we get formula (6.32), which shows the relation between the *symbol rate*, the *binary rate*, and the *size of the modulation alphabet*:

$$D = R\log_2(M) \tag{6.32}$$

6.2.2 8-phase shift keying (PSK)

We will start with the example of the 8-phase digital modulator that associates the portion of the sinusoidal signal $x(t) = A\cos(2\pi F_0 t + \Phi)$ lasting a duration of T with each 3-bit group. Because there are 8 ways of grouping three bits together, the values of Φ are chosen in the set comprised of the 8 phases regularly spread out between 0 and 2π. This set defined by $\{0,\ 2\pi/8,\ 4\pi/8,\ \ldots,\ 14\pi/8\}$ makes up the constellation.

This constellation is represented in Figure 6.8, which shows a coding example. Notice that the chosen coding is such that the codes of two adjacent symbols differ *by only one bit*. This is called a *Gray* code. We will see (equation (6.44)) what the point of such a coding is for the value of the bit error

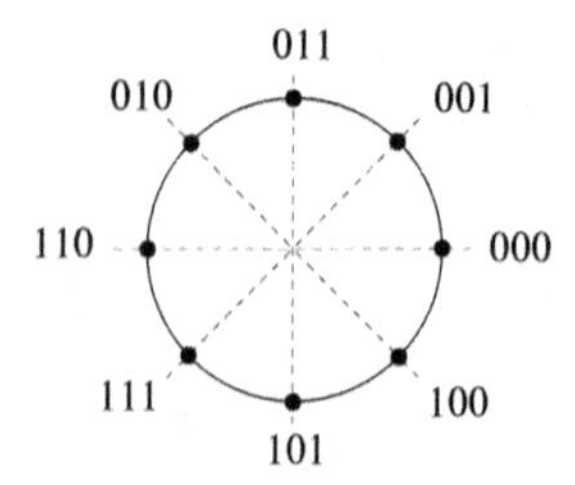

Figure 6.8 – 8-*PSK state constellation*

probability. Figure 6.9 shows the signal corresponding to a sequence of 15 bits.

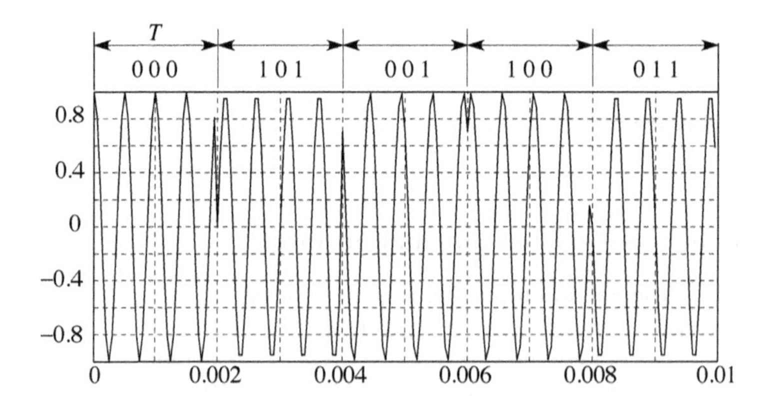

Figure 6.9 – 8-*PSK modulation signal*

Let us check that the complex envelope of a phase modulation is written:

$$\alpha(t) = A \sum_k a_k \text{rect}(t - kT)$$

where the $a_k = e^{j\Phi_k}$ are the complex symbols shown in Figure 6.8. Indeed, the signal $x_e(t)$ in the interval $(kT, (k+1)T)$ is written:

$$\begin{aligned} x_e(t) &= A\cos(2\pi F_0 t + \Phi_k) = A \times \text{Re}\left(e^{2j\pi F_0 t + j\Phi_k}\right) \\ &= \text{Re}\left(A a_k e^{2j\pi F_0 t}\right) = \text{Re}\left(\alpha(t) e^{2j\pi F_0 t}\right) \end{aligned}$$

By referring to expression (6.30), we can conclude that the complex envelope is equal to Aa_k in the interval $(kT, (k+1)T)$, which is the expected result.

Exercise 6.2 (Phase modulator) (see page 301) Consider an 8-PSK modulator and a binary rate of 1,500 bps.

1. Determine the symbol rate.

2. Determine a coding such that two neighboring points of the constellation differ by only one bit (*Gray code*).

3. Write a program that generates the signal transmitted, for a carrier frequency $F_0 = 2$ kHz (take a sampling frequency equal to 20 kHz for the display) and for the binary sequence [000 101 001 100 011].

6.2.3 PAM modulation

We now return to expression (6.27) of a baseband modulation, and we will consider that the sequence $\{a_k\}$ is a sequence with possible values in the constellation comprised of M real symbols and defined by:

$$a_k \in \{-(M-1) : 2 : +(M-1)\} \tag{6.33}$$

Most of the time, M is a power of 2. For instance, if $M = 8$, the alphabet is the set $\{-7, -5, -3, -1, +1, +3, +5, +7\}$. This modulation is called an M state *pulse amplitude modulation*, or M-PAM. Still in the case where $M = 8$, the association of 3-bit sequences with alphabet symbols can be done by using the following Gray code:

Sequence	000	001	011	010	110	111	101	100
Symbol	-7	-5	-3	-1	$+1$	$+3$	$+5$	$+7$

An example of the type of signal transmitted, when $h_e(t)$ is a rectangular impulse with a duration T, is shown in Figure 6.10.

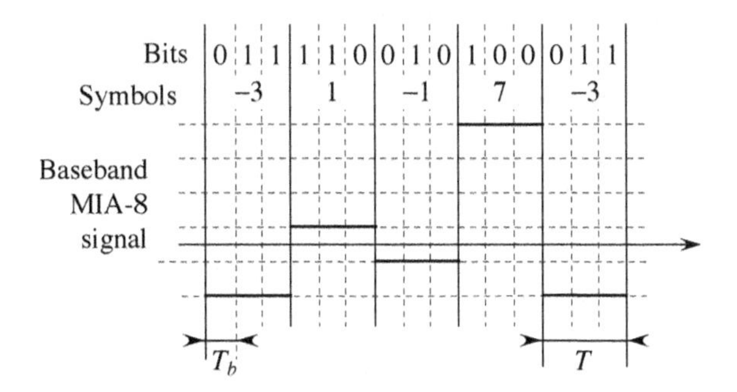

Figure 6.10 – *An example of 8-PAM modulation*

We will see, using formula (6.35), that this signal, in the case where $h_e(t)$ is a rectangular impulse, theoretically takes up an infinite amount of space in the spectrum, meaning that it cannot be transmitted through a B band limited channel without distortion, particularly if $B < 1/T$. We will see in section 6.2.5 that it is then preferable to choose $h_e(t)$ so as to satisfy a criterion better suited to the demodulation problem, called the *Nyquist criterion*.

In any case, the signal is deformed when it is transmitted through the channel. If we assume that the channel acts as a linear filter with the impulse response $h_c(t)$ added to a noise $w(t)$, the received signal is the following:

$$x_r(t) = \sum_{k=-\infty}^{+\infty} a_k h(t - kT) + w(t)$$

where the impulse $h(t) = (h_e \star h_c)(t)$ corresponds to the cascading of the emission filter $h_e(t)$ and $h_c(t)$.

Under the hypothesis that $w(t)$ is a white, centered, Gaussian noise, it can be shown [22] that in order to perform an optimal detection of the sequence of symbols $\{a_k\}$, and hence of the sequence of binary elements $\{d_k\}$, all we need to do is filter the signal $x_r(t)$ as it is received by the *matched* filter with the impulse response $h^*(-t)$, and to make the decision based only on the samples taken at the symbol rate $1/T$ from the matched filter's output.

Generally speaking, the transmission channel acts as a low-pass filter, which causes the signals to stretch out in time. In most cases, such as for instance the phone channel, $h_c(t)$ stretches out beyond T. As a consequence, the signal corresponding to the symbol a_k "overflows" onto the following time intervals. This is called *InterSymbol Interference*, or *ISI*.

This phenomenon is a nuisance as it makes it more difficult to retrieve the symbols from the matched filter's output samples. Even if we assume that the noise is Gaussian and white, the optimal receiver has to use a complex algorithm, called the *Viterbi algorithm*, to retrieve the most likely sequence of symbols (see section 6.3). However, as we will see, retrieving the symbols is very simple in the absence of ISI.

Error probability and signal-to-noise ratio

An important element for determining the performances parameters of the transmission system is the value of the *Bit Error Rate*, or *BER*. For the simplest modulations, it is possible to find an analytical expression [9]. However, most of the time, it is simply estimated using a simulation program that compares the sequence of emitted bits to the sequence of decided bits. Remember, while we are on the subject, that to obtain a 10% accuracy on the error probability P_e with a 70% confidence, we need a test sequence with a length $N \approx 100/P_e$. For $P_e = 10^{-2}$, this means $N = 10{,}000$, making it understandable that this can lead to a long simulation time, even for a fast computer.

The BER is usually plotted against the signal-to-noise ratio E_b/N_0 between the mean energy E_b necessary to send a bit and the quantity N_0, *where* $N_0/2$ *represents the* psd *of a white noise*. We wish to emphasize that N_0 is expressed in Joules, as it should be, because N_0 represents a power spectral density, and is therefore measured in Watt/Hz. The ratio E_b/N_0 can also be expressed as a function of the signal-to-noise ratio of the powers. If P_s refers to the power of the useful signal, and P_b refers to the noise power in the $(-B, +B)$ band, then:

$$P_s = \frac{E_b}{T_b} \text{ and } P_b = N_0 B$$

Hence the signal-to-noise ratio also has the expression:

$$\frac{E_b}{N_0} = \frac{P_s BT}{P_b} = \frac{P_s}{P_b}\frac{B}{R}$$

where R refers to the symbol rate (expression (6.29)).

6.2.4 Spectrum of a digital signal

Consider the digital signal:

$$x(t) = \sum_k a_k g(t - kT + U) \tag{6.34}$$

where $\{a_k\}$ is a *WSS* random sequence with possible values belonging to a finite alphabet. We will use the notations $m_a = \mathbb{E}\{a_n\}$ and $R_a(k) = \mathbb{E}\{a^c_{n+k}a^c_n\}$ where $a^c_n = a_n - m_a$. $g(t)$ is a modulation pulse, and U is a random variable uniformly distributed on $(0, T)$ and independent of the random variables $\{a_k\}$. T represents the time interval separating the transmission of two consecutive symbols. We are going to prove that the signal $x(t)$ is WSS stationary and that its power spectral density has the expression:

$$\begin{aligned} S_x(f) &= \frac{|G(f)|^2}{T}\sum_\ell R_a(\ell)e^{-2j\pi f\ell T} \\ &+ \frac{|m_a|^2}{T^2}\sum_{\ell\neq 0}\left|G\left(\frac{\ell}{T}\right)\right|^2 \delta(f - \ell/T) \end{aligned} \tag{6.35}$$

First, notice that the probability distribution of U has the probability density $p_U(u) = \mathbb{1}(u \in (0, T))/T$. Hence, because the random variables a_k and U are assumed to be independent (the product's expectation is equal to the product of the expectations), we have:

$$\mathbb{E}\{x(t)\} = m_a \sum_k \mathbb{E}\{g(t - kT + U)\} = \frac{m_a}{T}\sum_k \int_0^T g(t - kT + u)du$$

Making the variable change $v = t - kT + u$ and noticing that the integrals for $k \in \mathbb{Z}$ are defined on adjacent intervals, we get:

$$\mathbb{E}\{x(t)\} = \frac{m_a}{T}\sum_k \int_{t-kT}^{t-kT+T} g(v)dv = \frac{m_a}{T}\int_{-\infty}^{+\infty} g(v)dv = \frac{m_a}{T}G(0)$$

where $G(f)$ refers to the Fourier transform of $g(t)$.

We now calculate $\mathbb{E}\{x(t+\tau)x^*(t)\}$. Because a_k and U are independent:

$$\mathbb{E}\{x(t+\tau)x^*(t)\} = \sum_k \sum_n \mathbb{E}\{a_k a_n^*\} \mathbb{E}\{g(t+\tau-kT+U)g^*(t-nT+U)\}$$

If we make the variable change $v = t+\tau-kT+u$ and use the expression $p_U(u) = \mathbb{1}(u \in (0,T))/T$, we get:

$$\begin{aligned}\mathbb{E}\{g(t+\tau-kT+U)&g^*(t-nT+U)\} \\ &= \frac{1}{T}\int_0^T g(t+\tau-kT+u)g^*(t-nT+u)du \\ &= \frac{1}{T}\int_{t-kT}^{t-kT+T} g(v)g^*(v-\tau-(n-k)T)dv\end{aligned}$$

By defining $\ell = (n-k)$, by using the equality $\mathbb{E}\{a_k a_n^*\} = R_a(k-n)+|m_a|^2$ and by noticing that the integrals for $k \in \mathbb{Z}$ are defined on adjacent intervals, we have:

$$\mathbb{E}\{x(t+\tau)x^*(t)\} = \sum_\ell (R_a(\ell)+|m_a|^2)\frac{1}{T}\int_{-\infty}^{+\infty} g(v)g^*(v-\tau+\ell T)dv$$

If we denote by $h(\theta)$ the convolution of $g(\theta)$ with $g^*(-\theta)$, the integral can be written simply as $h(\tau-\ell T)$. This leads us to:

$$\mathbb{E}\{x(t+\tau)x^*(t)\} = \frac{1}{T}\sum_\ell (R_a(\ell)+|m_a|^2)h(\tau-\ell T)$$

which depends only on τ. $x(t)$ is therefore a WSS process. The Fourier transform of $h(\tau-\ell T)$ is $H(f)\exp(-2j\pi\ell T f)$. Because $h(\theta) = g(\theta) \star g^*(-\theta)$, we have $H(f) = G(f)G^*(f) = |G(f)|^2$. If we use the Poisson formula, we infer that:

$$\sum_\ell h(\tau-\ell T) = \frac{1}{T}\sum_m H(m/T)\exp(+2j\pi m\tau/T)$$

If we calculate the Fourier transform of the two sides of $\mathbb{E}\{x(t+\tau)x^*(t)\}$, we then get:

$$\Gamma(f) = \frac{|G(f)|^2}{T}\sum_\ell R_a(\ell)\exp(-2j\pi f\ell T) + \frac{|m_a|^2}{T^2}\sum_m |G(m/T)|^2\delta(f-m/T)$$

where $\delta(f-f_0)$ was used to denote the Fourier transform of $\exp(2j\pi f_0\tau)$. Remember that the psd is precisely $\Gamma(f)$ from which the peak $|\mathbb{E}\{x(t)\}|^2\delta(f)$ at the origin is subtracted, which leads us to the expected result.

As you can see, the spectrum depends, on the one hand on the chosen pulse, and on the other hand on the correlations introduced in the sequence $\{a_k\}$ by way of the expression:

$$S_a(f) = \sum_{\ell} R_a(\ell) e^{-2j\pi f \ell T}$$

This function is periodic with period $1/T$. Therefore all we have to do is calculate it on an frequency interval with a length of $1/T$, or by considering the normalized variable $u = fT$, on an interval with a length of 1. If the sequence $\{a_n\}$ is real, then $S_a(f)$ is an even function, and we can restrict our calculations to the positive frequencies. Theoretically, $S_x(f)$ has an infinite support. However, because of the multiplication of the periodic function $S_a(f)$ by the function $|G(f)|^2$, we can restrict the plotting of $S_x(f)$ to a few length $1/T$ intervals, since $|G(f)|^2$ usually decreases fast, typically like $1/f^2$ for the rectangular pulse. This is why in exercise 6.4, we only represented the function in the frequency interval $(0, 1/T)$, that is to say the interval $(0, 1)$ for the normalized variable $u = fT$.

Peaks can appear in multiples of $1/T$ in the case where $m_a \neq 0$. These peaks can be used to retrieve the symbol rate by filtering the signal after receiving it.

In the particular case where the sequence a_k is a sequence of uncorrelated centered variables with the same variance σ_a^2, $R_a(k) = \mathbb{E}\{a_{n+k}a_n\} = 0$ for $k \neq 0$ and the spectrum's expression comes down to:

$$S_x(f) = \sigma_a^2 \frac{|G(f)|^2}{T} \tag{6.36}$$

In the following exercises, we are going to study, through calculation, then through simulation the coding contribution for two codes of great practical importance: the AMI code and the HDB3 code. We will see in particular that these codes are such that $S_x(f)$ is null in $f = 0$. This is one of the important elements involved in the choice of a modulation, because many systems "pass" the spectrum's components very poorly around 0. Also, a zero gain in 0 can possibly add a continuous components, ensuring the system's power supply.

Exercise 6.3 (AMI code) (see page 303) In AMI coding, the sequence a_k is obtained using the following coding rule: if the bit $d_k = 0$, then $a_k = 0$ is transmitted, and if the bit $d_k = 1$, then $a_k = -1$ and $a_k = 1$ are *alternately transmitted.* AMI stands for Alternate Mark Inversion. We can check that $a_k \in \{-1, 0, 1\}$ is obtained from the $d_k \in \{0, 1\}$ by using the following relations:

$$\begin{cases} s_{k+1} = (1 - 2d_k)s_k \\ a_k = d_k s_k \end{cases}$$

1. Let $\{d_k\}$ be an sequence of random variables, i.i.d. in $\{0, 1\}$, with the probability $\Pr(d_k = 0) = \Pr(d_k = 1) = 1/2$. Calculate $\mathbb{E}\{a_k\}$ and $\mathbb{E}\{a_{n+k}a_n\}$. Use the result to find the expression:

$$S_a(f) = \sum_k R_a(k)e^{-2j\pi kfT}$$

2. Design a program that calculates the theoretical psd of $S_a(f)$ and compares it with the psd estimate obtained through simulation. Use the `welch` function to estimate the psd:

```
function [sf,gamma]=welch(x,lnwin,wtype,Lfft,beta)
%!========================================================!
%! SYNOPSIS: [sf,gamma]=WELCH(x,lnwin,wtype,Lfft,beta)  !
%!        x     = Input sequence                        !
%!        lnwin = analysis window length                !
%!        wtype = cindow: h(ham) or r(rect)             !
%!        Lfft  = FFT length                            !
%!        beta  = confidence parameter                  !
%!        gamma = confidence interval (100*beta%)       !
%!        sf    = spectrum                              !
%!========================================================!
x=x(:); N=length(x); sf=zeros(Lfft,1);
%===== suppression of the trend
[abid x]=tendoff(x);
Ks2=fix(lnwin/2); lnwin=2*Ks2; nbblocks=fix(N/Ks2)-1;
if (wtype(1)=='h')
    fen=0.54-0.46*cos(2*pi*(0:lnwin-1)/lnwin);
    elseif (wtype(1)=='r')
        fen=ones(1,lnwin);
    else
        disp('Unknown window'); return
end
wfen=(1/sqrt(fen*fen'))*fen';
for tt=1:nbblocks
    ti=(tt-1)*Ks2+1;tf=ti+lnwin-1; pt=x(ti:tf) .* wfen;
    apf=abs(fft(pt,Lfft)) .^2; sf=sf+apf;
end
sf=sf/nbblocks; gamma=sqrt(2)*erfinv(beta)/sqrt(nbblocks);
return
```

Exercise 6.4 (HDB3 code) (see page 304) In AMI coding, see exercise 6.3, the presence of a long sequence of zeros can cause the receiver to desynchronize. We then have to make sure never to transmit more than three consecutive zeros. In HDB3 coding (HDB stands for *High Density Bipolar*) solves this problem in the following way: when we encounter a sequence of four consecutive zeros, the fourth zero is coded as a “1”. To prevent any ambiguity when decoding, this 1 is coded by violating the alternation rule: this is called *bipolar violation*.

EXAMPLE: consider the sequence 101100000000000010. Its coding leads to:

bit:	+1	0	+1	+1	0	0	0	0	0	0	0	0	0	0	0	0	+1	0
symb.:	+1	0	−1	+1	0	0	0	+1	0	0	0	+1	0	0	0	+1	−1	0

Locally, this sequence has a mean different from 0. To avoid this local decentering phenomenon, an alternating rule is applied to the bipolar violation. In order to do this, we have to introduce a additional variable (p_v) that memorizes the bipolar violation.

If p_1 denotes the variable used to store the polarity of the last bit coded as "1", then we have the following algorithm (see the diagram in Figure 6.11):

1. If the last bit coded as 1 is transmitted with a + polarity, which is memorized as $p_1 = +1$, then the sequence 0000 is associated with either the sequence $0\,0\,0+1$, or the sequence $-1\,0\,0-1$, depending on whether the bit $p_v = -1$ or $+1$ respectively. Then the polarity of p_v is changed, and we redefine $p_1 = p_v$.

2. In the other case, $p_1 = -1$, the sequence 0000 is associated with either $+1\,0\,0+1$, or $0\,0\,0-1$, depending on whether the bit $p_v = -1$ or $+1$ respectively. Then the polarity of p_v is changed, and we redefine $p_1 = p_v$.

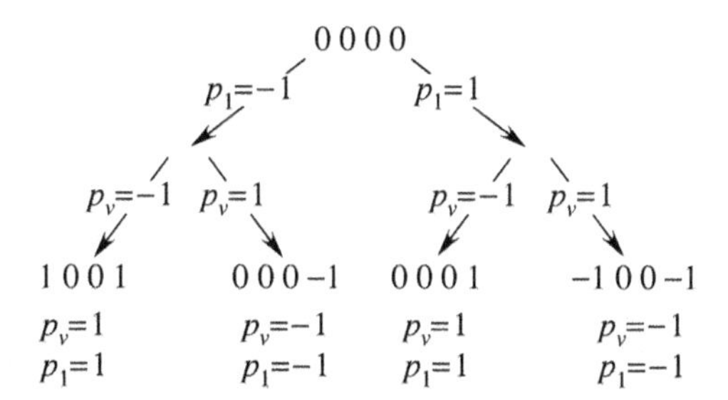

Figure 6.11 – *HDB3 coding: processing the four consecutive zeros and updating the bipolar violation bits*

It can be shown that the resulting sequence is centered. The decoding is particularly simple: if a bipolar violation is encountered, the last 4 bits are set to zero.

1. Starting with the initial conditions $p_v = +1$ and $p_1 = -1$, apply HDB3 coding to the binary sequence:

   0 1 1 1 0 0 0 0 1 0 0 0 0 0 1 0

2. Write a MATLAB® function that performs the HDB3 coding of any sequence of bits.

3. With the use of the `welch` function (see exercise 6.3), perform the signal's spectrum estimation based on a sequence of 10,000 symbols and for a rectangular pulse with a duration equal to the symbol rate.

Exercise 6.5 (Linear equalization of a communications channel) (see page 306) Consider the double side band modulation with the complex envelope:

$$\alpha(t) = \sum_k a_k h_e(t - kT)$$

The carrier frequency value F_0 is not specified since all of the processings will apply to the complex envelopes of the signals encountered in different places along the transmission line.

Let $a_k = u_k + jv_k$ where the possible values of u_k and v_k belong to $\{-3, -1, +1, +3\}$. This type of modulation is called Quadrature Amplitude Modulation (QAM).

The received signal is sampled at a rate $1/T$. Let $\{x_k\}$ be the sequence of the samples. We will assume that the channel introduces by filtering an intersymbol interference affecting $L = 2$ symbols, meaning that:

$$x_k = h_0 a_k + h_1 a_{k-1} + w_k$$

where w_k refers to a white, centered, complex, Gaussian, additive noise. A simple, but not very effective idea is to invert the filter $H(z) = h_0 + h_1 z^{-1}$.

1. Represent the constellation associated with this code. Find a Gray code associated with this constellation.

2. Determine a causal approximation of $G(z) = 1/H(z)$ using a 21 coefficient FIR filter in the following two cases: $h_0 = 1, h_1 = -0.6$ and $h_0 = 1, h_1 = -1.6$.

3. Write a program that represents, in the complex plane, the samples that are emitted, received, and processed by the inverse filter .

6.2.5 The Nyquist criterion in digital communications

Expressing the Nyquist criterion for PAM modulation

Given the digital signal $\sum_k a_k h_e(t - kT)$, corresponding to a PAM modulation, we are going to try to determine the conditions that the pulse $h_e(t)$ has to satisfy so that the emitted signal, once it has travelled through a B *band limited ideal low-pass channel* (its complex gain has the expression $H_c(f) = \mathbb{1}(-B \leq f \leq$

B)), then gone through the matched filter with the impulse response $h_e^*(-t)$, leads an output with *no intersymbol interference at the sampling times.*

Since the channel is B band, we can choose a B band limited signal without being any less general. In that case the received signal can be written:

$$x_r(t) = \sum_k a_k h_e(t - kT) + w(t)$$

where a_k refers to a sequence of symbols from the alphabet $\{-(M-1), \cdots, -3, -1, +1, +3, \cdots, +(M-1)\}$. If we assume that $p(t) = h_e(t) \star h_e^*(-t)$, and calculate the Fourier transform, we get:

$$P(f) = |H_e(f)|^2$$

Hence, if $P(f)$ is B band limited, that is if $P(f) = 0$ for $|f| > B$, then $h_e(t)$ is B band limited itself. This condition precisely expresses the constraint that has to be satisfied if the channel is B band limited. If $y(t)$ now refers to the output signal of the matched filter[2] with the impulse response $h_e^*(-t)$, then we can write:

$$y(t) = \sum_k a_k p(t - kT) + b(t)$$

where $b(t)$ represents the noise $w(t)$ after it has been filtered. The samples taken from the sampler's output at the times nT then have the expression:

$$\begin{aligned} y(nT) &= \sum_k a_k p(nT - kT) + b(nT) \\ &= a_n p(0) + \underbrace{\sum_{k \neq n} a_k p((n-k)T)}_{\text{ISI}} + b(nT) \end{aligned} \tag{6.37}$$

The first term is directly related to the symbol a_n emitted at the time nT. The second one represents the contribution, at the sampling times, of all the symbols emitted other than a_n. This term is called the *InterSymbol Interference.*

A situation of particular practical importance is the one where $p(t)$ satisfies the two following conditions, called the *Nyquist criterion*:

1. $P(f) = 0$ for $|f| > B$, where B refers to the channel's bandwidth;

2. $p(kT) = 0$ for $k \neq 0$.

[2] All of the results in this section are still true in the case of side band modulations with the carrier frequency F_0 if we replace $x_r(t)$ with its complex envelope with respect to F_0.

According to the Poisson formula, condition 2 is equivalent to:

$$\sum_{k=-\infty}^{+\infty} P\left(f - \frac{k}{T}\right) = T \sum_{m=-\infty}^{+\infty} p(mT)\exp(-2\pi jmfT) = Tp(0)$$

This expression means that the algebraic sum of the spectra shifted by $1/T$, $2/T$, etc. is equal to a constant. If we introduce the symbol rate $R = 1/T$, we can determine a necessary condition for the Nyquist criterion to be satisfied on a B band channel. This condition is expressed:

$$B > \frac{R}{2} \tag{6.38}$$

Let us examine how the Nyquist criterion leads to a simplified decision rule. If $p(t)$ satisfies the Nyquist criterion, then the expression (6.37) giving $y(nT)$ can be simplified and we have:

$$y(nT) = a_n p(0) + b(nT) \tag{6.39}$$

Because $y(nT)$ only depends on one symbol, and because it can be shown (you can do it as an exercise) that the noise samples are independent, since they are uncorrelated and Gaussian, the decision can be taken *symbol by symbol* simply by comparing $y(nT)$ to thresholds.

What we did is start out with a transmission system designed for a pulse $h_e(t)$ such that $p(t) = h_e(t) \star h_e^*(-t)$ verifies the Nyquist criterion, and in the end, the reception set is composed of:

- a matched filter with the impulse response $h_e^*(-t)$;
- a sampler at the rate T;
- and a symbol-by-symbol decision-making system that compares the sample with thresholds.

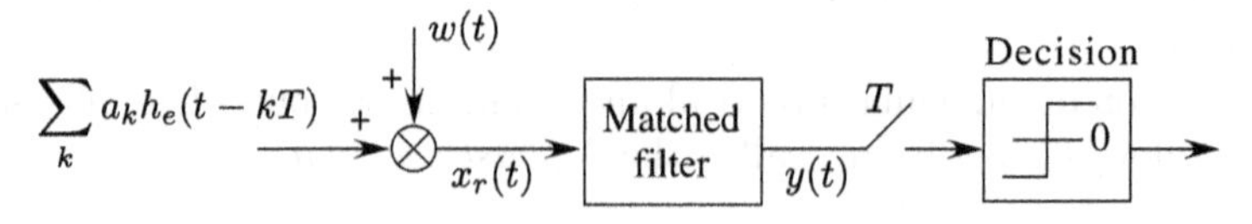

Figure 6.12 – *Diagram of the receiver*

For example, with an PAM modulation with four symbols $\{-3, -1, +1, +3\}$, the decision rule is as follows:

Observation	Decision
$y(nT) < -2p(0)$	$\widehat{a}_n = -3$
$-2p(0) \leq y(nT) < 0$	$\widehat{a}_n = -1$
$0 \leq y(nT) < 2p(0)$	$\widehat{a}_n = +1$
$2p(0) \leq y(nT)$	$\widehat{a}_n = +3$

When the ISI is low but not completely non-existent, we can still use symbol-by-symbol decision. To explain what a low ISI means, we have to go back to expression (6.37). The most unfavorable case concerning the use of symbol-by-symbol decision occurs when all the symbols other than a_n interfere destructively with the amplitude $a_n p(0)$. For example, if $a_n = 1$ and $p(0) > 0$, all the other symbols assume the value $\pm(M-1)$ causing each contribution to be negative. This leads us to the following definition on how to measure the level of ISI, also called *the maximum ISI*:

$$D = \frac{(M-1)\sum_{k\neq 0}|p(k)|}{|p(0)|} \tag{6.40}$$

If D is close to 0, the ISI is low and the symbol-per-symbol detector will yield excellent results. Otherwise, optimal processing requires a more complex algorithm called the Viterbi algorithm (see page 211) that takes into account all of the received symbols. In section 6.2.6, we will discuss another tool for evaluating the ISI called the *eye pattern*.

The raised cosine Nyquist function

We are now going to define a set of real functions $h_e(t)$ depending on one parameter, and such that $p(t) = h_e(t) \star h_e(-t)$ satisfies the Nyquist criterion. These functions play a major practical role. Consider the function:

$$h_e(t) = \frac{1}{\pi\sqrt{T}} \frac{4\alpha\dfrac{t}{T}\cos\left((1+\alpha)\pi\dfrac{t}{T}\right) + \sin\left((1-\alpha)\pi\dfrac{t}{T}\right)}{\dfrac{t}{T}\left(1-16\alpha^2\dfrac{t^2}{T^2}\right)} \tag{6.41}$$

where the parameter $\alpha \in (0,1)$ is called the *roll-off factor*.

It can be shown by way of a long calculation that $p(t) = h_e(t) \star h_e(-t)$ satisfies the Nyquist conditions, meaning that $p(kT) = 0$ for $k \neq 0$ and $P(f) = 0$ for $|f| > B$ with:

$$B = \frac{1}{2T}(1+\alpha) \tag{6.42}$$

Theoretically, $h_e(t)$ has an infinite duration. However, the function quickly becomes evanescent, and we can maintain satisfactory properties if we truncate the function to keep about a dozen lobes around 0.

The `racnyq.m` function generates the samples of $h_e(t)$ based on the roll-off factor `alpha`, the number of lobes `nblobes` remaining to the right and left of 0, as well as the number of samples `Npts` on the interval with a duration T. Save this function as `racnyq.m`:

```
function hracNyq=racnyq(alpha,nblobes,Npts);
%!=======================================================!
%! RACNYQ: Square root raised cosine response            !
%! SYNOPSIS: hracNyq=RACNYQ(alpha,nblobes,Npts)          !
%!      alpha   = Roll-off                               !
%!      nblobes = Response length (even symbol number)   !
%!      Npts    = Number of points per symbol            !
%!      hracNyq = Response                               !
%!=======================================================!
deminblobes=nblobes/2; tsurT=(1:deminblobes*Npts)/Npts;
a4tsurT=4*alpha*tsurT;
gamma1=pi*(1+alpha)*tsurT; gamma2=pi*(1-alpha)*tsurT;
%=====
num= a4tsurT  .* cos(gamma1) + sin(gamma2);
den=(1- a4tsurT .* a4tsurT) .* tsurT;
%===== den(t0)=0 if tsurT=1/(4*alpha)
t0=find(abs(den) <=  2.2204e-14);
lh=length(num);
if isempty(t0)
   hracNyq=num ./ den;
else
   C1=pi*(1+alpha)/(4*alpha);
   hnul=(-0.5*cos(C1)+(pi/4)*sin(C1))*4*alpha;
   hracNyq=[num(1:t0-1) ./ den(1:t0-1) hnul ...
          num(t0+1:lh) ./ den(t0+1:lh)];
end
h0=4*alpha+pi*(1-alpha);
hracNyq=[hracNyq(lh:-1:1) h0 hracNyq]';
hracNyq=hracNyq/norm(hracNyq);
```

A long but not at all difficult calculation shows that:

$$P(f) = \begin{cases} T & \text{for } |f| < \dfrac{1-\alpha}{2T} \\ \dfrac{T}{2}\left[1 - \sin\left(\dfrac{\pi T}{\alpha}(f - 1/2T)\right)\right] & \text{for } \dfrac{1-\alpha}{2T} < |f| < \dfrac{1+\alpha}{2T} \\ 0 & \text{for } |f| > \dfrac{1+\alpha}{2T} \end{cases} \tag{6.43}$$

with $\alpha \in (0, 1)$. $P(f)$ is called a *raised cosine* function.

Let us now check numerically that the function $p(t)$ satisfies the Nyquist criterion. Let $R = 1/T = 1{,}000$ symbols per second and $B = 600$ Hz (this way we have $B > R/2$). Hence, according to (6.42), $\alpha = 2B/R - 1 = 0.2$. To perform the computation, we will take 10 points per symbol time T and truncate $h_e(t)$ down to about 20 lobes around 0. The convolution $p(t) = h_e(t) \star h_e(-t)$ is obtained using the MATLAB® function `conv`. $p(t)$ and its Fourier transform are plotted for 1,024 frequency points. To do this, type:

```
%===== critnyq.m
clear
R=1000; T=1/R; B=600;
%=====
alpha=(2*B/R-1);
NpS=10; Fe=NpS*R; Te=1/Fe; nblobes=20;
h=racnyq(alpha,nblobes,NpS); lh=length(h);
p=conv(h,h(lh:-1:1));
%===== temporal response
lpm1=length(p)-1;
subplot(211); plot(Te*(-lpm1/2:lpm1/2),p)
set(gca,'xtick',(-30:30)*T)
grid ; set(gca,'xlim',[-5*T 5*T]);
%===== spectral response
Lfft=1024; fq=Fe*(0:Lfft-1)/Lfft;
subplot(212); plot(fq,abs(fft(p,Lfft))); grid
set(gca,'xlim',[0 R]);
B=(1+alpha)*R/2; text(B,2,'B');
```

The results are shown in Figure 6.13. As you can see, the function $p(t)$ is null for every multiple of T, except at the origin, and the spectrum is B band limited, meaning that $P(f) = 0$ for $|f| > B = (1 + \alpha)/2T$.

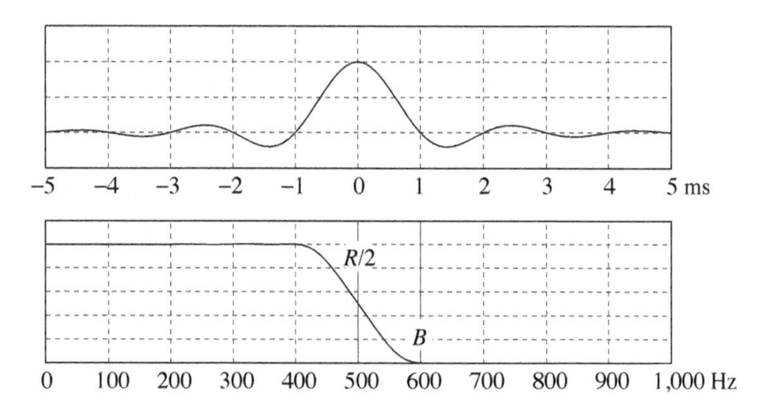

Figure 6.13 – *A raised cosine function satisfying the Nyquist criterion*

6.2.6 The eye pattern

As we have said before, the receiver shown in Figure 6.12 is so simple because of the absence of ISI. This is why it is essential to have access to a practical tool for measuring its level. We have already given on page 198 a quantitative definition, with equation (6.40), of the maximum ISI, and we are now going to present another very important tool, called the *eye pattern*, for reasons that will be obvious if you refer to Figure 6.14.

The eye pattern is obtained by superimposing a large number of trajectories, with durations $2T$, of the matched filter's output signal. It can be displayed on an oscilloscope by synchronizing with the symbol rate $1/T$. The afterglow of the screen makes it possible for the superposition to persist. In the presence of noise, as we are going to see, the wider the eye is vertically, the lower the error probability. Therefore, we have to choose the decision time where the eye is vertically "widest".

In the following exercise, we are going to generate a complete simulation line for 2-PAM modulation, in the form of five different programs. We will examine the eye pattern and choose the optimal sampling time. All the computed values are saved as we go along, so they can be used as input data for the next program.

Exercise 6.6 (2-PAM modulation) (see page 308) Consider a PAM modulation with two levels $\{-1, +1\}$. The binary rate is equal to 1,000 bps. The signal's expression is:

$$x_e(t) = \sum_k a_k h_e(t - kT) = \sum_k a_k \text{rect}(t - kT)$$

To display the signal as if it were time-continuous, choose $F_s = 20$ kHz as the sampling frequency.

1. Write a program that computes the samples `xe` of the signal $x_e(t)$ for a random sequence of 300 bits. Display part of the chronogram of `xe` (`zoom xon`). Save all of the values.

2. Write a program that generates and displays the received signal `xr`. To simulate a transmission channel, use a low-pass filter $h_c(t)$ the impulse response of which is obtained by `hc=rif(lhsT*NT-1,bc)` with `lhsT=3.5` and `bc=0.06`, and where `NT` refers to the number of points corresponding to the time interval between two symbols. Therefore the channel's response stretches out over 3.5 symbols.

3. Let $h(t) = (h_c \star h_e)(t)$. The receiver's matched filter therefore has the impulse response $h(-t)$. The matched filter's output signal is denoted by `xa`. Write a program that superimposes the "sections" of `xa` lasting a duration of two "symbol periods" so as to display the *eye pattern*. The program will have to be designed so that the time coordinate of the place

where the eye is vertically "widest" can be interactively defined as input data. You can use the function `ginput`.

4. Write a program that displays the matched filter's output signal `xa` and superimposes the values taken from the sampler's output.

5. Write a program that, based on the value of the signal-to-noise ratio, expressed in dB:

   - adds a white, centered, Gaussian noise with the power P_b to the received signal `xc`;
   - decides that the bit is equal to 1 if its value at the sampler's output is positive and 0 if it is not;
   - and evaluates the error probability.

In exercise 6.6, the overall impulse response ensures that the ISI is low enough for the symbol-by-symbol decision to yield good results. The low level of interference is clearly shown by the eye pattern without noise represented in Figure 6.14. The trajectories almost converge to the same point at multiples of T. Therefore, if the sampling is done at these times, the values located around 1 are likely to correspond to the transmission of a bit 1.

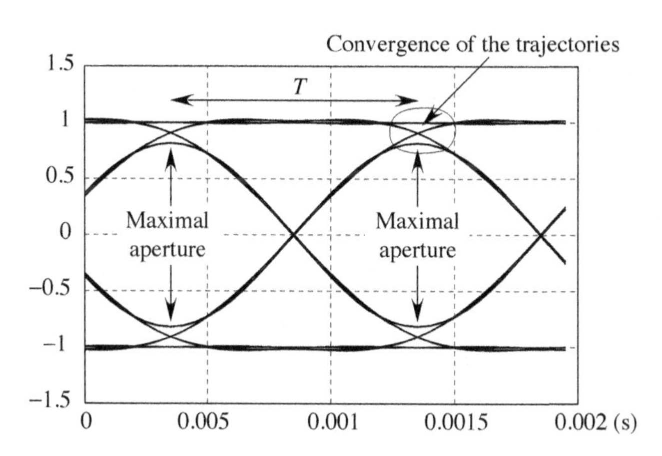

Figure 6.14 – *PAM-2: eye pattern without noise*

6.2.7 PAM modulation on the Nyquist channel

We are now going to present a sequence of programs designed to simulate a PAM modulation on the Nyquist channel. The following sections describe in this order:

- the generation of symbol sequences;
- the generation of the emitted digital signal;
- the addition of noise by the channel;
- matched filtering and the examination of the eye pattern;
- symbol-by-symbol decision.

Generating the sequence of symbols

The gsymb.m function generates a sequence of N symbols taken from the alphabet $\{-(M-1), \ldots, -3, -1, +1, +3, \ldots, (M-1)\}$ with $M = 2^m$. Save this function as gsymb.m:

```
function [symb,Esymb]=gsymb(m,N)
%!=====================================================!
%! Generates a random sequence with values in an MIA  !
%! M=2^m symbols alphabet with a uniform distribution !
%! SYNOPSIS: [symb,Esymb]=GSYMB(m,N)                  !
%!     m       such as M=2^m                          !
%!     N     = sequence length                        !
%!     symb  = random sequence                        !
%!     Esymb = mean energy                            !
%!=====================================================!
M=2^m; alphabet=2*(0:M-1)+1-M;
la=length(alphabet);
ind=fix(rand(1,N)*la)+1;
symb=alphabet(ind);
Esymb=alphabet*alphabet'/M;
```

Generating the receiver filter's output signal

We are going to try to numerically compute the sample at the rate $T_s = T/N_T$ of the signal $x_e(t) = \sum_k a_k h_e(t - kT)$. N_T represents the number of samples per symbol duration T. These samples can be used for creating the continuous-time signal using a digital-to-analog converter. In this case, these samples will be used for the display. They have the expression:

$$x_e(nT_s) = \sum_k a_k h_e(nT_s - kN_T T_s) = \sum_k a_k h_e((n - kN_T)T_s)$$

The samples of $h_e(t)$ taken at the rate T_s can be obtained with the use of the racnyq.m function. The function $h_e(t)$ can be truncated down to about 30 lobes around 0. If we choose a high enough value for N_T, the resulting plot of $x_e(t)$ is "almost continuous". To calculate the samples of $x_e(t)$, we can use the diagram explaining the principle, from Figure 6.15, drawn for $N_T = 4$.

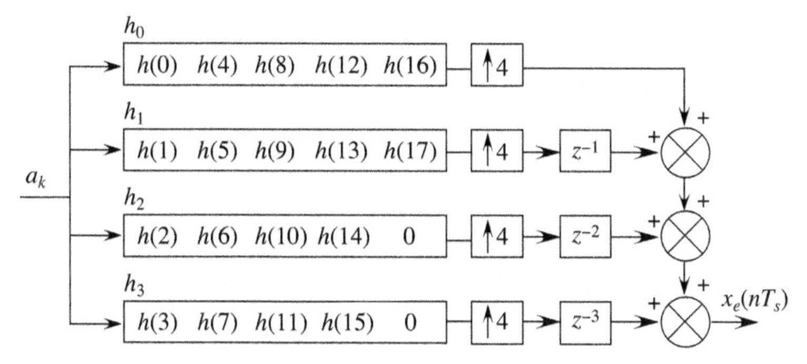

Figure 6.15 – *Constructing the signal transmitted based on the symbols and the polyphase components of the emission filter in the case where $N_T = 4$*

If we let $n = \ell N_T + r$, where $r = 0, \ldots, N_T - 1$, then we indeed have:

$$x_e((\ell N_T + r)T_s) = \sum_k a_k h_e(((\ell - k)N_T + r)T_s)$$

If we let $\widetilde{x}_r(\ell) = x_e((\ell N_T + r)T_s)$ and $\widetilde{h}_r(\ell) = h_e((\ell N_T + r)T_s)$, then we can write:

$$\widetilde{x}_r(\ell) = \sum_k a_k \widetilde{h}_r(\ell - k)$$

The component $\widetilde{x}_r(\ell)$ is called the r-th N_T-*polyphase component*. It is obtained by filtering the sequence a_k by the filter with the impulse response $\widetilde{h}_r(\ell)$ obtained by undersampling the sequence $h_e(mT_s)$ by a factor N_T. In MATLAB®, given a sequence **he** of the coefficients of $h_e(t)$, the filter with the impulse response $\widetilde{h}_r(\ell)$ is obtained simply by typing:

```
hr = he(r:NT:length(he));
```

In Figure 6.15, we chose a response $h_e(t)$ with 18 non-zero coefficients. The number of samples per symbol is $N_T = 4$. The transmitted signal is obtained by superimposing the 4 shifted components.

The roll-off expression of α is obtained with relations (6.32) and (6.42):

$$\alpha = 2\frac{B}{R} - 1 = 2\frac{B}{\log_2(M)D} - 1$$

Once the samples of $x_e(t)$ have been generated, a noise is added by setting the signal-to-noise ratio. The noised signal is filtered by the filter with the impulse response $h_e(-t)$. $y(t)$ is the resulting signal. The following program computes the samples of $y(t)$ and displays the result of the superimposing the trajectories with a duration of $2T$:

```
%===== NyquistEye.m
% Eye pattern for a Nyquist channel
% Uses: GSYMB, RACNYQ, GSIG
clear
N=5000;          % sequence length (bits) of the source
NbTraj=200;      % number of trajectories
%===== binary rate
Db=1000; tt=sprintf('Binary rate: %4.0f bits/s',Db);
disp(tt); nbbps=input('Number of bits per symbol: ');
M=2^nbbps;
%===== modulation alphabet
alphabet=2*(0:M-1)+1-M;
%===== Baud rate
Dsymb=Db/nbbps;
Bmin=Dsymb/2;                % alpha = 0
Bmax=Dsymb;                  % alpha = 1
tt=sprintf('Band of the channel B between %3.1f and %3.1f :',...
            Bmin,Bmax);
disp(tt); Bc=input('B = ');
if (Bc<Bmin)
   disp('ISI=0 impossible: you must increase the bandwidth.');
   return;
end
if (Bc>Bmax),
   disp('The bandwidth must be decreased.'); return;
end
alpha=(2*Bc/Dsymb)-1; nblobes=30;
[symb, Esymb]=gsymb(nbbps,N);        % generating of symbols
%===== square root Nyquist
NpS=10; he=racnyq(alpha,nblobes,NpS);
lh=length(he); xe=zeros(NpS,N);
%===== emitted signal (polyphase components in the columns)
for ii=1:NpS,hr=he(ii:NpS:lh); xe(ii,:)=filter(hr,1,symb); end
xe=xe(:,1:N); st=zeros(NpS*N,1); st(:)=xe;
%===== mean energy per symbol
Es_sim=(st'*st)/N;           % estimated value
Es=Esymb*(he'*he);
%===== mean energy per bit
Eb=Es/nbbps; RSB=input('Ratio Eb/N0 (dB) = ');
PB=Eb*10^(-RSB/10); sigmab=sqrt(PB/2);
xt=st+sigmab*randn(NpS*N,1); % adding of noise
%===== square root Nyquist
htfa=he(lh:-1:1);            % matched filter
stfa=filter(htfa,1,xt);      % output to be sampled
%===== eye pattern
nbtraces=(N-nblobes)/2;
seye=zeros(NpS*2,nbtraces); seye(:)=stfa(nblobes*NpS+1:N*NpS);
MaxLev=M-1+4*sigmab+4; t0=NpS+1;
```

```
plot([t0 t0],[-MaxLev MaxLev],':'); hold on
%===== some trajectories
for tt=1:NbTraj, plot((1:2*NpS),seye(:,tt)); end
%===== sampling time
t0=NpS+1; plot([t0 t0],[-MaxLev MaxLev],':'); hold off
set(gca,'xlim',[1 2*NpS])
save eyediagr
```

By superimposing portions of trajectories with a duration $2T$, that is to say $2N_T$ points, we get the eye pattern shown in Figure 6.16. If M refers to the alphabet size, then there are $(M-1)$ eye apertures. In the case of a non-ideal channel, the trajectories no longer have the shapes shown here. In particular, they no longer perfectly converge in at the multiples of T, which is imposed by the Nyquist criterion.

Figure 6.16 represents the eye pattern for $M = 4$ (number of bits per symbol $= 2$), a rate of 1,000 bps, and a spectrum support of 300 Hz, hence $\alpha = 0.2$.

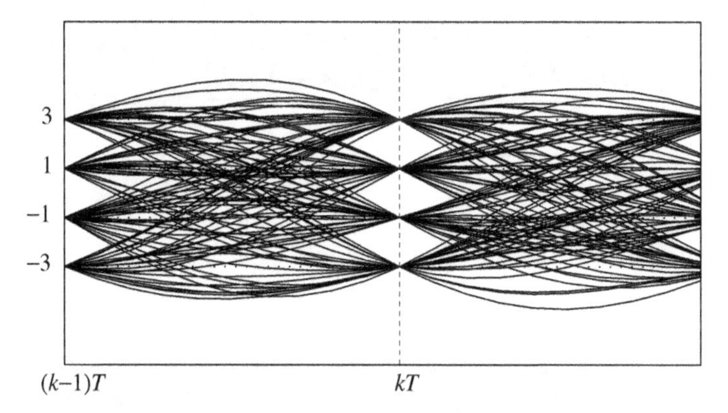

Figure 6.16 – *The eye pattern for $M = 4$. The noise is null. Spectrum support $B = 300$ Hz, rate 1,000 bps*

Figure 6.17 shows the same eye diagram for $M = 4$, but for a spectrum support of 500 Hz, hence $\alpha = 1$. The vertical opening at the times kT is the same as before. In terms of probability, the performances are the same. However, if the sampling time is slightly shifted, the vertical opening becomes wider at that time for the 500 Hz. The error probability will be smaller. To put it more simply, the vertical opening guarantees a better resistance when the sampler is desynchronized.

Figures 6.18 and 6.19 show the eye pattern for a signal-to-noise ratio of 15 dB.

If we “place ourselves” in the center of the eye, and set the thresholds to the values -2, 0 and 2, the error probability has to remain very small, which is checked with the following program. Type:

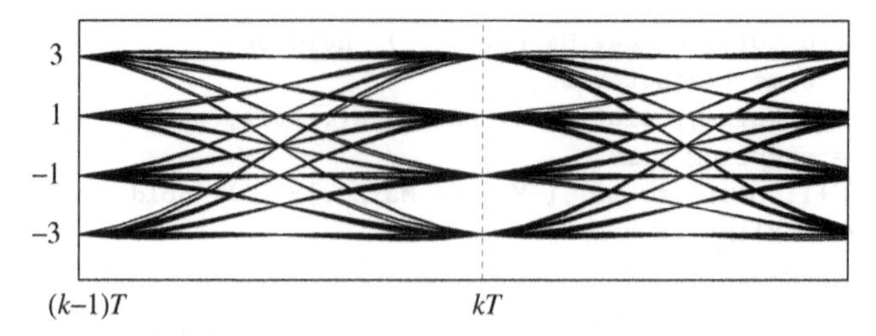

Figure 6.17 – *Eye pattern for $M = 4$. The noise is null. Spectrum support $B = 500$ Hz, rate 1,000 bps*

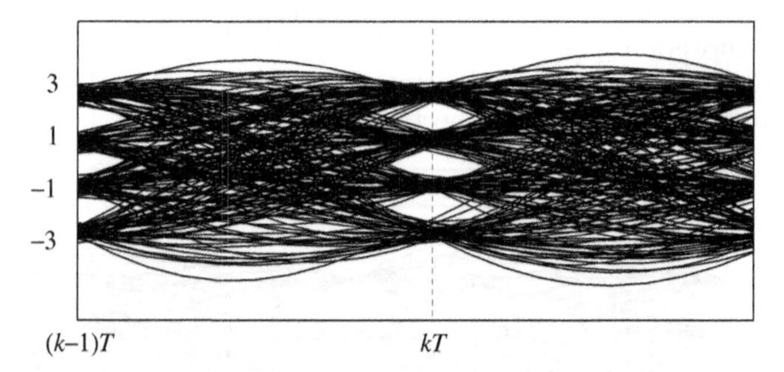

Figure 6.18 – *Eye pattern for $M = 4$. The signal-to-noise ratio is equal to 15 dB, the spectrum support is $B = 300$ Hz and the rate 1,000 bps*

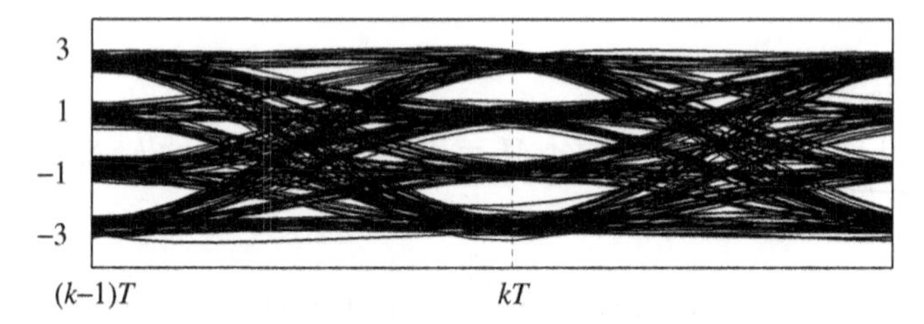

Figure 6.19 – *Eye pattern for $M = 4$. The signal-to-noise ratio is equal to 15 dB, the spectrum support is $B = 500$ Hz and the rate 1,000 bps*

```
%===== peestim.m
% Sampling the output of the matched filter
clear; load eyediagr
%===== thresholds
thold=2*(0:M-2)+1-M+1; thold=[-inf thold inf];
stfaech=stfa(t0:NpS:N*NpS); lsfa=length(stfaech);
%===== decision symb/symb by the "nearest" rule
```

```
aux=stfaech(NpS:lsfa); laux=length(aux);
deci=zeros(1,laux); lalphabet=length(alphabet);
for jj=1:lalphabet
    v=find(aux>thold(jj) & aux<=thold(jj+1));
    deci(v)=alphabet(jj)*ones(1,length(v));
end
decal=nblobes-NpS; deci=deci(decal+1:length(deci));
ldiff=length(deci); diff=deci - symb(1:ldiff);
nbe=length(find(diff~=0)); pe=nbe/length(diff);
TEB=pe/nbbps;
disp('********************************************')
tt=sprintf('*Binary rate: %d bits/s',Db); disp(tt)
tt=sprintf('*Symbol rate: %3.1f bauds, M=%i',Dsymb,M);
disp(tt)
tt=sprintf('*Channel band: %3.0f Hz, alpha: %3.1f',Bc,alpha);
disp(tt)
tt=sprintf('*Eb/N0=%ddB        ==> TEB=%5.2d ', SNR,TEB);
disp(tt)
tt=sprintf('*Nb errors/symbol= %i pour N=%i',nbe,N); disp(tt)
disp('********************************************')
```

The program estimates the *error probability P_s per symbol.* If *the signal-to-noise ratio is high*, we can consider that the only errors are caused by a decision in favor of one of the two symbols adjacent to the symbol actually transmitted. Therefore, if we are using a Gray code, these adjacent symbols only cause one *false bit over the $m = \log_2(M)$ emitted bits*, leading us to the following formula, which gives us the relation between the probability per symbol P_s and the *bit error rate* BER:

$$\text{BER} \approx \frac{P_s}{\log_2(M)} \tag{6.44}$$

With this program, we can also test the error probability increase when the sampling time is shifted. All we need to do is modify the variable `t0`. We have to check that this increase becomes greater as the chosen band becomes narrower. However, when we are right in the center of the eye (where the trajectories converge in the absence of noise), the error probability is independent of the value of B.

6.3 Linear equalization and the Viterbi algorithm

Consider the case of the *binary baseband modulation* that associates the continuous-time signal $x_e(t) = \sum_k a_k h_e(t - kT)$ with the sequence $\{d_k\}$ of binary elements $\{0, 1\}$ according to the following rule: if $d_k = 0$, then $a_k = -1$

and if $d_k = 1$, then $a_k = 1$. The signal $h_e(t)$ refers to the impulse response of the emission filter. The signal $x_e(t)$ is filtered by the transmission channel and is subjected to additive noise. The received signal can then be written $x_r(t) = \sum_k a_k h(t - kT) + w(t)$ where $h(t)$ is obtained by convoluting the filter $h_e(t - kT)$ with the channel filter, and $w(t)$ is the noise. When the signal is received, it is filtered by the matched filter then sampled. The result is a sequence of values $x_a(n)$, the expressions of which linearly depend on the sequence a_k of the type:

$$x_a(n) = \sum_k a(k)g(n-k) + b(n) \tag{6.45}$$

The term $b(n)$ accounts for the noise, which is assumed to be Gaussian. Without being any less general, we can assume that $b(n)$ is white. If it is not, we know that there exists a causal filter that can make it white. This filter changes the values of the coefficients $g(k)$, but the general form of expression (6.45) stays the same. The variance of $b(n)$ is denoted by σ_b^2.

The coefficients $g(n)$ take into account the emission filter, as well as the channel filter and the receiver matched filter. From now on, the sequence $g(n)$ is causal and has a *finite* duration L. This hypothesis is actually quite realistic. The only problem is that the complexity of the processes increases with L. Remember that we have already discussed the case where $k \neq 0$: this is the case where the ISI is equal to 0.

In the end, we can describe the digital transmission line as a "black box", with the sequence of symbols $a(n)$ ($\in \{-1, 1\}$) as the input, and as the output the sequence of real values given by the expression:

$$x_a(n) = g_0 a(n) + g_1 a(n-1) + \cdots + g_{L-1} a(n - L + 1) + b(n) \tag{6.46}$$

The choice of the likeliest decision sequence is based on the observations $x_a(n)$ for $n \in \{1, ..., N\}$. In the example we are going to discuss, we have chosen $L = 3$.

In practice, the emission and reception filters are known, whereas the channel filter is not. We have already said that in order to measure the quantities $g(k)$, we could then use a set sequence of symbols called the *training sequence.* From now on, we will assume that the values of g_0, g_1, ..., g_{L-1} are known. If we then use the Gaussian distribution of the sequence $b(n)$ and expression (6.46), the probability density of the observed sequence $\{x_a(1), \ldots, x_a(N)\}$ can be written:

$$p_X(\boldsymbol{x}; a_1, ..., a_N) = \frac{1}{(\sigma_b\sqrt{2\pi})^N} \exp\left(-\frac{1}{2\sigma_b^2} A\right) \tag{6.47}$$

$$\text{with } A = \sum_{n=1}^{N} [x_a(n) - g_0 a(n) - g_1 a(n-1) - g_2 a(n-2)]^2$$

where we have assumed that $a(0) = a(-1) = -1$. Based on the N observations $\{x_a(1), \ldots, x_a(N)\}$, the *maximum likelihood rule* consists of finding the sequence $\{a(1), \ldots, a(N)\}$ that maximizes $p_X(\boldsymbol{x}; a_1, ..., a_N)$, that is to say the sequence that minimizes the quantity:

$$\ell(\{x_a\}, \{a\}) = \sum_{n=1}^{N} [x_a(n) - g_0 a(n) - g_1 a(n-1) - g_2 a(n-2)]^2 \qquad (6.48)$$

Theoretically, there are 2^N possible sequences and we could just bluntly compute the 2^N values of $\ell(\{x_a\}, \{a\})$ for the observed sequence $\{x_a(1), \ldots, x_a(N)\}$, and choose the sequence a that leads to the minimum.

The Viterbi algorithm makes it possible to cut down to $N \times 2^L$ the number of values that have to be calculated. Before we present this algorithm, we are going to study a simpler, but suboptimal process.

6.3.1 Linear equalization

The first idea consists of using the filter with the impulse response $w(n)$ that inhibits the effect of the channel $\{g_0, g_1, g_2\}$. This is called equalizing the channel and the filter $w(n)$ is called an *equalizer.*

There are two common approaches:

- The equalizer is chosen so that, in the absence of noise, it completely eliminates the intersymbol interference. This is called *Zero Forcing.*

- The equalizer is chosen so as to minimize, in the presence of noise, the square deviation between the original signal and the signal after it has been equalized. This is called the Wiener equalization. It requires for the noise's variance σ_b^2 to be known. The expression of its complex gain is given by:

$$W(f) = \frac{G^*(f)}{|G(f)|^2 + \sigma_b^2/\sigma_x^2}$$

Obviously, in the absence of noise ($\sigma_b = 0$), the two solutions coincide.

Exercise 6.7 ("Zero Forcing" linear equalization) (see page 309) In the absence of noise, the Zero Forcing equalizer eliminates the ISI. Its transfer function is $W(z) = 1/G(z)$ with $G(z) = g_0 + g_1 z^{-1} + g_2 z^{-2}$. If $G(z)$ has its zeros inside the unit circle, the solution is stable and causal. Otherwise, the stable solution has a non-causal impulse response. It is possible to implement it by approximating it with a finite delay (see exercise 6.5).

Because $G(z)$ is a polynomial, that is to say a filter with a finite impulse response, the series expansion of $W(z)$ has an infinite number of coefficients and hence the filter has an infinite impulse response. In practice, it is often approximated by a filter with a long enough finite impulse response. Here we are going to design the filter $W(z) = 1/G(z)$ in its exact form using the MATLAB® function `filter`, which is possible only if $G(z)$ has all its zeros strictly inside the unit circle.

1. Using equation (6.46), show that the output signal $y(n)$ of the equalizer $W(z)$ can be expressed as $y(n) = a(n) + u(n)$, where $u(n)$ is a noise.

2. Let $g_0 = 1$, $g_1 = -1.4$ and $g_2 = 0.8$. Notice that the zeros are inside the unit circle, and therefore the equalizer $w(n)$ is stable and causal. Determine the variance of $u(n)$. You can use expression (6.49):

$$\left(\begin{bmatrix} 1 & g_1 & g_2 \\ g_1 & g_2 & 0 \\ g_2 & 0 & 0 \end{bmatrix} + \begin{bmatrix} 1 & 0 & 0 \\ g_1 & 1 & 0 \\ g_2 & g_1 & 1 \end{bmatrix}\right) \begin{bmatrix} R(0)/2 \\ R(1 \\ R(2) \end{bmatrix} = \begin{bmatrix} \sigma^2 \\ 0 \\ 0 \end{bmatrix} \tag{6.49}$$

3. Use the result to find the probability distribution of $y(n)$ when $a(n) = -1$ and when $a(n) = +1$.

4. Given the previous, we have come up with the following rule: if $y(n)$ is positive, then the decision is that $a(n) = 1$, and otherwise the decision is that $a(n) = -1$. Determine the expression of the error probability.

5. Write a program that:

   - generates a random binary sequence, made up of -1 and $+1$, with a length N;
   - filters the resulting sequence by the filter with the impulse response `gc=[1 -1.4 0.8];`
   - adds a Gaussian noise with a variance σ^2 such that the signal-to-noise ratio is equal to R dB;
   - passes the obtained signal through a filter with the transfer function $W(z) = 1/G(z)$;
   - compares the equalizer's output to the threshold 0 to decide what symbol was transmitted;
   - evaluates the number of errors (bear in mind that for the evaluation to be relevant, about a hundred errors have to be counted, hence N must be chosen high enough);
   - compares the results with the theoretical plot.

Exercise 6.8 (Wiener equalization) (see page 311) Consider again equation (6.46):

$$x(n) = g_0a(n) + g_1a(n-1) + \cdots + g(L-1)a(n-L+1) + b(n) \tag{6.50}$$

$a(n)$ is assumed to be a sequence of equally distributed i.i.d. random variables with possible values in $\{-1, +1\}$. This means that $\mathbb{E}\{a(n)\} = 0$ and that $\mathbb{E}\{a(n+k)a(n)\} = \delta(k)$.

We wish to find a filter with a finite impulse response $w(n)$, with a length N, that minimizes the square deviation:

$$\mathbb{E}\left\{|a(n-d) - \widehat{a}(n)|^2\right\} \tag{6.51}$$

where $\widehat{a}(n) = x(n) \star w(n)$ refers to this filter's output. This filter is an example of the Wiener filter. d is a positive integer that accounts for the fact that a delay is required, because if the filter $h(n)$ is not minimum phase, then we know that the stable inverse is not causal. In practice, a delay must therefore be introduced to obtain a proper causal approximation.

1. Determine, as a function of the autocovariance function $R_{aa}(k)$ of the sequence $a(n)$ and of the covariance function $R_{ax}(k)$ between $a(n)$ and $x(n)$, the filter $w(n)$ that minimizes (6.51).

2. Determine the expression of $R_{ax}(k)$ as a function of $R_{aa}(k)$, and the expressions of $g(k)$ and of the autocovariance function $R_{bb}(k)$ of the noise $b(n)$.

3. Write the problem in matrix form. Determine the solution's expression.

4. Write a program that performs the equalization.

5. Write a program that uses the equalized output and applies symbol-by-symbol decision. Compare the results, in terms of error probability, with those obtained with the Zero Forcing equalizer.

6.3.2 The soft decoding Viterbi algorithm

As we said, based on the N observations, we have to calculate the 2^N quantities:

$$\ell(\{x_a\}, \{a\}) = \sum_{n=1}^{N} [x(n) - g_0 a(n) - g_1 a(n-1) - g_2 a(n-2)]^2$$

corresponding to the 2^N length N binary sequences. Let us assume that, initially, $a(0) = -1$ et $a(-1) = -1$. Let $s(n) = [a(n-1) \quad a(n-2)]^T$ be the size 2 vector constructed by concatenating two consecutive symbols. In our case, $s(n)$ can only assume four different values, denoted symbolically by {00, 01, 10, 11}.

Let us calculate the probabilities for $s(n+1)$ to be equal to 00, 01, 10 and 11, respectively, knowing that $s(n) = 00$. If $s(n) = 00$, the only possible states for $s(n+1)$ are 00 if $a(n) = -1$ and 10 if $a(n) = +1$. Therefore:

$$\begin{aligned}
\Pr(s(n+1) = 00|s(n) = 00) &= 1/2 \\
\Pr(s(n+1) = 01|s(n) = 00) &= 0 \\
\Pr(s(n+1) = 10|s(n) = 00) &= 1/2 \\
\Pr(s(n+1) = 11|s(n) = 00) &= 0
\end{aligned}$$

The probabilities of $s(n+1)$ with respect to the three other values of $s(n)$ can be calculated in the same way and represented by Figure 6.20.

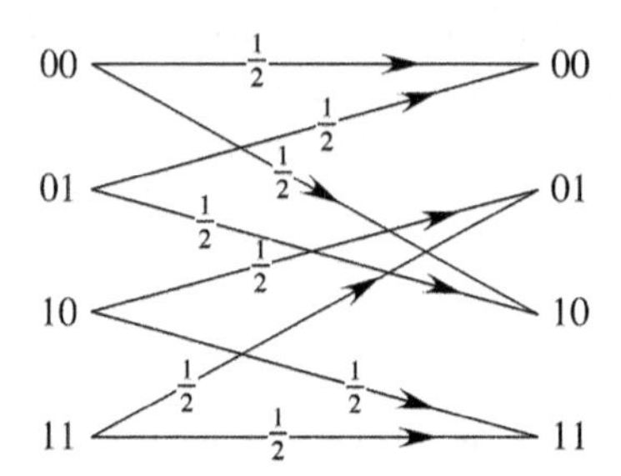

Figure 6.20 – *Transition probability graphs for the four states depending on the input symbol*

The probabilities can be grouped together in a *transition matrix*:

$$\mathscr{M} = \begin{bmatrix} 1/2 & 0 & 1/2 & 0 \\ 1/2 & 0 & 1/2 & 0 \\ 0 & 1/2 & 0 & 1/2 \\ 0 & 1/2 & 0 & 1/2 \end{bmatrix}$$

representing the state transition probability for two consecutive times. This description of the evolution of $s(n)$ is called a *Markov*, or *Markovian* model. The Markovian property is what makes it possible to use the Viterbi algorithm. Consider again the expression we wish to minimize, and let $\boldsymbol{g} = [g_1 \quad g_2]^T$. We are going to use the definition of $s(n)$. If we stop the computation at step p, we get:

$$d(s(p), p) = \sum_{n=1}^{p} (x_a(n) - g_0 a(n) - \boldsymbol{g}^T s(n))^2$$

From now on, the quantity $d(s(p), p)$ will be called the *path metric* $(a(1), \ldots, a(p))$ at step p.

Let us assume that at step p, we still have four sequences $(a_1, \ldots, a_p)$ competing each other ending with $s(p)$ equal to 00, 01, 10 or 11 respectively and with the metric $d(00, p)$, $d(01, p)$, $d(10, p)$ and $d(11, p)$ respectively.

In order to calculate the metric of the sequence $(a(1), \ldots, a(p+1))$, we then have to write the positive term $(x(p+1) - g_0 a(p+1) - \boldsymbol{g}^T s(p+1))^2$ for the two possible values of $a(p+1)$, that is to say -1 and $+1$. Hence each state has *two possible children*. We will therefore have to calculate, as functions of the observation $x_a(p+1)$ at the time $(p+1)$, 8 values that lead to one of the four possible states. These calculations are summed up in the following table:

$s(p)$		$s(p+1)$	
00	$\rightarrow$	00	$v(00,00) = d(00,p) + (x_a(p+1) - (-g_0 - g_1 - g_2))^2$
	$\searrow$	10	$v(00,10) = d(00,p) + (x_a(p+1) - (+g_0 - g_1 - g_2))^2$
01	$\rightarrow$	00	$v(01,00) = d(01,p) + (x_a(p+1) - (-g_0 - g_1 + g_2))^2$
	$\searrow$	10	$v(01,10) = d(01,p) + (x_a(p+1) - (+g_0 - g_1 + g_2))^2$
10	$\rightarrow$	01	$v(10,01) = d(10,p) + (x_a(p+1) - (-g_0 + g_1 - g_2))^2$
	$\searrow$	11	$v(10,11) = d(10,p) + (x_a(p+1) - (+g_0 + g_1 - g_2))^2$
11	$\rightarrow$	01	$v(11,01) = d(11,p) + (x_a(p+1) - (-g_0 + g_1 + g_2))^2$
	$\searrow$	11	$v(11,11) = d(11,p) + (x_a(p+1) - (+g_0 + g_1 + g_2))^2$

Notice that there are two ways of ending up in each state. For example we end up in $s(p+1) = 10$ either by starting in $s(p) = 00$, if we choose $a(p+1) = 1$, either by starting in $s(p) = 01$, if we choose $a(p+1) = 1$. These two possibilities correspond to the two values $v(00,10)$ and $v(01,10)$ respectively. It is useless to keep the largest one of the two, since any further value will have a higher metric. Hence $d(10, p+1) = \min(v(00,10), v(01,10))$.

In the end, the algorithm calculates, at step $p+1$:

$$\begin{aligned}
d(00,p+1) &= \min\{d(00,p) + [x_a(p+1) - (-g_0 - g_1 - g_2)]^2, \quad (6.52)\\
&\qquad d(01,p) + [x_a(p+1) - (-g_0 - g_1 + g_2)]^2\}\\
d(01,p+1) &= \min\{d(10,p) + [x_a(p+1) - (-g_0 + g_1 - g_2)]^2, \quad (6.53)\\
&\qquad d(11,p) + [x_a(p+1) - (-g_0 + g_1 + g_2)]^2\}\\
d(10,p+1) &= \min\{d(00,p) + [x_a(p+1) - (+g_0 - g_1 - g_2)]^2, \quad (6.54)\\
&\qquad d(01,p) + [x_a(p+1) - (+g_0 - g_1 + g_2)]^2\}\\
d(11,p+1) &= \min\{d(10,p) + [x_a(p+1) - (+g_0 + g_1 - g_2)]^2, \quad (6.55)\\
&\qquad d(11,p) + [x_a(p+1) - (+g_0 + g_1 + g_2)]^2\}
\end{aligned}$$

At each step, all the *parents* leading to the smallest metric have to be memorized. For example if the minimum of $d(00, p+1)$ is obtained for $d(01,p) + (x_a(p+1) - (-g_0 - g_1 + g_2)^2)$, it will be memorized as the state 00 at the time p that leads to the state 01 at the time $p+1$. The same is done for the three other states. The initial values are calculated based on the fact that we have assumed $a(0) = a(-1) = -1$, that is to say $s(0) = 00$. This means that when the first symbol is transmitted, either the state 00 or the state 10 is reached, and therefore:

$$\begin{aligned} d(00,1) &= [x_a(1) - (-g_0 - g_1 - g_2)]^2 \\ d(10,1) &= [x_a(1) - (hg_0 - g_1 - g_2)]^2 \end{aligned}$$

We then infer that:

$$\begin{aligned} d(00,2) &= d(00,1) + [x_a(2) - (-g_0 - g_1 - g_2)]^2 \\ d(01,2) &= d(10,1) + [x_a(2) - (-g_0 + g_1 - g_2)]^2 \\ d(10,2) &= d(00,1) + [x_a(2) - (+g_0 - g_1 - g_2)]^2 \\ d(11,2) &= d(10,1) + [x_a(2) - (+g_0 + g_1 - g_2)]^2 \end{aligned}$$

Obviously, the intermediate metrics do not need to be memorized. Only the four current ones have to be. Hence, formulae (6.52) to (6.55) have to be used as updating formulae for the four states reached at the considered state. The algorithm stops at the end of the length N observation sequence. Then the sequence considered to be the most likely is the one that leads to the state with the smallest metric. The sequence of states is determined by going back up the table of parents corresponding to this sequence. Once the sequence of states has been obtained, we simply compute the symbol sequence. The `viterbi.m` program determines the emitted sequence using the Viterbi algorithm. The results are given in Figure 6.21. The plot ('o') represents the error probability when using the Viterbi algorithm. The plot ('×') reproduces the results provided by the program from exercise 6.5 (a Zero Forcing equalization followed by a symbol-by-symbol threshold detection) by typing, after running it:

```
hold on; semilog(RSBdB, PeTheo,'x'); hold off
```

As you can see, the Viterbi results are much better than the ones obtained by linear equalization.

```
%===== viterbi.m
N=5000; hc=[1 -1.4 0.8]; SNRdB=(5:17); lgsnr=length(SNRdB);
PeVi=zeros(lgsnr,1); ak=sign(randn(1,N));
ak(1)=-1; ak(2)=-1; sk=filter(hc,1,ak); vs=sqrt(sk*sk'/N);
%=====
dec=NaN*ones(4,4); dec(1,1)=-1; dec(2,1)=1; dec(3,2)=-1;
dec(4,2)=1; dec(1,3)=-1; dec(2,3)=1; dec(3,4)=-1; dec(4,4)=1;
%=====
for jj=1:lgsnr
    SNR=10^(SNRdB(jj)/20); sigma_b=vs/SNR;
    bk=sigma_b*randn(1,N); xk=sk+bk;
    %===== indexing four states (with the metric d2)
    %     1 for 00, 2 for 01, 3 for 10, 4 for 11
    %===== asc is the parents' sequence
    d2=zeros(4,1); asc=zeros(4,N);
    %==== initialization
```

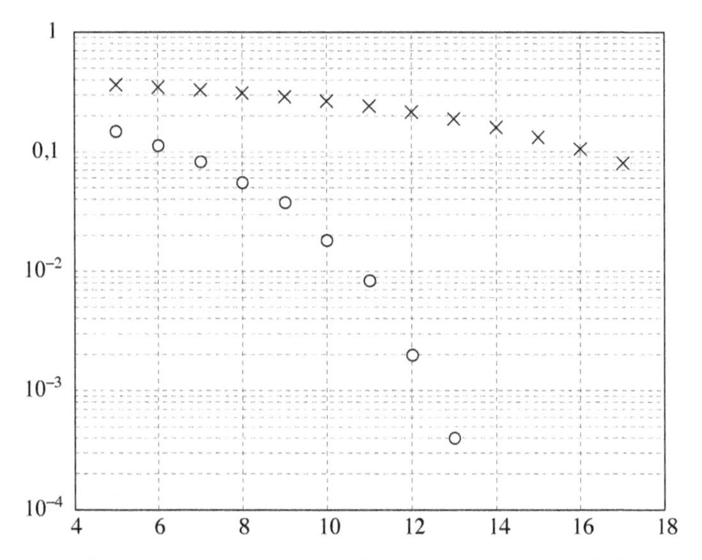

Figure 6.21 – *Comparing the error probabilities as functions of the signal-to-noise ratio in dB. Plot ('×'): zero forcing linear equalization. Plot ('○'): Viterbi algorithm. Results obtained through a simulation with a length of* 5,000. *The channel filter has the finite impulse response* $(1 \; -1.4 \; 0.8)$

```
    d21=(xk(3)-(-hc(1)-hc(2)-hc(3)))^2;
    d23=(xk(3)-(hc(1)-hc(2)-hc(3)))^2;
    asc(1,3)=1; asc(3,3)=1;
    %=====
    d2(1)=d21+(xk(4)-(-hc(1)-hc(2)-hc(3)))^2;
    d2(2)=d23+(xk(4)-(-hc(1)+hc(2)-hc(3)))^2;
    d2(3)=d21+(xk(4)-(hc(1)-hc(2)-hc(3)))^2;
    d2(4)=d23+(xk(4)-(hc(1)+hc(2)-hc(3)))^2;
    asc(:,4)=[1;3;1;3]; d2kk=zeros(4,1); ind=zeros(4,1);
    for kk=5:N   %======
  [d2kk(1) ind(1)]=min([d2(1)+(xk(kk)-(-hc(1)-hc(2)-hc(3)))^2;...
         d2(2)+(xk(kk)-(-hc(1)-hc(2)+hc(3)))^2]);
  [d2kk(2) ind(2)]=min([d2(3)+(xk(kk)-(-hc(1)+hc(2)-hc(3)))^2;...
         d2(4)+(xk(kk)-(-hc(1)+hc(2)+hc(3)))^2]);
  [d2kk(3) ind(3)]=min([d2(1)+(xk(kk)-(hc(1)-hc(2)-hc(3)))^2;...
         d2(2)+(xk(kk)-(hc(1)-hc(2)+hc(3)))^2]);
  [d2kk(4) ind(4)]=min([d2(3)+(xk(kk)-(hc(1)+hc(2)-hc(3)))^2;...
         d2(4)+(xk(kk)-(hc(1)+hc(2)+hc(3)))^2]);
  asc(:,kk)=[1*(ind(1)==1)+2*(ind(1)==2);...
  3*(ind(2)==1)+4*(ind(2)==2);...
     1*(ind(3)==1)+2*(ind(3)==2);3*(ind(4)==1)+4*(ind(4)==2)];
       d2=d2kk;
    end          %======
```

```
        [metN indN]=min(d2); akVi=zeros(1,N-2); akVi(1)=-1; akVi(2)=-1;
        for kk=N:-1:3
            EI=asc(indN,kk); EF=indN; akVi(kk)=dec(EI,EF); indN=EI;
        end
        nbe=sum(abs(ak(1:N-2)-akVi(3:N))/2); PeVi(jj)=nbe/N;
end
semilogy(SNRdB, PeVi,'o'); grid
```

6.4 Compression

6.4.1 Scalar Quantization

The scalar quantization of a random quantity X consists of partitioning $\mathbb{R}$ in n sub-intervals:

$$I_1 = (a_0 = -\infty, a_1), \ldots, I_k = (a_{k-1}, a_k), \ldots, I_n = (a_{n-1}, a_n = +\infty)$$

and of defining a sequence $\{\mu_1, \ldots, \mu_n\}$ of real values. From then on:

- the *coding* process associates x, the value assumed by the random variable X, with the index k of the interval to which the word belongs, that is associates k with x if $x \in I_k$;
- the *decoding* process associates the index k with the value μ_k.

The value μ_k can be called either the *code word* or the *representative element* of the interval I_k, or simply the *reconstruction value*. The set of code words is called a *codebook*.

Hence you can see that the coding consists of associating the value x with the adress of one of the words in the codebook. In practices, the codebook size n is often chosen equal to a power of 2, that is $n = 2^N$, and the memory addresses are then written using N bits.

The linear quantization of a signal is the simplest case of scalar quantization: all of the n intervals, except for the first one and the last one, are chosen with the same length q, and the representative elements are taken from the middle of the interval. Coding amounts to testing whether x belongs to one of the n intervals of the type $(kq - q/2, kq + q/2)$ and to transmit k.

Mathematically, these two operations, coding and decoding, are summarized by the application:

$$X \to \mu(X)$$

What this actually means is that no difference is made between all the elements of the interval I_k and the code word μ_k. In this context, coding improves as the distorsion caused by these operations grows fainter, for a given number n of intervals. Of course, for the elements close to μ_k, the error is small. On the other hand, for the elements far away, the error is high. Hence it is best to have intervals of small length for the most frequent values of x and to let the intervals be longer for the less probable values. We are now going to give a mathematical expression for this relation.

The problem is to find, based on a probability distribution of X known beforehand, the best partitioning and the best representative elements with respect to a given criterion of distorsion between X and $\mu(X)$. In the search for a solution, the pioneers are undoubtedly Lloyd and Max, and the reader can look up the famous reference [13, 15].

We will give here the solution in the case where the criterion that has to be minimized is the square deviation defined by $\mathbb{E}\{(X - \mu(X)\}^2)$, and the expression of which is:

$$J(\{a_k\}, \{\mu_k\}) = \mathbb{E}\left\{(X - \mu(X))^2\right\} = \sum_{k=1}^{n} \int_{a_{k-1}}^{a_k} (x - \mu_k)^2 p_X(x) dx \quad (6.56)$$

where $p_X(x)$ refers to the probability density of X. The problem consists of determining the $(2n - 1)$ values a_1, ..., a_{n-1}, μ_1, ..., μ_n that minimize $J(\{a_k\}, \{\mu_k\})$. This solution can be obtained by setting to zero the partial derivatives of J with respect to the a_k and the μ_k. This leads to a system of equations which, once simplified, can be written:

$$\begin{cases} a_k = \dfrac{\mu_k + \mu_{k+1}}{2} & k = 1, \ldots, (n-1) \\ \\ \mu_k = \dfrac{\int_{a_{k-1}}^{a_k} x p_X(x) dx}{\int_{a_{k-1}}^{a_k} p_X(x) dx} & k = 1, \ldots, n \end{cases} \quad (6.57)$$

The first expression simply means that the interval I_k is the set of points closest to μ_k. And according to the second one, μ_k can be interpreted as the *barycenter* of the interval I_k weighted by $p_X(x)$.

Unfortunately, the expressions 6.57 generally do not lead to a simple analytical form for the quantities a_1, ..., a_{n-1} and μ_1, ..., μ_n. Numerical methods can then be used, such as the gradient algorithm, the general form of which is:

$$\theta_p = \theta_{p-1} - \lambda \left.\frac{\partial J}{\partial \theta}\right|_{\theta=\theta_{p-1}}, \qquad \lambda > 0 \quad (6.58)$$

where θ refers to the parameter for which we want to determine the numerical value that minimizes the criterion $J(\theta)$. The index p refers in this case to the p-th iteration. The *positive* number λ is the *gradient step*.

In our case, the parameter we have to determine is $\theta = (a_1, \ldots, a_{n-1}, \mu_1, \ldots, \mu_n)^T$, which is comprised of $(2n-1)$ values. Using 6.56, we can infer the $(2n-1)$ component expression of the gradient:

$$\frac{\partial J}{\partial \theta} = \begin{bmatrix} \dfrac{\partial J}{\partial a_1} & \cdots & \dfrac{\partial J}{\partial a_{n-1}} & \dfrac{\partial J}{\partial \mu_1} & \cdots & \dfrac{\partial J}{\partial \mu_n} \end{bmatrix}^T \quad (6.59)$$

with:

$$\begin{cases} \dfrac{\partial J}{\partial a_k} = ((a_k - \mu_k)^2 - (a_k - \mu_{k+1})^2)p_X(a_k) & k = 1, \ldots, (n-1) \\ \dfrac{\partial J}{\partial \mu_k} = -2 \int_{a_{k-1}}^{a_k} (x - \mu_k)p_X(x)dx & k = 1, \ldots, n \end{cases} \tag{6.60}$$

If $p_X(x)$ is a Gaussian distribution, the second relation of the system 6.60 can be numerically evaluated using MATLAB® with the `erf` function. The following program implements the gradient algorithm 6.58:

```
%===== lloyd.m
usrpi=1/sqrt(2*pi); rc2=sqrt(2);
%===== 7 parameters to be calculated
n=4; Ga=zeros(n-1,1); Gmua=zeros(n,1); Gmub=zeros(n,1);
a=[-3; 0; 3];           % initialization
mu=[-4; -2; 2; 4];
lambda=0.1;             % gradient step
for jj=1:2000
    mu1=mu(1:n-1); mu2=mu(2:n);
    expa=usrpi*exp(-(a .^2)/2);
    Ga=((a-mu1) .^2 - (a-mu2) .^2) .* expa;
    Gmua=-2*([0;expa]-[expa;0]);
    Gmub=mu .* (erf([a/rc2;+inf])-erf([-inf;a/rc2]));
    Gmu=Gmua+Gmub; mu=mu-lambda*Gmu;
    a=a-lambda*Ga;
end
%===== estimated parameters
[a;mu]'
```

For $n = 4$ and $\sigma = 1$, the program returned:

$$\begin{cases} a_3 = -a_1 = -0.9816 \quad a_2 = 0 \\ \mu_4 = -\mu_1 = 1.5104 \quad \mu_3 = -\mu_2 = 0.4528 \end{cases} \tag{6.61}$$

In the case of a centered, Gaussian random process with the variance σ^2, all we have to do is multiply these values by σ. As we have already said, the gradient method converges slowly and the calculation time is long. However, it can be shown that, if the function $\log(p_X(x))$ is strictly concave, the function J has only one minimum.

6.4.2 Vector Quantization

Introductory Example

The vector quantization problem is formulated in the same way as the scalar quantization. The difference is that we now wish to code length m vectors instead of real scalars. Hence we have to determine the best way of partitioning $\mathbb{R}^m$ in n regions and associate each one of these regions with the best representative element.

The simplest idea consists of coding the m components *separately*, using scalar quantization. Consider, for example, the case where $m = 2$ and where four bits are used to code a point of $\mathbb{R}^2$. We can allocate two bits to each of the vector's two components and then code each of the two components using two-bit scalar quantization. But we can also try to directly code the point in $\mathbb{R}^2$ using four bits. We are going to see, with an example first, that this second method can be better.

To obtain a sequence of correlated vectors we consider the AR-1 process defined by the equation $s(n)+as(n-1) = w(n)$, where $|a| < 1$ and where $w(n)$ is a centered, Gaussian, random sequence with the variance σ^2. We are going to use the following representation: based on the signal $s(n)$, we construct the two component vector $\boldsymbol{s}_2(p) = \begin{bmatrix} s(2p-1) & s(2p) \end{bmatrix}^T$ obtained by grouping together the two consecutive samples of $s(n)$. The following program displays a sequence of 1,000 values of $\boldsymbol{s}_2(p)$ in the form of a scatter in $\mathbb{R}^2$. The results are shown in Figure 6.22. The correlation found in the AR-1 process reveals itself by the elliptical shape of the scatter.

```
%===== partqv2ar1.m
N=1000; a=0.8; N2=N/2;
w=randn(N,1); s=filter(1,[1 a],w);
s2=zeros(2,N/2); s2(:)=s; s2=s2';
plot(s2(:,1),s2(:,2),'x'); axis('square')
%===== covariance matrix
%C=s2'*s2/(N/2);        % estimation of C
C=[1 -a;-a 1]/(1-a*a); % theoretical value of C
%===== confidence ellipsis
alpha=.9; s=-2*log(1-alpha);
hold on; ellipse([0 0],inv(C),s);
ellipse([0 0],inv(C),s); hold off, grid
```

with [2]:

```
function ellipse(X0,E,c)
%!==================================================!
%! Drawing an ellipse                                !
%! SYNOPSIS: ELLIPSE(X0,E,c)                         !
%!    X0 = coordinates of the ellipse's center (2x1) !
%!    E  = a positive (2x2) matrix                   !
%!    c  = scale Factor                              !
%!==================================================!
N=100; theta = (0:N) * (2*pi) ./ N ;
Y = sqrt(c)*[cos(theta);sin(theta)];
Fm1=sqrtm(E);
X = diag(X0)*ones(2,N+1)+Fm1\Y;
```

```
plot(X(1,:),X(2,:)); set(gca,'DataAspectRatio',[1 1 1])
```

Coding the vector $\boldsymbol{s}_2(p) \in \mathbb{R}^2$ using four bits means that we have to partition $\mathbb{R}^2$ in 16 regions. In the case where the two components of $\boldsymbol{s}_2(p)$ are coded separately using two bits, this partition is made of rectangular regions with their sides parallel to the axes. This is shown in Figure 6.22(a). The positions of the boundary lines are determined by the optimal values of the 2-bit scalar quantization of a Gaussian random variable, provided by 6.61.

Figure 6.22(b) also shows a partition based on 16 other regions, which takes more into account the elliptical distribution of the points in the plane. The regions are delimited first by the ellipse's major axis and second by the eight lines parallel to the minor axis located in scalar positions indicated by the 3-bit Lloyd-Max algorithm.

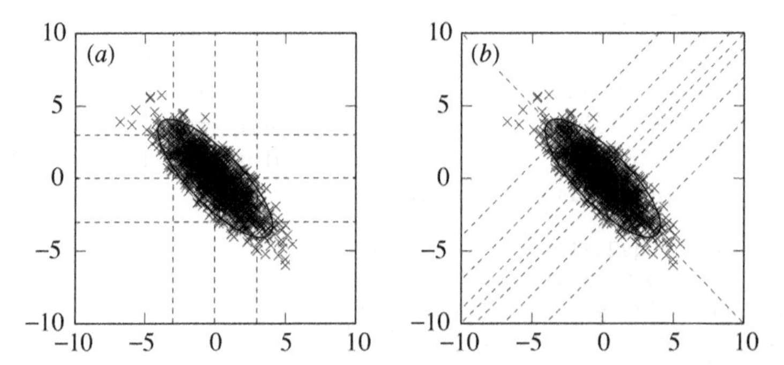

Figure 6.22 – *Two partitions of the plane in 16 regions (four bits are used to code each representative element of the corresponding region)*

Another idea consists of whitening the two components of the sequence $\boldsymbol{s}_2(p)$ so as to transform the elliptical scatter into a circular scatter. This operation is achieved by applying a square root of the inverse of the covariance matrix:

$$\boldsymbol{R} = \begin{bmatrix} R(0) & R(1) \\ R(1) & R(0) \end{bmatrix} = \frac{\sigma^2}{1-a^2} \begin{bmatrix} 1 & -a \\ -a & 1 \end{bmatrix}$$

to the sequence of vectors $\boldsymbol{s}_2(p)$. Note that the operation suggested here does not decorrelate the vectors that are transformed, only their two components. In plainer terms, if we use the notation $\boldsymbol{t}_2(p) = \boldsymbol{M}\boldsymbol{s}_2(p)$, where $\boldsymbol{M}$ is the modal matrix, $\mathbb{E}\left\{\boldsymbol{t}_2(p)\boldsymbol{t}_2^T(p)\right\}$ is diagonal but $\mathbb{E}\left\{\boldsymbol{t}_2(p)\boldsymbol{t}_2^T(p+k)\right\} \neq \boldsymbol{0}$.

In practice, the matrix $\boldsymbol{R}$ can be estimated, from the length N sequence $s(n)$ (and therefore the length of $\boldsymbol{s}_2(p)$ is equal to $N/2$), as follows:

$$\widehat{\boldsymbol{R}} = \frac{1}{N/2} \sum_{p=1}^{N/2} \boldsymbol{s}_2(p)\boldsymbol{s}_2^T(p)$$

We then obtain a sequence of vectors $\widehat{\boldsymbol{R}}^{-1/2} \boldsymbol{s}_2(p)$ the two components of which are uncorrelated. We can then perform the 2-bit quantization of each of the two components. You can check by typing:

```
%===== partqv2w.m
N=5000; a=0.8; N2=N/2;
w=randn(N,1); s=filter(1,[1 a],w);
s2=zeros(2,N/2); s2(:)=s; s2=s2';
subplot(121),plot(s2(:,1),s2(:,2),'x');
grid; axis('square')
%===== decorrelation
R2est=s2'*s2/N/2; s2t=s2*sqrtm(inv(R2est));
subplot(122); plot(s2t(:,1),s2t(:,2),'x');
grid; axis('square')
```

As an exercise, you can perform a simulation on 1,000 values and compare the square deviation of these three quantization rules by choosing as representative elements the points located in the center of these regions.

The last method which, after estimating the covariance matrix, whitens the two components, can be extended to a higher number of components. This method can also be generalized to any form of transformation that tends to perform a decorrelation of the data. This is the case for example of the discrete Fourier transform, of the discrete cosine transform. The coding can then be performed component by component in the transformed region: in this context, this is called transform coding [19].

Voronoi Regions and Centroids

We now come back to the problem of determining the best partition and the best representative elements in vector quantization. Let:

$$\mathcal{R} = \{R_1, \ldots, R_n\} \tag{6.62}$$

be the partition of the observation space $\mathbb{R}^m$ in n disjoint regions and:

$$\mathcal{C} = \{\mu_1, \ldots, \mu_n\}$$

the codebook of the n representative elements (Figure 6.23) where μ_k is the element in $\mathbb{R}^m$ that replaces every point of the region R_k.

Let us assume that the probability distibution of the vector observation X, with possible values in $\mathbb{R}^m$, has a probability density denoted by $p_X(x)$. Then the mean square deviation has the expression:

$$J(\mathcal{R}, \mathcal{C}) = \sum_{k=1}^{n} \int_{R_k} \|x - \mu_k\|^2 p_X(x) dx \tag{6.63}$$

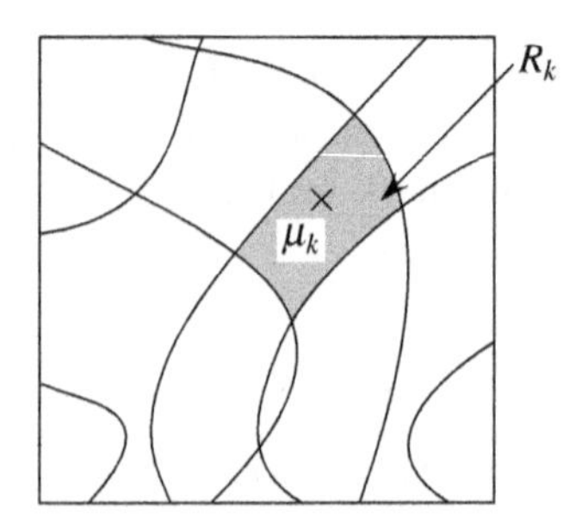

Figure 6.23 – *Partition $\mathcal{R}$ of the plane*

where $\|v\| = \sqrt{v^T v}$ represents the euclidean-norm of the vector v. The goal is to minimize $J(\mathcal{R}, \mathcal{C})$ with respect to $\mathcal{R}$ and $\mathcal{C}$. This can be performed numerically, by repeating the two following operations:

1. We start with n representative elements $\mathcal{C}$, and find the n regions that minimize $J(\mathcal{R}, \mathcal{C})$;

2. Once the n regions have been determined, we find the n new representatives that minimize $J(\mathcal{R}, \mathcal{C})$.

After each step, $J(\mathcal{R}, \mathcal{C})$ becomes smaller. The two operations are iterated until $J(\mathcal{R}, \mathcal{C})$ reaches a value deemed small enough. It is unfortunately possible with this algorithm to end up at a local minimum. Let us examine these two operations in detail:

1. If $\mathcal{C}$ is set, minimizing $J(\mathcal{R}, \mathcal{C})$ with respect to $\mathcal{R}$ amounts, according to expression 6.63, to assigning to R_k all of the points of $\mathbb{R}^m$ such that:

$$R_k = \{\ x \in \mathbb{R}^m :\ \|x - \mu_k\| < \|x - \mu_j\|\ \forall j \neq k\ \} \tag{6.64}$$

   The regions defined by such a partition are called *Voronoi regions*. This concept can of course be extended to other distances, not just the basic euclidean distance.

   Figure 6.24 shows the Voronoi regions for a set of points of $\mathbb{R}^2$ when the distance used in the euclidean distance. In this case, we start with the set of representative elements, and the lines delimiting the regions are simply defined by the perpendicular bisectors of the line segments formed by the pairs of these points. The points belonging to the perpendicular bisectors can indifferently be assigned to either one of the adjacent regions.

2. If $\mathcal{R}$ is set, and under sufficient regularity conditions, $J(\mathcal{R}, \mathcal{C})$ can be minimized with respect to $\mathcal{C}$ by setting to zero the partial derivative of

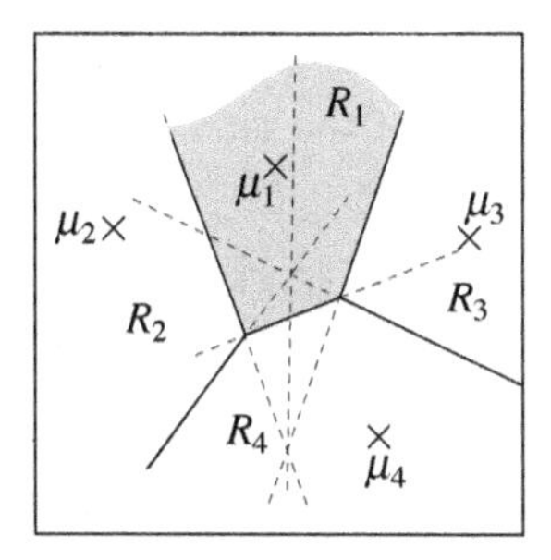

Figure 6.24 – *$\mathcal{R}$ partition by the perpendicular bisectors*

$J(\mathcal{R}, \mathcal{C})$ with respect to μ_k. This leads to:

$$\mu_k = \frac{\int_{R_k} x p_X(x) dx}{\int_{R_k} p_X(x) dx} \tag{6.65}$$

The point in $\mathbb{R}^m$ defined by the expression 6.65 is called the *centroid* of the region R_k for the ditribution with the probability distribution $p_X(x)$.

In practice, the probability distribution of the observation X we wish to code is unknown, and therefore it probability density $p_X(x)$ even more so. However, the latter can be estimated based on a set of outcomes. This leads us to the algorithm presented in the following section.

LBG Algorithm

The *LBG* algorithm, named after its creators *Linde, Buzzo* and *Gray* (see reference [12]), allows us to determine the n best representative elements of an unknown distribution based on a set of N outcomes called a *training set*:

$$\mathcal{D} = \{x_1, \ldots, x_N\} \quad \text{where} \quad x_i \in \mathbb{R}^m$$

In practice, $N \gg n$. Replacing the unknown probability distribution $p_X(x)$ with the histogram of the values is equivalent to assigning the same probability $1/N$ to every observation. If we then replace this in 6.65, we get:

$$\mu_k = \frac{\sum_{R_k} \frac{1}{N} x_i}{\sum_{R_k} \frac{1}{N}} = \frac{1}{N_k} \sum_{\{i:x_i \in R_k\}} x_i \tag{6.66}$$

where N_k refers to the number of points in the region R_k. Using this result, we can implement the following algorithm. We start with an initial codebook $\mathcal{C}$ of n centroids $\{\mu_1, \ldots, \mu_n\}$ belonging to $\mathbb{R}^m$, then successively perform the following operations:

1. construction of the n Voronoi regions associated with the n centroids, that is to say the partition of the set $\mathcal{D}$ in n subsets containing the points closest to the μ_k;

2. computation, based on expression 6.66, of the n new corresponding centroids;

3. stopping criterion: if the n new centroids are "close" enough to the previous ones, the algorithm stops, otherwise, it picks up at step 1.

It can be shown that the estimated square deviation, defined by:

$$\mathcal{E} = \frac{1}{N}\sum_{k=1}^{n}\sum_{i=1}^{N}\|x_i - \mu_k\|^2 \mathbb{1}(x_i \in R_k) \tag{6.67}$$

decreases after each iteration and that the algorithm converges to a local minimum.

On the other hand, this algorithm poses two problems related to the initialization: first, we are not sure that the algorithm converges to the global minimum, and second, empty spaces can appear while the algorithm is running. To avoid these problems, the LBG algorithm has the following efficient initialization procedure:

1. the centroid is calculated for the set of the whole training sequence;

2. based on each centroid, two points are constructed by a small shift in two opposite directions. As a consequence, the number of points is doubled;

3. the Voronoi regions, associated with the set of previously obtained points, are determined, as well as their respective centroids;

4. the algorithm goes back to step 2 until a set of n centroids is obtained. The resulting codebook size is a power of 2.

The `voronoi.m` Function

The `voronoi.m` function determines the n Voronoi regions associated with the codebook `C`. It returns the vector `indR` that contains, for each element of the training sequence `x`, the number of the region to which it belongs.

```
function indR=voronoi(x,C)
%!=============================================!
%! SYNOPSIS: indR=VORONOI(x,C)                  !
%!      x    = (m x T) training sequence        !
%!      C    = n code words (m x n) array       !
%!      indR = length T vector associating a point !
%!             x to the region it belongs to    !
```

```
%!==============================================!
[m, T]=size(x); [m, n]=size(C);
%===== euclidean norms(^2) of the representatives
norC=ones(1,m)*(C .* C);
%===== searching the minimal distance for each x
indR = zeros(T,1);
for tt=1:T
    d2=norC - 2*x(:,tt)'*C; [bid,ind]=min(d2);
    indR(tt)=ind(1);
end
```

The centroides.m Function

The `centroides.m` function determines the n centroids (matrix `C`) based on the training sequence along with the number of the region to which each of its elements belong. It returns the value of the square deviation `E` given by expression 6.67.

```
function [C,E] = centroides(x,n,indR)
%!==============================================!
%! Computing the n centroids of n regions       !
%! SYNOPSIS : [C,E]=CENTROIDES(x,n,indR)        !
%!    x    = training sequence                  !
%!    n    = number of centroids                !
%!    indR = index of the region x belongs to   !
%!    C    = centroids                          !
%!    E    = square error                       !
%!==============================================!
[m,T]=size(x); C=zeros(m,n);
E=0;
for jj=1:n     % for each region
    indRep=find(indR==jj); % indices of points
    % belonging to region jj
    xjj=x(:,indRep);
    [m, lxjj]=size(xjj);
    if lxjj~=0
        % estimated mean
        C(:,jj)=(xjj*ones(lxjj,1))/lxjj;
        d2=(C(:,jj)*ones(1,lxjj)-xjj) .^2;
        E=E+sum(sum(d2));
    end
end
E=E/T; % estimated Error
```

The initlbg.m Function

The `initlbg.m` function determines an initial size n codebook based on the training sequence `x`. The value `delta` sets the shifts in the direction $(1, 1, \ldots, 1)$ and in the direction $(-1, -1, \ldots, -1)$.

```
function C0 = initlbg(x,n)
%!=============================================================!
%! Initialization of LBG: determine n centroids by dichotomy !
%! SYNOPSIS: C0=INITLBG(x,n)                                   !
%!      x  = learning sequence (mxT array)                     !
%!      n  = number of representatives (power of 2)            !
%!      C0 = n representatives (mxn array)                     !
%!=============================================================!
delta=0.01; [m,T]=size(x);
C0=x*ones(T,1)/T;     % first representative
indR=ones(1,T);       % initial array
jj=1; vv=zeros(m,n);
while (jj<n)
   for kk=1:jj
       vv(:,2*kk-1)=C0(:,kk)-delta;
       vv(:,2*kk)=C0(:,kk)+delta;
   end
   vv=vv(:,1:2*jj);
   jj=jj*2; C0=centroides(x,jj,indR);
end
```

Implementing the LBG algorithm

The `lbg.m` function implements the LBG algorithm and returns a size n codebook, based on the training sequence `x`. This function also returns the initial codebook produced by the function `initlbg.m` as well as the decreasing sequence of values of the square deviation. The length of `E` gives the number of iterations needed to reach the minimum.

```
function [Cf,Ci,E] = lbg(x,n)
%!=====================================================!
%! Calculating the dictionary using the LBG algorithm !
%! SYNOPSIS: [Cf,Ci,E]=LBG(x,n)                        !
%!     x  = (mxT) training sequence with:              !
%!         m  dimension of the observations            !
%!         T  number of observations                   !
%!     n  = number of representatives (power of 2)     !
%!     Cf = dictionary                                 !
%!     Ci = initial dictionary                         !
%!     E  = square error based on the iteration step   !
%! Uses: voronoi.m, centroides.m initlbg.m             !
%!=====================================================!
[m, T]=size(x); N=log2(n); Cn=zeros(m,n);
indR=zeros(1,T); E(1)=0; ncv=1; epsilon=0.001;
%===== initialization
Ci=initlbg(x,n); Cf=Ci;
%===== iterations
rep=0;
while ncv
```

```
    rep=rep+1;
    %===== searching the Voronoi regions corresponding
    %      to the n centroids. indR gives each point
    %      of x the number the region it belongs to.
    indR = voronoi(x,Cf);
    %===== calculating of new centroids and distorsion E
    [Cn,E(rep)]=centroides(x,n,indR);
    if max(abs(Cn-Cf))<epsilon
       ncv=0; Cf=Cn;
    else
       Cf=Cn;
    end
end
```

Example 6.3 (Applying the LBG function) We are going to use the `lbg.m` function to find the four representative elements of a centered Gaussian distribution with the variance 1. Type:

```
x=randn(1,4000);
[Cf,Ci,E]=lbg(x,4);
Cf
```

Through simulation, we found, values for `Cf` such as:

```
-1.4899   -0.4369   0.4565   1.5036
```

This result is in agreement with the values $\mu_4 = -\mu_1 = 1.5104$ and $\mu_3 = -\mu_2 = 0.4528$ obtained by numerical resolution of the Lloyd-Max equation (see page 222).

The Codebook Size Problem

Let us assume that we have to code the elements of $\mathbb{R}^m$ using 20 bits. A direct application of vector quantization leads to the creation of a $2^{20} \approx 1\,000\,000$ word codebook. If this number is too large for the considered application, we can, at the cost of an acceptable performance loss (see exercise 6.9), break the 20 bits down to two or more smaller sub-codebooks. For instance, if the 20 bits are broken down to two sets of 10 bits, two codebooks have to be created, each one containing $2^{10} \approx 1000$ words. An illustration of this method is given by the vector quantization of the prediction coefficients in a speech coder.

The *MELP coder* (*Mixed Excitation Linear Prediction*) converts speech sampled at 8,000 Hz into a sequence of binary symbols flowing at a rate of 2,400 bits per second. As the name implies, the coder performs a linear prediction analysis on 10 coefficients every 22.5 ms. Therefore, it has at its disposal $2400 \times 22.5\,10^{-3} = 54$ bits to code a frame. 25 of these 54 bits are assigned to coding the 10 prediction coefficients. A direct application of vector quantization would lead to producing a $2^{25} \approx 32{,}000{,}000$ word codebook in $\mathbb{R}^{10}$.

Obviously, this is infeasible. Instead, the solution consists of creating several nested codebooks. The principle can be explained, without being any less general, by first considering a vector coding using 6 bits. A direct construction would then consist of using a dictionary of $2^6 = 64$ elements. Instead, and at the cost of an acceptable performance loss, we can construct a first dictionary of 4 representative elements, therefore coded using 2 bits, then code the distance between the vector to be coded and the representative element using the 4 remaining bits with a single dictionary of $2^4 = 16$ representative elements. Thus, the set of two dictionnaries is now only comprised of $16+4 = 20$ representative elements instead of 64. To get a better idea, imagine a first partition of the space in 4 regions, then each region is partitioned *in an identical way* in 16 sub-regions. All we need to do when decoding is concatenate the value of the value of the representative element associated with the first two bits and the value of the representative element associated with the last 4 bits. The solution chosen for the MELP coder uses 4 sub-codebooks with the sizes $2^7 = 128$, $2^6 = 64$, $2^6 = 64$ and $2^6 = 64$, respectively, for a total of only 320 words to code the vector of the 10 prediction coefficients using 25 bits.

Exercise 6.9 (Performances with two sub-codebooks) (see page 315)
Consider the AR-1 process $s(n)$ defined by the recursive equation:

$$s(n) + as(n-1) = w(n)$$

where $a = 0.9$ and where $w(n)$ is a centered, Gaussian random sequence with a variance equal to 1. The "vector" signal $\boldsymbol{s}_2(p) = [s(2p-1)\ s(2p)]^T$ is reconstructed from the signal $s(n)$ by grouping together two consecutive samples of $s(n)$. We wish to code $\boldsymbol{s}_2(p)$ using 6 bits.

1. Write a program that generates the length $N = 5000$ sequence $\boldsymbol{s}_2(p)$ then constructs a codebook of $2^6 = 64$ representative elements with the use of the `lbg.m` function.

2. Using the same sequence, construct two sub-codebooks with $2^2 = 4$ and $2^4 = 16$ representative elements, respectively. Compare the performances. Bear in mind that the computation can take up to a few minutes on a standard computer. However, this usually is not a problem since the computation is done once and for all before the use of the codebooks.

Image Applications

To illustrate the performances of the LBG algorithm, we are now going to discuss the example of the compression of an image defined in levels of gray. We are going to start with the size (256×256) image of Lena, and cut it up into T "thumbnails" of m *pixels*. If we want to code each thumbnail using b

bits (hence b/m bits per pixel), we have to find 2^b representative elements. The following program applies the `lbg.m` function to the T thumbnails.

```
%===== imagette.m
clear
b=3;      % number of bits per representative element
n=2^b;    % number of  representative elements
ml=4;     % number of pixels per row
mc=4;     % number of pixels per column
m=ml*mc; % number of pixels
nbbit_pixel=b/m;
%=====
load lena; % reading the image pixc and colormap cmap
[lig col]=size(pixc); moy_pixc=mean(mean(pixc));
pixc_centre=pixc-moy_pixc;
blocl=lig/ml;            % number of blocks per row
blocc=col/mc;            % number of blocks per column
nbimagette=lig*col/m; % number of images
%===== the image is divided in (m x 1) vector thumbnails
imaget=zeros(m,nbimagette);
if m==1,
imaget(:)=pixc_centre;
else
   %=====  T column vectors
   for ll=0:blocl-1
       il=ll*ml+1;
       for cc=0:blocc-1
           ic=cc*mc+1; ii=ic+il-1;
           aux=pixc_centre(il:il+ml-1,ic:ic+mc-1);
           imaget(:,ii)=aux(:);
       end
   end
end
%===== searching the n representative elements
[Cf,Ci,E]=lbg(imaget,n);          % LBG
norC=ones(1,m)*( Cf .* Cf );
d2=zeros(n,1); auxcol=zeros(m,1);
%===== coding the thumbnails
imag0=zeros(ml,mc);               % initialization
pixc_cod_centre=zeros(lig,col);
for ll=0:blocl-1
    il=ll*ml+1;
    for cc=0:blocc-1
        ic=cc*mc+1; ii=ic+il-1;
        aux=pixc_centre(il:il+ml-1,ic:ic+mc-1);
        auxcol=aux(:);
        for nn=1:n
            d2(nn)=-2*auxcol'*Cf(:,nn) + norC(nn);
        end
        [bid, ind]=min(d2); Ccod=Cf(:,ind); imag0(:)=Ccod;
```

```
            %===== Reconstruction
            pixc_cod_centre(il:il+ml-1,ic:ic+mc-1)=imag0;
        end
    end
    pixc_cod=pixc_cod_centre+moy_pixc;
    subplot(121); imagesc(pixc); colormap(cmap); axis('square');
    subplot(122); imagesc(pixc_cod); colormap(cmap); axis('square');
```

If $m = 1$ and $b = 1$, this is equivalent to coding each pixel using 1 bit. The result has to be a "finely detailed" image but with 2 levels of gray. If, on the contrary, $m = 4$ and $b = 4$, hence still 1 bit per pixel, the image definition will not be as good, with $2^4 = 64$ thumbnails to represent the image. Figures 6.25 and 6.26 illustrate the differences obtained depending on the values of m and b.

Figure 6.25 – *Original image on the left and the result of the LBG compression with* $m = 16$ *and* $b = 3$

Figure 6.26 – *LBG compression:* $(m = 4, b = 2)$ *on the left,* $(m = 4, b = 4)$ *on the right*

Chapter 7

Hints and Solutions

H1 Mathematical concepts

H1.1 (δ-method) (see page 20) We shall use the hypothesis $\text{cov}(X_1, X_2) = \sigma^2 \boldsymbol{I}_2$. The Jacobian, here noted $\partial \boldsymbol{g}$, of $g : (X_1, X_2) \to (R, \theta)$ is deduced from the Jacobian $\partial \boldsymbol{h}$ of $h : (R, \theta) \to (X_1, X_2)$ following:

$$\partial \boldsymbol{h} = \begin{bmatrix} \cos(\theta) & -R\sin(\theta) \\ \sin(\theta) & R\cos(\theta) \end{bmatrix} \Rightarrow \partial \boldsymbol{g} = \partial \boldsymbol{h}^{-1} = \frac{1}{R} \begin{bmatrix} R\cos(\theta) & R\sin(\theta) \\ -\sin(\theta) & \cos(\theta) \end{bmatrix}$$

We have:

$$\text{cov}\left(\begin{bmatrix} R \\ \theta \end{bmatrix}\right) = \begin{bmatrix} \text{cov}(R, R) & \text{cov}(R, \theta) \\ \text{cov}(R, \theta) & \text{cov}(\theta, \theta) \end{bmatrix} = \begin{bmatrix} \text{var}(R) & \text{cov}(R, \theta) \\ \text{cov}(R, \theta) & \text{cov}(\theta, \theta) \end{bmatrix}$$

and, from (1.51):

$$\text{cov}\left(\begin{bmatrix} R \\ \theta \end{bmatrix}\right) \approx \partial \boldsymbol{g}\, \text{cov}(X_1, X_2)\, \partial^T \boldsymbol{g} = \sigma^2 \boldsymbol{I}_2$$

we deduce:

$$\text{var}(R) = \begin{bmatrix} 1 & 0 \end{bmatrix} \sigma^2 \begin{bmatrix} 1 \\ 0 \end{bmatrix} = \sigma^2 \tag{7.1}$$

This result is therefore correct on the condition that $(\mu_1^2 + \mu_2^2)/\sigma^2$ is large. This result may appear weak; note, however, that the δ-method allows us to calculate an approximation even in cases where $\text{cov}(X_1, X_2)$ is different from $\sigma^2 \boldsymbol{I}_2$.

Let $\nu = \sqrt{\mu_1^2 + \mu_2^2}$. Program `deltaMethodRice.m` estimates the variance of R by carrying out a simulation using 100000 draws. If $\nu/\sigma > 4$, then the approximation is acceptable. Note that the function $\boldsymbol{g}$ under consideration is not differentiable at point $(0, 0)$, associated with the case where $\mu_1 = \mu_2 = 0$. Type:

```
%=== deltaMethodRice
clear all
nu = 8; sigma = 2; a = 2*rand*pi;
mu1 = nu*cos(a); mu2 = nu*sin(a);
Lruns = 100000;
X = ones(Lruns,1)*[mu1 mu2] + sigma * randn(Lruns,2);
R = sqrt( (X .^2)*ones(2,1));
sprintf('true value = %4.2f, simulation value = %4.2f',...
    sigma,std(R))
```

H1.2 (Asymptotic confidence interval) (see page 22)

1. From the hypotheses, we deduce that, for any k, $\mathbb{E}\{X_k\} = p$ and $\text{var}(X_k) = p(1-p)$. According to the central limit theorem 1.10, when N tends toward infinity we have:

$$\sqrt{N}\,(\widehat{p} - p) \xrightarrow{d} \mathcal{N}(0, p(1-p))$$

2. From this, we deduce:

$$\mathbb{P}\left\{-\epsilon \leq \sqrt{N}(\widehat{p} - p) \leq \epsilon\right\} \approx \int_{-\epsilon}^{\epsilon} \frac{1}{\sqrt{2\pi p(1-p)}} e^{-\frac{u^2}{2p(1-p)}} du$$

Taking $w = u/\sqrt{p(1-p)}$ and $\delta = \epsilon/\sqrt{p(1-p)}$, we obtain:

$$\mathbb{P}\left\{-\delta\sqrt{p(1-p)} \leq \sqrt{N}(\widehat{p} - p) \leq \delta\sqrt{p(1-p)}\right\} \approx \int_{-\delta}^{\delta} \frac{1}{\sqrt{2\pi}} e^{-w^2/2} dw$$

3. Solving the double inequality in terms of p, we deduce the confidence interval at $100\,\alpha\%$:

$$\text{CI}_{100\alpha\%} = \left(\frac{\widehat{p} + \frac{\delta^2}{2N} - \sqrt{\Delta}}{1 + \frac{\delta^2}{N}}, \frac{\widehat{p} + \frac{\delta^2}{2N} + \sqrt{\Delta}}{1 + \frac{\delta^2}{N}}\right) \quad (7.2)$$

where $\Delta = \widehat{p}(1 - \widehat{p})\frac{\delta^2}{N} + \frac{\delta^4}{4N^2}$ and where δ is linked to value α by

$$\int_{-\delta}^{\delta} \frac{1}{\sqrt{2\pi}} e^{-w^2/2} dw = \alpha$$

Typically, for $\alpha = 0.95$, $\delta = 1.96$.

4. Type the program:

```
%==== ICpercent.m
clear all
alpha = 0.95;
delta = norminv(1-(1-alpha)/2);
p = 0.2; N = 300;
Lruns = 1000; CI = zeros(2,Lruns);
for ir=1:Lruns
   X = rand(N,1) < p;
   hatp = mean(X);
   Delta = hatp*(1-hatp)*(delta^2)/N+(delta^4)/4/N/N;
   CI(1,ir) = (hatp+(delta^2)/2/N-sqrt(Delta))/(1+delta^2/N);
   CI(2,ir) = (hatp+(delta^2)/2/N+sqrt(Delta))/(1+delta^2/N);
end
% percent of the true value inside the confidence interval
X=100*sum(and(CI(1,:)<p, CI(2,:)>p))/Lruns;
disp('Part of true values inside the confidence interval')
[num2str(X),'%']
```

H2 Statistical inferences

H2.1 (ROC curve and AUC for two Gaussians) (see page 33)

1. The significance level α represents the probability that, under H_0, the statistic $\Phi(X)$ will be higher than the threshold ζ. Consequently, to estimate α for the threshold ζ, we need to count how many values of $\Phi(x)$ are greater than ζ in the database H_0. The same approach should be taken for the power; to obtain the ROC curve, we alter ζ across a range dependent on the minimum and maximum values of $\Phi(X)$. One simple approach is to rank the set of values of $\Phi(X)$ in decreasing order, then calculate the cumulative sum of the indicators associated with the two hypotheses. This is written:

$$\begin{cases} \widehat{\alpha}_j & = & \dfrac{1}{N_0}\sum_{i_0=1}^{N_0} \mathbb{1}(\Phi_{0,i_0} > T_j) \\ \widehat{\beta}_j & = & \dfrac{1}{N_1}\sum_{i_1=1}^{N_1} \mathbb{1}(\Phi_{1,i_1} > T_j) \end{cases}$$

where T_j is the sequence of values ranked in decreasing order $\{T_{0,i_0}\} \cup \{T_{1,i_1}\}$, where $j = 1$ to $N_t = N_0 + N_1$.

2. Type the program:

```
%===== roc2simplesHypot.m
clear all
```

```
N0 = 1000; N1 = 1200; Nt = N0+N1;
n = 10; m0 = 0; m1 = 0.5;
%===== data bases H0 and H1
X0 = m0+randn(N0,n); X1 = m1+randn(N1,n);
%===== statistics Phi(X)
Phi0 = mean(X0,2); Phi1 = mean(X1,2);
%===== Experimental ROC curve estimate
c0 = zeros(Nt,1); c1 = zeros(Nt,1);
[Tsort,idx] = sort([Phi0;Phi1],'descend');
c0(idx<=N0) = 1; hatalpha = cumsum(c0)/N0;
c1(idx>N0) = 1; hatbeta = cumsum(c1)/N1;
hatalpha = [0;hatalpha];
hatbeta = [0;hatbeta];
plot(hatalpha,hatbeta), grid
zeta = linspace(0,10,200);
%===== theoretical ROC curve
alpha = 1-normcdf(zeta,m0,1/sqrt(n));
beta = 1-normcdf(zeta,m1,1/sqrt(n));
hold on; plot(alpha,beta,'.--'); hold off
%===== AUC
W = 0;
for i0=1:N0
    for i1=1:N1
        W = W+(Phi1(i1)>Phi0(i0))+0.5*(Phi1(i1)==Phi0(i0));
    end
end
eauc = W/(N0*N1);
tt = sprintf('Experimental AUC = %5.2f',eauc);
text(0.3,0.7,tt,'fontsize',18,'fontw','b')
```

H2.2 (Student distribution for H_0) (see page 35) In the following program, the variance and the mean have been chosen at random, and we see that they have no effect on the distribution of $V(X)$.

Type:

```
%===== studentlawmdiffm0.m
% Uses tpdf from the stats toolbox
clear all
sigma   = rand*100;
L       = 10000;
n       = 100;
m0      = 10*randn;
X       = randn(n,L)+m0;
num     = (mean(X,1)-m0)*sqrt(n);
Xc      = X - ones(n,1)*mean(X,1);
denum   = sqrt(sum(Xc .^2,1)/(n-1));
V       = num ./ denum;
[hb,xb] = hist(V,100);
testim  = hb/L/(xb(2)-xb(1));
%=====
```

```
bar(xb,testim);
ttheo = tpdf(xb,n-1);
hold on, plot(xb,ttheo,'.-r'), hold off
```

H2.3 (Unilateral mean testing) (see page 36) The log-likelihood is written:

$$\mathcal{L}(\theta) = -\frac{n}{2}\log(2\pi) - \frac{n}{2}\log(\sigma^2) - \frac{1}{2\sigma^2}\sum_{k=1}^{n}(X_k - m)^2$$

Firstly, maximizing with respect to σ^2, we obtain:

$$\widetilde{\mathcal{L}}(m) = -\frac{n}{2}\log(2\pi) - \frac{n}{2}\log\sum_{k=1}^{n}(X_k - m)^2$$

The maximum for H_0 is obtained for $m = n^{-1}\sum_{k=1}^{n} X_k$ on the condition that $n^{-1}\sum_{k=1}^{n} X_k \geq m_0$, otherwise it is obtained for m_0. The maximum for H_0 is therefore written:

$$\begin{aligned} -\frac{n}{2}\log(2\pi) \quad &- \quad \frac{n}{2}\log\sum_{k=1}^{n}(X_k - \widehat{m})^2\mathbb{1}(\widehat{m} \geq m_0) \\ &- \quad \frac{n}{2}\log\sum_{k=1}^{n}(X_k - m_0)^2\mathbb{1}(\widehat{m} < m_0) \end{aligned}$$

The maximum for Θ is expressed:

$$-\frac{n}{2}\log(2\pi) - \frac{n}{2}\log\sum_{k=1}^{n}(X_k - \widehat{m})^2$$

The log-GLRT, noted A, is written:

$$\begin{aligned} A \quad = \quad & -\frac{n}{2}\log\sum_{k=1}^{n}(X_k - \widehat{m})^2\left(\mathbb{1}(\widehat{m} \geq m_0) + \mathbb{1}(\widehat{m} < m_0)\right) \\ & +\frac{n}{2}\log\sum_{k=1}^{n}(X_k - \widehat{m})^2\mathbb{1}(\widehat{m} \geq m_0) \\ & +\frac{n}{2}\log\sum_{k=1}^{n}(X_k - m_0)^2\mathbb{1}(\widehat{m} < m_0) \end{aligned}$$

$$\begin{aligned}
A &= -\frac{n}{2}\left(\log\frac{\sum_{k=1}^{n}(X_k-\widehat{m})^2}{\sum_{k=1}^{n}(X_k-m_0)^2}\right)\mathbb{1}(\widehat{m}<m_0)\\
&= -\frac{n}{2}\left(\log\frac{\sum_{k=1}^{n}(X_k-\widehat{m})^2}{\sum_{k=1}^{n}(X_k-\widehat{m}+(\widehat{m}-m_0))^2}\right)\mathbb{1}(\widehat{m}<m_0)\\
&= -\frac{n}{2}\left(\log\frac{\sum_{k=1}^{n}(X_k-\widehat{m})^2}{\sum_{k=1}^{n}(X_k-\widehat{m})^2+n(\widehat{m}-m_0)^2}\right)\mathbb{1}(\widehat{m}<m_0)\\
A &= \frac{n}{2}\log(1+T(X))\,\mathbb{1}(\widehat{m}<m_0) \qquad (7.3)
\end{aligned}$$

taking:

$$T(X) = \frac{(\widehat{m}-m_0)^2}{n^{-1}\sum_{k=1}^{n}(X_k-\widehat{m})^2}$$

Taking the exponential of the two members with $n \neq 0$, we obtain:

$$\mathrm{GLRT}^{\frac{2}{n}} = (1+T(X))\,\mathbb{1}(\widehat{m}<m_0) + \mathbb{1}(\widehat{m}\geq m_0) = T(X)\,\mathbb{1}(\widehat{m}<m_0)$$

and the test:

$$V(X) = T(X)\,\mathbb{1}(\widehat{m}<m_0) \underset{H_0}{\overset{H_1}{\gtrless}} \eta$$

If $\eta < 0$, we always decide in favor of H_1.

$$U(X) = \frac{\sqrt{n}(m_0-\widehat{m})}{(n-1)^{-1/2}\sqrt{\sum_{k=1}^{n}(X_k-\widehat{m})^2}}$$

We then verify that V is an increasing monotone function of U. Comparing V to a threshold is therefore equivalent to comparing U to a threshold. We know that, for H_0, the statistic $U(X)$ follows a Student distribution with $(n-1)$ degrees of freedom. The test is therefore written:

$$\frac{\sqrt{n}(m_0-\widehat{m})}{(n-1)^{-1/2}\sqrt{\sum_{k=1}^{n}(X_k-\widehat{m})^2}} \underset{H_0}{\overset{H_1}{\gtrless}} T_{n-1}^{[-1]}(1-\alpha)$$

H2.4 (Mean equality test) (see page 38)

1. The model is constituted by $\{\text{i.i.d. } \mathcal{N}(n_1; m_1, \sigma^2)\}\otimes\{\text{i.i.d. } \mathcal{N}(n_2; m_2, \sigma^2)\}$ where $\theta = (m_1, m_2, \sigma) \in \Theta = \mathbb{R}\otimes\mathbb{R}\otimes\mathbb{R}^+$. Hypothesis $H_0 = \{\theta \in \Theta : s.t. \quad m_1 = m_2\}$.

2. Let $n = n_1 + n_2$. The maximization of the probability density for Θ leads us to take $\widehat{m}_1 = n_1^{-1}\sum_{k=1}^{n_1} X_{1,k}$, $\widehat{m}_2 = n_2^{-1}\sum_{k=1}^{n_2} X_{2,k}$ and $\widehat{\sigma}_1^2 = n^{-1}\sum_{k=1}^{n_1}(X_{1,k} - \widehat{m}_1)^2 + n^{-1}\sum_{k=1}^{n_1}(X_{2,k} - \widehat{m}_2)^2$.

   Maximization of the probability density for H_0 leads us to take $\widehat{m}_0 = n^{-1}\sum_{k=1}^{n_1} X_{1,k} + n^{-1}\sum_{k=1}^{n_2} X_{2,k}$ and $\sigma_0^2 = n^{-1}\sum_{k=1}^{n_1}(X_{1,k} - \widehat{m}_0)^2 + n^{-1}\sum_{k=1}^{n_2}(X_{2,k} - \widehat{m}_0)^2$. From this, we deduce:

$$\widehat{\sigma}_0^2 = \widehat{\sigma}_1^2 + \frac{n_1}{n}(\widehat{m}_0 - \widehat{m}_1)^2 + \frac{n_2}{n}(\widehat{m}_0 - \widehat{m}_2)^2$$

   and the GLRT:

$$\Lambda(X) \quad \propto \quad \frac{\widehat{\sigma}_0^2}{\widehat{\sigma}_1^2} = 1 + \frac{\frac{n_1}{n}(\widehat{m}_0 - \widehat{m}_1)^2 + \frac{n_2}{n}(\widehat{m}_0 - \widehat{m}_2)^2}{n^{-1}\sum_{k=1}^{n_1}(X_{1,k} - \widehat{m}_1)^2 + n^{-1}\sum_{k=1}^{n_1}(X_{2,k} - \widehat{m}_2)^2}$$

   Comparing the GLRT to a threshold can be shown to be equivalent to comparing the following statistic $W(X)$ to a threshold:

$$W(X) = \frac{\sqrt{n_H}\,|\widehat{m}_1 - \widehat{m}_2|}{S/\sqrt{n-2}}$$

   where $n_H^{-1} = n_1^{-1} + n_2^{-1}$ et $S^2 = \sum_{k=1}^{n_1}(X_{1,k} - \widehat{m}_1)^2 + \sum_{k=1}^{n_2}(X_{2,k} - \widehat{m}_2)^2$.

3. Based on property 2.2 (section 2.3.3), we deduce that $\widehat{m}_1$, $\widehat{m}_2$, $e_1 = X_1 - \widehat{m}_1$ and $e_2 = X_2 - \widehat{m}_2$ are jointly independent and Gaussian. Hence, for H_0:

$$\frac{U}{\sigma} = \frac{\sqrt{n_H}(\widehat{m}_1 - \widehat{m}_2)}{\sigma} \sim \mathcal{N}(0,1)$$

   We also deduce that $S^2/\sigma^2 \sim \chi_{n-2}^2$ and that S^2 is independent of U. Hence, in accordance with (1.57):

$$V(X) = \frac{U(X)/\sigma}{S(X)/\sigma\sqrt{n-2}} \sim \frac{\mathcal{N}(0,1)}{\sqrt{\chi_{n-2}^2/(n-2)}} = T_{n-2}$$

   where T_{n-2} is a Student variable with $(n-2)$ degrees of freedom. This distribution is symmetrical around the value 0. The test therefore takes the form:

$$W(X) \quad \underset{H_0}{\overset{H_1}{\gtrless}} \quad \eta$$

   where $\eta = T_{n-2}^{[-1]}(1 - \alpha/2)$ and where α is the confidence level.

4. The p-value is written:

$$p\text{-value} = 2\int_{T(X_1,X_2)}^{+\infty} T_{n-2}(t)dt$$

5. The program below provides a p-value of 0.63, so the hypothesis of mean equality is accepted.

```
%===== T-test
clear all
data = [1    1    1    1    2    2    2; ...
        51.0 53.3 55.6 51.0 55.5 53.0 52.1];
d1 = data(2,data(1,:)==1);
d2 = data(2,data(1,:)==2);
N1=length(d1); N2=length(d2); N=N1+N2;
m1=mean(d1); m1c = d1-m1;
m2=mean(d2); m2c = d2-m2;
NH = 1/(1/N1+1/N2);
U  = abs(m1-m2)* sqrt(NH);
S2 = m1c*m1c'+m2c*m2c';
W  = U/sqrt(S2/(N-2));
pvalue = 2*(1-tcdf(W,N-2));
if pvalue<0.05, dec='H0 false'; else dec='H0 true'; end
fprintf('\tp-value of H0={m1=m2} = %4.2f ==> %s\n',pvalue,dec)
```

Program `studentlawdiffm1m2.m` shows by simulation that W follows a Student distribution with $n-2$ degrees of freedom (see appendix A.2.3).

```
%===== studentlawdiffm1m2.m
% verification of the distribution under H0
sigma = rand*100;
L     = 10000;
n1  = 100; n2 = 150;
n   = n1+n2; nh = 1/(1/n1+1/n2);
m0  = 10*randn;
X1  = randn(n1,L)+m0; X2 = randn(n2,L)+m0;
m1  = mean(X1,1);
m2  = mean(X2,1);
Xc1 = X1 - ones(n1,1)*m1;
Xc2 = X2 - ones(n2,1)*m2;
S2  = sum(Xc1 .^2,1)+sum(Xc2 .^2,1);
U   = (m1-m2)*sqrt(nh);
W   = U ./ sqrt(S2/(n-1));
[hb, xb] = hist(W,100);
testim   = hb/L/(xb(2)-xb(1));
%=====
bar(xb, testim);
ttheo      = tpdf(xb,n-2);
hold on, plot(xb, ttheo,'.-r'), hold off
```

H2.5 (Correlation test) (see page 38)

1. The log-likelihood is given by (2.83) with $d = 2$. Parameter $\theta = (m_1, m_2, \sigma_1, \sigma_2, \rho) \in \Theta = \mathbb{R} \times \mathbb{R} \times \mathbb{R}^+ \times \mathbb{R}^+ \times (-1, 1)$. Hypothesis $H_0 = \mathbb{R} \times \mathbb{R} \times \mathbb{R}^+ \times \mathbb{R}^+ \times \{(-\rho_0, \rho_0)\}\}$.

   The maximum log-likelihood in relation to m_1 and m_2 is given by expression (2.84):

$$\widetilde{\ell}(C) = -n\log(2\pi) - \frac{n}{2}\log\det\{C\} - \frac{n}{2}\mathrm{trace}\left\{C^{-1}\widehat{C}\right\}$$

   where $\widehat{C}$ is obtained using expressions (2.88) and (2.89), and where

$$C = \begin{bmatrix} \sigma_1^2 & \rho\sigma_1\sigma_2 \\ \rho\sigma_1\sigma_2 & \sigma_2^2 \end{bmatrix}$$

2. The maximum log-likelihood for all parameters together is obtained using expression (2.90). Let $\gamma_i = \sigma_i^{-1} > 0$. Given that the log-likelihood is an increasing function of ρ_0^2, maximization for H_0 may be carried out using the expression:

$$\begin{aligned}\widetilde{\ell}(C) &= -n\log(2\pi) + n\log\gamma_1 + n\log\gamma_2 - \frac{n}{2}\log(1-\rho_0^2) \\ &\quad -\frac{n}{2(1-\rho_0^2)}\left(\gamma_1^2\widehat{C}_{11} + \gamma_2^2\widehat{C}_{22} - 2\rho_0\gamma_1\gamma_2\widehat{C}_{12}\right)\end{aligned}$$

   Canceling the derivatives with respect to γ_i and after a long but simple calculation process, we obtain the following GLRT test:

$$\widehat{\rho}^2 \underset{H_0}{\overset{H_1}{\gtrless}} \eta\rho_0^2 \quad \text{where} \quad \widehat{\rho}^2 = \frac{\widehat{C}_{12}^2}{\widehat{C}_{11}\widehat{C}_{22}}$$

   Threshold η is determined based on the distribution of $\widehat{\rho}^2$ for H_0 and from the significance level, or, more precisely, from the approximate distribution of $0.5\log((1+\widehat{\rho})/(1-\widehat{\rho}))$.

3. The following program gives a p-value of 0.0176. We can therefore reject H_0. This indicates that the correlation is likely to have a modulus greater than 0.7.

```
%===== testcorrelationWH.m
clear all
Hcm = [162;167;167;159;172;172;168];
N = length(Hcm);
```

```
Wkg = [48.3;50.3;50.8;47.5;51.2;51.7;50.1];
Xc=Hcm-mean(Hcm);Yc=Wkg-mean(Wkg);
hatcorr = abs(Xc'*Yc) ./ sqrt((Xc'*Xc)*(Yc'*Yc));
hatf = 0.5*log((1+hatcorr)/(1-hatcorr));
rho0 = 0.7; f0 = 0.5*log((1+rho0)/(1-rho0));
pvalue = 2*(1-normcdf(hatf,f0,1/sqrt(N-3)));
pvalue
```

The following program shows that the distribution of the Fisher transformation of $\widehat{\rho}$, for H_0, approximately follows a Gaussian distribution of mean f_0 and variance $1/(N-3)$. The generation process uses the square root of the theoretical covariance matrix. The means, variances and the number of samples N may be modified.

```
%===== correlationdistribution.m
clear all
m1=2*randn;m2=3*randn;s1=5*rand;s2=2*rand;
rho0=0.6; N = 20;
Corr = [s1^2 rho0*s1*s2;rho0*s1*s2 s2*s2];
Lruns=3000;
hatf=zeros(Lruns,1);
for ir=1:Lruns
    Z=randn(N,2); U=Z*sqrtm(Corr);
    X=U(:,1)+m1;Y=U(:,2)+m2;
    Xc=X-mean(X);Yc=Y-mean(Y);
    hatcorr = (Xc'*Yc) ./ sqrt((Xc'*Xc)*(Yc'*Yc));
    hatf(ir) = 0.5*log((1+hatcorr)/(1-hatcorr));
end
hatf0 = 0.5*log((1+rho0)/(1-rho0));
[hh,dd] = hist(hatf,30);
hatpdff = hh/Lruns/(dd(2)-dd(1));
figure(1); bar(dd,hatpdff)
hold on
plot(dd,normpdf(dd,hatf0,1/sqrt(N-3)),'.-')
hold off
```

H2.6 (CUSUM) (see page 39)

1. The statistical model is a family of probability distributions dependent on parameter m with values in $\{1,\dots,n\}$ and a log-density which is written:

$$\ell(x;m) = \begin{cases} \displaystyle\sum_{k=1}^{m} \log p(x_k;\mu_0) + \sum_{k=m+1}^{n} \log p(x_k;\mu_1) & \text{if} \quad m \in \{1,\dots,n-1\} \\ \displaystyle\sum_{k=1}^{n} \log p(x_k;\mu_0) & \text{if} \quad m = n \end{cases}$$

Using this notation, the hypothesis to test is $H_0 = \{n\}$.

2. For the sample of length n, the test function of the GLRT is written:

$$\begin{aligned} T_n(X) &= \max_{1\leq m\leq n-1} \ell(X;m) - \ell(X;n) \\ &= \sum_{k=1}^{n} s_k - \min_{1\leq m\leq n} \sum_{k=1}^{m} s_k \end{aligned} \quad (7.4)$$

taking $s_k = \log p(x_k;\mu_1)/p(x_k;\mu_0)$. Hence, $T_n(X) \geq 0$.

3. Assuming that we know the test function for a sample of size $n-1$ and that we are observing a new value s_n, $C(n)$ is either greater or less than $\min_{1\leq m\leq n-1}\sum_{k=1}^{m} s_k$. If $C(n) > \min_{1\leq m\leq n-1}\sum_{k=1}^{m} s_k$, then $\min_{1\leq m\leq n}\sum_{k=1}^{m} s_k = \min_{1\leq m\leq n-1}\sum_{k=1}^{m} s_k$ and $T_n = C(n) - \min_{1\leq m\leq n-1}\sum_{k=1}^{m} s_k = T_{n-1} + s_n$. If $C(n) \leq \min_{1\leq m\leq n-1}\sum_{k=1}^{m} s_k$, then $T_n = C(n) - C(n) = 0$. As $T_n \geq 0$, we deduce that:

$$T_n = \max\{T_{n-1}, 0\} \quad (7.5)$$

The following program verifies that the direct formula (7.4) gives the same values as the recursive expression of the CUSUM, (7.5).

```
%===== CUSUMrecursiveformula
clear all
n=160; s=randn(n,1); C=cumsum(s);
Tdirect = zeros(n,1); Trecursive = zeros(n,1);
for is=2:n
    Tdirect(is)=C(is)-min(C(1:is));
    Trecursive(is)=max([Trecursive(is-1)+s(is),0]);
end
plot([Tdirect Trecursive])
```

4. The following program implements the CUSUM test for a change in the means of two Gaussians with the same variables. The test works better as the difference between the means increases in relation to the standard deviation.

```
%===== CUSUMtest.m
n=120; m=32; mu0=1; mu1=5;
x=zeros(n,1);
x(1:m)=randn(m,1)+mu0;
x(m+1:n)=randn(n-m,1)+mu1;
T=zeros(n,1);
T(1)=(x(1)-mu1)^2-(x(1)-mu0)^2;
for in=2:n
    sn=(x(in)-mu1)^2-(x(in)-mu0)^2;
    T(in)=max([T(in-1)+sn,0]);
end
plot(T); hold on; plot([m m],[0 max(T)],'--'); hold off
```

H2.7 (Proof of (2.26)) (see page 40)

1. $\mathbb{E}\{N_j\} = Np_j$, using the fact that

$$\mathbb{E}\{\mathbb{1}(X_k \in \Delta_j)\} = \mathbb{P}\{X_k \in \Delta_j\}$$

In the same way, $\mathbb{E}\{N_j N_m\} = np_j\delta(j,m) + n(n-1)p_j p_m$. Therefore:

$$C = D - PP^T$$

where $D = \text{diag}(P)$.

2. We have:

$$\begin{aligned} C &= D - D^{1/2}D^{-1/2}PP^TD^{-1/2}D^{1/2} \\ &= D - D^{1/2}VV^TD^{1/2} \\ &= D^{1/2}(I - VV^T)D^{1/2} \end{aligned}$$

where $V = D^{-1/2}P = \begin{bmatrix}\sqrt{p_1} & \cdots & \sqrt{p_g}\end{bmatrix}^T$. And thus:

$$\Gamma = D^{-1/2}CD^{-1/2} = I - VV^T$$

We note that $V^TV = 1$ and that VV^T is the projector onto V with rank 1. Γ is therefore a projector of rank $(g-1)$. A unit matrix U therefore exists such that

$$\Gamma = U\begin{bmatrix} I_{g-1} & 0 \\ 0 & 0 \end{bmatrix}U^T \tag{7.6}$$

3. Let:

$$\widehat{P} = \begin{bmatrix}\frac{N_1}{N} & \cdots & \frac{N_g}{N}\end{bmatrix}^T$$

The central limit theorem 1.10 tells us that the random variable

$$Y = \sqrt{n}\left(\widehat{P} - P\right) \xrightarrow{d} \mathcal{N}(0, C)$$

converges in distribution toward a Gaussian of mean vector 0 and covariance matrix C. The random vector $Z = D^{-1/2}Y$ therefore converges in distribution following:

$$Z = D^{-1/2}Y \xrightarrow{d} \mathcal{N}(0, \Gamma)$$

Using equation (7.6), $Z^T Z$ appears asymptotically as the sum of the squares of $(g-1)$ Gaussian, independent, centered random variables with variance 1. $Z^T Z$ therefore converges in distribution toward a variable of χ^2 with $(g-1)$ degrees of freedom. This demonstrates expression (2.26).

H2.8 (Chi2 fitting test) (see page 41) Type the program:

```
%===== chi2test.m
% Uses chi2pdf from the stats toolbox
clear all
g = 8; n = 202;
m = 1; Lruns = 1;%3000;
Tchi2 = zeros(Lruns,1);
for irun=1:Lruns
    sigma    = 3*rand; x = sigma*randn(n,1);
    hatsigma = std(x); xsort = sort(x);
    nbe      = fix(n / g); rest = rem(n,g);
    hbin     = zeros(g,1); edges = zeros(g+1,1);
    edges(1) = -inf; id1 = 1;
    for ig=1:g
        if ig<=rest
            id2 = id1+nbe;
        else
            id2 = id1+nbe-1;
        end
        edges(ig+1) = mean(xsort(id2-1:id2));
        hbin(ig) = id2-id1+1;
        id1 = id2+1;
    end
    cdfj        = normcdf(edges,0,hatsigma);
    npj         = diff(cdfj)*n;
    X2          = sum((((hbin-npj).^2) ./ npj));
    Tchi2(irun) = X2;
end
[hval,xval] = hist(Tchi2,50);
X2distrib   = hval / Lruns / (xval(2)-xval(1));
%=====
figure(1), bar(xval,X2distrib), hold on
xvallin = linspace(0,xval(50),100);
plot(xvallin,chi2pdf(xvallin,g-1),'m','linew',1.5), hold off
```

H2.9 (Decomposition of Z) (see page 48)

1. As $\Pi_{\mathbb{1}}$ and Π_{X_c} are orthogonal projectors, $(\Pi_{\mathbb{1}} + \Pi_{X_c})$ is a projector. To show that $\Pi_Z = \Pi_{\mathbb{1}} + \Pi_{X_c}$, we must simply demonstrate that $(\Pi_{\mathbb{1}} + \Pi_{X_c})Z = Z$. Using the fact that $Z = \begin{bmatrix}\mathbb{1}_N & X\end{bmatrix}$ we obtain, successively:

$$\begin{aligned}(\Pi_{\mathbb{1}} + \Pi_{X_c})Z &= (\Pi_{\mathbb{1}} + \Pi_{X_c})\begin{bmatrix}\mathbb{1}_N & X\end{bmatrix}\\ &= \begin{bmatrix}\Pi_{\mathbb{1}}\mathbb{1}_N & \Pi_{\mathbb{1}}X\end{bmatrix} + \begin{bmatrix}\Pi_{X_c}\mathbb{1}_N & \Pi_{X_c}X\end{bmatrix}\\ &= \begin{bmatrix}\mathbb{1}_N & \Pi_{\mathbb{1}}X\end{bmatrix} + \begin{bmatrix}0 & X_c\end{bmatrix}\\ &= Z\end{aligned}$$

   where, following equations (2.45) and (2.46), we use the fact that $X_c = \Pi_{X_c}X_c = \Pi_{X_c}X + 0$.

2. As $\Pi_{\mathbb{1}}$, Π_{X_c} and Π_Z are projectors and $\Pi_Z = \Pi_{\mathbb{1}} + \Pi_{X_c}$, we have, for any $v \in \mathbb{C}^N$:

$$v^H v \geq v^H \Pi_Z v = v^H \Pi_{\mathbb{1}} v + v^H \Pi_{X_c} v \geq v^H \Pi_{\mathbb{1}} v$$

   In other words $\boldsymbol{I}_N \geq \Pi_Z \geq \Pi_{\mathbb{1}}$.

3. We have $u_n^T \Pi_{\mathbb{1}} u_n = 1/N$, $u_n^T u_n = 1$ and $u_n^T \Pi_Z u_n = h_{n,n}$.

4. Type and run program `leverageeffect.m`. Changing the test value, 0 or 1, we observe the effect of significant leverage.

```
%===== leverageeffect.m
clear all
N=100; P=2; X=randn(N,P);
if 0
    X(N,:) = mean(X(1:N-1,:),1);
else
    X(N,:) = 100*mean(X(1:N-1,:),1);
end
Z = [ones(N,1) X]; PiZ = Z*((Z'*Z)\Z');
plot(X(:,1),X(:,2),'x')
hold on, plot(X(N,1),X(N,2),'or'); hold off
title(sprintf('%4.2f < h_N = %4.2f =< 1',1/N,PiZ(N,N)))
```

H2.10 (Car consumption) (see page 54) Type and run the following program:

```
%===== regcar.m
load carsmall
%===== suppress any nan
indnotnan = not(or(or(isnan(MPG), isnan(Weight)),...
```

```
                   isnan(Horsepower)));
Z1 = Weight(indnotnan); Z2 = Horsepower(indnotnan);
y  = MPG(indnotnan);
Z  = [ones(length(Z1),1) Z1 Z2]; iZTZ = inv(Z'*Z);
n  = length(y); p = size(Z,2)-1;
%===== scater plot
hatbeta = Z \ y; haty = Z*hatbeta;
e = y-haty;
hatsigma2 = e'*e/(n-p-1); hatsigma = sqrt(hatsigma2);
%===== test
T0y = hatbeta(1)/sqrt(iZTZ(1,1))/hatsigma;
    pvalue(1) = 2*(1-tcdf(abs(T0y),n-p-1));
T1y = hatbeta(2)/sqrt(iZTZ(2,2))/hatsigma;
    pvalue(2) = 2*(1-tcdf(abs(T1y),n-p-1));
T2y = hatbeta(3)/sqrt(iZTZ(3,3))/hatsigma;
    pvalue(3) = 2*(1-tcdf(abs(T2y),n-p-1));
txt = cell(p+1,1);
for i1=1:p+1
    txt{i1} =...
        sprintf('%i\t%4.3f\t%4.3E',i1-1,hatbeta(i1),pvalue(i1));
end
disp('*************')
fprintf('\tbeta\tpval\n'); disp(txt)
%=====
figure(1)
for i1=1:p
    subplot(2,2,2*i1-1); plot(Z(:,i1+1),y,'.')
    hold on, plot(Z(:,i1+1),haty,'xr'), hold off
end
subplot(1,2,2), plot3(Z1,Z2,y,'.')
hold on, plot3(Z1,Z2,haty,'xr'), hold off, grid on
set(gca,'xlim',[min(Z1) max(Z1)],'ylim',[min(Z2) max(Z2)],...
    'zlim',[min(haty) max(haty)])
```

We see that the p-values are all very small. We can therefore be confident in rejecting the idea that the coefficients β_k are null.

Figure H2.1 shows the N observed points and the predicted points in the space (x_1, y), (x_2, y) and (x_1, x_2, y). Note that the predicted points $\widehat{y}_n$ do not lie in a straight line. The predicted solutions $\widehat{y}_n$ are of the form $y_n = \beta_0 + \beta_1 x_{i,1} + \beta_2 x_{i,2}$; they therefore belong to the same plane of dimension 2, but there is no reason for them to lie on the same straight line. The same applies to their projections onto planes (Ox_1y) and (Ox_2y).

We obtain $R^2 = 0.7521$ and $\frac{(N-r)R^2}{(r-1)(1-R^2)} = 136.5$, with a p-value ≈ 0. This allows us to reject the hypothesis that all of the coefficients are null.

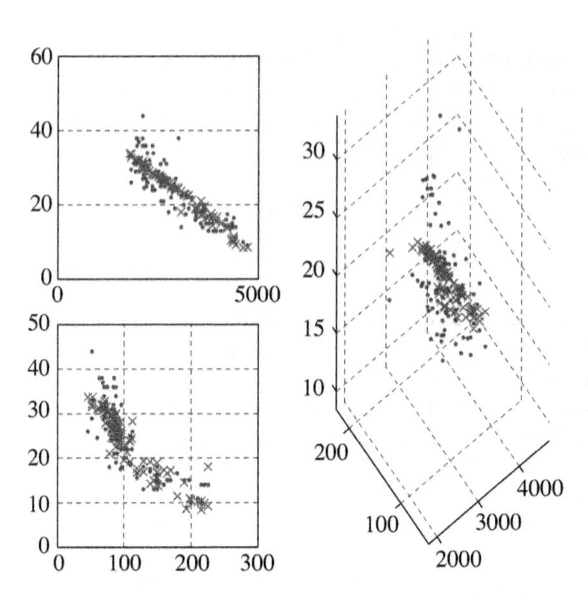

Figure H2.1 – *Top left: the N points $(x_{n,1}, y_n)$ and $(x_{n,1}, \widehat{y}_n)$ in the space Ox_1y. The points ('.') are observed values and ('x') the predicted values. Bottom left: the N points $(x_{n,2}, y_n)$ and $(x_{n,2}, \widehat{y}_n)$ in the space Ox_2y. Right: the N points $(x_{n,1}, x_{n,2}, y_n)$ and $(x_{n,1}, x_{n,2}, \widehat{y}_n)$ in the space Ox_1x_2y*

H2.11 (Central limit for a moment estimator) (see page 57)

Let $\widehat{m}_1 = N^{-1}\sum_{n=1}^{N} X_n$ and $\widehat{m}_2 = N^{-1}\sum_{n=1}^{N} X_n^2$. Expression (2.73) may be rewritten $\begin{bmatrix}\widehat{\alpha} & \widehat{\lambda}\end{bmatrix}^T = g(\widehat{m}_1, \widehat{m}_2)$ where

$$g : \begin{bmatrix} m_1 \\ m_2 \end{bmatrix} \longrightarrow \begin{bmatrix} k \\ \lambda \end{bmatrix} = \begin{bmatrix} \dfrac{m_1^2}{m_2 - m_1^2} \\ \dfrac{m_2 - m_1^2}{m_1} \end{bmatrix} \tag{7.7}$$

$$\Leftrightarrow g^{-1} : \begin{bmatrix} k \\ \lambda \end{bmatrix} \longrightarrow \begin{bmatrix} m_1 \\ m_2 \end{bmatrix} = \begin{bmatrix} k\lambda \\ \dfrac{k(1+k)}{\lambda^2} \end{bmatrix}$$

The central limit theorem 1.10 states that, when N tends toward infinity:

$$\sqrt{N} \begin{bmatrix} \widehat{m}_1 - k\lambda \\ \widehat{m}_2 - k(1+k)k^2 \end{bmatrix} \rightarrow \mathcal{N}(0, C(\alpha, \lambda))$$

where

$$C(k,\lambda) = \begin{bmatrix} \operatorname{cov}(X_n, X_n) & \operatorname{cov}(X_n, X_n^2) \\ \operatorname{cov}(X_n, X_n^2) & \operatorname{cov}(X_n^2, X_n^2) \end{bmatrix}$$

Readers may wish to determine the expression of $C(k,\lambda)$ as a function of k and λ. The continuity theorem 1.11, applied to function g, states that, when N tends toward infinity:

$$\sqrt{N}\begin{bmatrix} \widehat{k} - k \\ \widehat{\lambda} - \lambda \end{bmatrix} \rightarrow \mathcal{N}(0, \Gamma(\alpha,\lambda))$$

where $\Gamma(k,\lambda) = JCJ^T$ with

$$J = \begin{bmatrix} \dfrac{\partial g_1}{m_1} & \dfrac{\partial g_1}{m_2} \\ \dfrac{\partial g_2}{m_1} & \dfrac{\partial g_2}{m_2} \end{bmatrix}$$

This may be calculated in terms of (k,λ) using (7.7). Clearly, the performance depends on the parameter values. In the context of a practical problem, if we wish to calculate a confidence interval, the values of the parameter (k,λ) may be replaced by an estimate.

H2.12 (Estimating the proportions of a mixture) (see page 58)

1. The statistical model has a density of:

$$p(x_1,\dots,x_N) = \prod_{n=1}^{N}\left(\frac{\alpha}{\sigma_1\sqrt{2\pi}}\, e^{-(x_n-m_1)^2/2\sigma_1^2} + \frac{1-\alpha}{\sigma_2\sqrt{2\pi}}\, e^{-(x_n-m_2)^2/2\sigma_2^2}\right)$$

   where the parameter $\alpha \in (0,1)$. We deduce that $\mathbb{E}\{X_n\} = \alpha m_1 + (1-\alpha)m_2$ and $\mathbb{E}\{X_n^2\} = \alpha(m_1^2+\sigma_1^2) + (1-\alpha)(m_2^2+\sigma_2^2)$.

2. We have $S(\alpha) = \mathbb{E}\{S(X)\} = \alpha m_1 + (1-\alpha)m_2 \Rightarrow \alpha = (S(\alpha)-m_2)/(m_1-m_2)$. Hence, $\widehat{\alpha}_1 = (S(X)-m_2)/(m_1-m_2)$.

3. $S(\alpha) = \mathbb{E}\{S(X)\}$. Therefore:

$$S(\alpha) \;=\; \begin{bmatrix} \alpha m_1 + (1-\alpha)m_2 \\ \alpha(m_1^2+\sigma_1^2) + (1-\alpha)(m_2^2+\sigma_2^2) \end{bmatrix}$$

Let $\widehat{S}(X) = \begin{bmatrix}\widehat{S}_1(X) & \widehat{S}_2(X)\end{bmatrix}$, and $J(\alpha) = \|S(\alpha) - \widehat{S}(X)\|^2$. From this, we deduce the value $\widehat{\alpha}_2$ which minimizes $J(\alpha)$, canceling the derivative of $J(\alpha)$.

4. Using the fact that for a Gaussian, centered, random variable U of variance σ^2, we have $\mathbb{E}\left\{U^3\right\} = 0$ and $\mathbb{E}\left\{U^4\right\} = 3\sigma^4$, we obtain:

$$C(\alpha) = \alpha \begin{bmatrix}\sigma_1^2 & 0 \\ 0 & 3\sigma_1^4\end{bmatrix} + (1-\alpha)\begin{bmatrix}\sigma_2^2 & 0 \\ 0 & 3\sigma_2^4\end{bmatrix}$$

We obtain the estimator of α_3 by digitally minimizing the function $K(\alpha) = (\widehat{S}(X) - S(\alpha))^T C^{-1}(\alpha)(\widehat{S}(X) - S(\alpha))$ in relation to α.

5. Type and run the following program:

```
%=== MMmixture.m
% Uses boxplot from the stats toolbox
% Uses fgmm
clear all
alpha=0.5; m1=10; m2=10.1; s1=0.2; s2=0.5;
N=100; Lruns=1000;
hatalpha=zeros(Lruns,3);
La = 100;
for ir=1:Lruns
    U=rand(N,1)<alpha;
    X=(m1+s1*randn(N,1)) .* U + (m2+s2*randn(N,1)) .*(1-U);
    %===== we use only the mean statistic
    S1=mean(X);
    F11=m1; F12=m2;
    hatalpha(ir,1)=(S1-F12)/(F11-F12);
    %===== we use the two first statistics
    S2=sum(X .^2)/N; S=[S1;S2];
    F1=[F11;m1^2+s1^2]; F2=[F12;m2^2+s2^2];
    hatalpha(ir,2)= (F1-F2) \ (S-F2);
    hatalpha(ir,3)=fgmm(S,m1,m2,s1,s2,La);
end
std(hatalpha-alpha)
%=====
boxplot(hatalpha)
hold on, plot([0 10],[1 1]*alpha,':r'), hold off
```

where the function fgmm.m is:

```
function [alphaopt,Calphaopt] = fgmm(S,m1,m2,s1,s2,La)
%!====================================================!
%! SYNOPSIS:                                          !
%!    [alphaopt,Calphaopt]= FGMM(S,m1,m2,s1,s2,La)    !
```

```
%!    S          =                                              !
%!    m1, m2     =                                              !
%!    s1, s2     =                                              !
%!    La         =                                              !
%!    alphaopt   =                                              !
%!    Calphaopt  =                                              !
%!=============================================================!
F1=[m1;m1^2+s1^2]; W1=[s1^2 0;0 3*s1^4];
F2=[m2;m2^2+s2^2]; W2=[s2^2 0;0 3*s2^4];
alphalist=linspace(0,1,La);
valt = zeros(La,1);
for ia=1:La
    alpha=alphalist(ia);
    W = alpha*W1+(1-alpha)*W2;
    M = alpha*F1+(1-alpha)*F2;
    valt(ia)=(S-M)'*(W\(S-M));
end
[minb, iamin]=min(valt);
alphaopt=alphalist(iamin);
Calphaopt = alphaopt*W1+(1-alphaopt)*W2;
```

In this case, the minimization applies to the scalar variable $\alpha \in (0, 1)$. The minimization operation may therefore be carried out using a value grid. Modifying the parameters m_i, σ_i, we see that the reduction of the mean squares error obtained with $\widehat{\alpha}_2$ and $\widehat{\alpha}_3$ in relation to $\widehat{\alpha}_1$ depends on a significant extent on the choice of these values.

H2.13 ($\Gamma(k, \lambda)$ distribution) (see page 62)

1. The $\Gamma(1, \lambda)$ distribution is an exponential distribution of parameter λ. The log-likelihood is expressed $\mathcal{L}(\lambda) = -N\log(\lambda) - \frac{1}{\lambda}\sum_{n=1}^{N} X_n$. Canceling the derivative with respect to λ, we obtain:

$$\widehat{\lambda} = \frac{1}{N}\sum_{n=1}^{N} X_n \tag{7.8}$$

The Fisher information is written:

$$F = -\mathbb{E}\left\{N/\lambda^2 - 2S/\lambda^3\right\} = N/\lambda^2$$

Hence, following (2.82):

$$\sqrt{N}(\widehat{\lambda} - \lambda) \xrightarrow{d} \mathcal{N}(0, \lambda^2)$$

Type and run the following program:

```
%===== censoredsimul.m
clear all
lambda=2; N=100;
Lruns=300; hatlambda=zeros(Lruns,1);
for ir=1:Lruns
    Y = -lambda*log(rand(N,1));
    hatlambda(ir) = mean(Y);
end
[std(hatlambda-lambda), lambda/sqrt(N)]
```

This program uses the property of inversion of the cumulative distribution function $F(x) = 1 - e^{-x/\lambda}$, and thus $x = -\lambda \log(1 - F)$, as presented in section 3.3.1. Finally, if F is a uniform r.v., then $(1 - F)$ is also a uniform r.v. This explains the generation of an exponential law using `Y=-lambda*log(rand(N,1));`.

2. The log-likelihood is written:

$$\mathcal{L}(\theta) = -Nk \log \lambda - N \log \Gamma(k) + \sum_{n=1}^{N} (k-1) \log X_n - \sum_{n=1}^{N} \frac{X_n}{\lambda}$$

Canceling the Jacobian with respect to (k, λ), we obtain

$$\begin{cases} -\dfrac{Nk}{\lambda} + \dfrac{1}{\lambda^2} \sum_{n=1}^{N} X_n & = \quad 0 \\ -N \log(\lambda) - N \dfrac{\Gamma'(k)}{\Gamma(k)} + \sum_{n=1}^{N} \log X_n & = \quad 0 \end{cases}$$

From the first equation, let $\widehat{\lambda} = \frac{1}{Nk} \sum_{n=1}^{N} X_n = S/k$. Applying this to the second equation gives us:

$$k - \frac{\Gamma'(k)}{\Gamma(k)} = \log S - T$$

where $S = N^{-1} \sum_{n=1}^{N} X_n$ and $T = N^{-1} \sum_{n=1}^{N} \log X_n$. Function $\frac{\Gamma'(k)}{\Gamma(k)}$ is known as the "digamma" function and noted ψ. We have:

$$k - \psi(k) = \log S - T \tag{7.9}$$

The solution to (7.9) has no simple analytic expression, but may be calculated using a numerical procedure such as the Newton-Raphson algorithm.

H2.14 (Singularity in the MLE approach) (see page 62) We can assume, without loss of generality, that the observations verify $x_n \neq x_1$ for any $n \geq 2$. The log-likelihood is expressed $\ell(\theta) = \sum_{n=1}^{N} \log p_{X_n}(x_n; \theta)$. Choosing $m_1 = x_1$, the log-likelihood is written:

$$\begin{aligned}\ell(\widehat{\theta}) &= \log\left(\frac{1}{2\sqrt{2\pi}\sigma_1} + \frac{1}{2\sqrt{2\pi}\sigma_2} e^{-(x_1-m_2)^2/2\sigma_2^2}\right)\\ &+ \sum_{n=2}^{N} \log\left(\frac{1}{2\sqrt{2\pi}\sigma_1} e^{-(x_n-x_1)^2/2\sigma_1^2} + \frac{1}{2\sqrt{2\pi}\sigma_2} e^{-(x_n-m_2)^2/2\sigma_2^2}\right)\end{aligned}$$

Choosing any $m_2 \in \mathbb{R}$ and $\sigma_2 \in \mathbb{R}^+$, let us make σ_1 tend toward 0. As $x_n - x_1 \neq 0$, the second term tends toward a finite value. ℓ therefore tends toward infinity. To avoid this situation, a constraint must be introduced:

- either by requiring the variances to be above a certain threshold, which is equivalent to restricting the domain by replacing Θ with $\widetilde{\Theta} \subset \Theta$, and writing:

  $$\widehat{\theta} = \arg\max_{\widetilde{\Theta}} \ell(\theta)$$

- or, using a Bayesian approach, considering a probability distribution for θ with density $f(\theta)$ and writing:

  $$\widehat{\theta} = \arg\max_{\Theta}\ \ell(\theta) + \log f(\theta)$$

- or by imposing a constraint on $\widehat{\theta}$. In the case of this specific example, we can impose that the unknown means must be computed with at least two unequal observations.

H2.15 (Parameters of a homogeneous Markov chain) (see page 63)

1. Using Bayes rule and the Markov property, we have:

$$\begin{aligned}\mathcal{L} &= \mathbb{P}\{X_1 = x_1, \ldots, X_N = x_N\}\\ &= \mathbb{P}\{X_N = x_N | X_{N-1} = x_{N-1}, \ldots, X_1 = x_1\}\\ &\qquad \mathbb{P}\{X_{N-1} = x_{N-1}, \ldots, X_1 = x_1\}\\ &= \mathbb{P}\{X_N = x_N | X_{N-1} = x_{N-1}\}\, \mathbb{P}\{X_{N-1} = x_{N-1}, \ldots, X_1 = x_1\}\end{aligned}$$

Reiterating, we obtain:

$$\mathcal{L} = \sum_{n=2}^{N}\sum_{s=1}^{S}\sum_{s'=1}^{S} p_{s|s'}\mathbb{1}(x_n = s, x_{n-1} = s') \sum_{s=1}^{S} \alpha_s \mathbb{1}(x_1 = s)$$

Taking the log, we have:

$$\log \mathbb{P}\{X_1 = x_1, \cdots, X_N = x_N\} = \sum_{n=2}^{N}\sum_{s=1}^{S}\sum_{s'=1}^{S} \log p_{s|s'}\mathbb{1}(x_n = s, x_{n-1} = s') + \sum_{s=1}^{S} \log \alpha_s \mathbb{1}(x_1 = s)$$

2. The estimator of the maximum likelihood of α_s is the solution to

$$\begin{cases} \max \sum_{s=1}^{S} \log \alpha_s \mathbb{1}(x_1 = s) \\ \sum_s \alpha_s = 1 \end{cases}$$

Hence $\alpha_s = \sum_{s=1}^{S} \mathbb{1}(x_1 = s)$. To clarify, the values of α_s are all null except for that for which the position is the observed value of X_1.

The estimator of the maximum likelihood of $p_{s|s'}$ is the solution to

$$\begin{cases} \max \sum_{n=2}^{N} \sum_{s=1}^{S} \sum_{s'=1}^{S} \log p_{s|s'} \mathbb{1}(x_n = s, x_{n-1} = s') \\ \forall s', \sum_{s=1}^{S} p_{s|s'} = 1 \end{cases}$$

The Lagrangian is written $\sum_{n=2}^{N} \sum_{s=1}^{S} \sum_{s'=1}^{S} \log p_{s|s'} \mathbb{1}(x_n = s, x_{n-1} = s') + \sum_{s'=1}^{S} \lambda_{s'} (\sum_{s=1}^{S} p_{s|s'} - 1)$. Canceling the derivative of the Lagrangian with respect to $p_{s|s'}$, we obtain:

$$\frac{1}{p_{s|s'}} \sum_{n=2}^{N} \mathbb{1}(x_n = s, x_{n-1} = s') + \lambda_{s'} = 0$$

and:

$$\widehat{p}_{s|s'} = \frac{\sum_{n=2}^{N} \mathbb{1}(x_n = s, x_{n-1} = s')}{\sum_{s=1}^{S} \sum_{n=2}^{N} \mathbb{1}(x_n = s, x_{n-1} = s')}$$

The meaning of this expression is evident: we count the number of ordered pairs of form (s', s) and we divide the result by the total number of pairs beginning with s'. In practice, we simply need to calculate the numerator for every (s', s) then apply $\sum_{s=1}^{S} \widehat{p}_{s|s'} = 1$.

3. Type:

```
%===== estimMarkovchain.m
clear all
S = 3; P = rand(S);
P = P ./ (ones(S,1)*ones(1,S)*P);
alpha = rand(S,1); alpha = alpha/sum(alpha);
cumsumalpha = cumsum(alpha);
cumsumP = cumsum(P);
N = 10000; X = zeros(N,1);
%===== X(1) under alpha
U = rand; X(1) = find(cumsumalpha>U,1);
for n=2:N
    Pn=cumsumP(:,X(n-1));
    %===== X(n) under Pn
    U = rand; X(n)=find(Pn>=U,1);
end
%===== estimation of P
hatP = zeros(S);
for n=2:N
    hatP(X(n),X(n-1))=hatP(X(n),X(n-1))+1;
end
%===== normalization
hatP = hatP ./ (ones(S,1)*ones(1,S)*hatP);
[P;hatP]
```

H2.16 (Logistic regression) (see page 65) The following function estimates the parameters of a logistic model, and is based on the Newton-Raphson algorithm.

```
function [alpha,loglike,Covalpha]=logisticNR(Z,Y,tol,ITERMAX)
%!============================================================!
%! Newton-Raphson algorithm                                    !
%! SYNOPSIS                                                    !
%!    [alpha,loglike,Covalpha]=LOGISTICNR(X,Y,tol,ITERMAX)     !
%! Inputs:                                                     !
%!    Z       = array N x (p+1), explicative variables         !
%!              (first column of 1)                            !
%!    Y       = array N x 1, response in {0,1}                 !
%!    tol     = relative gap, typically 1e-8                   !
%!    ITERMAX = maximal iteration number, typically 100        !
%! Outputs:                                                    !
%!    alpha    = regresssion coefficients (p+1) x 1            !
%!    loglike  = log-likelihood                                !
%!    Covalpha = covariance of the alpha estimates             !
%!============================================================!
Yp = Y'; Z = Z';
p = size(Z,1)-1; N = size(Z,2);
sumZY = sum(Z  .* ((ones(p+1,1)*(Yp==1))),2);
relativediff  = +inf;
```

```
ell_old      = -inf;
alpha = eye(p+1,1);
loglike = zeros(ITERMAX,1);
kNR = 0;
while and(relativediff>tol, kNR<ITERMAX)
    kNR = kNR+1;
    linkp = ones(p+1,1)*(1 ./ (1+exp(alpha'*Z)));
    dell = -sumZY+sum(Z .* linkp,2);
    link2p = exp(alpha'*Z) ./ (1+exp(alpha'*Z) .^2);
    d2ell = zeros(p+1);
    for in=1:N
        d2ell = d2ell + link2p(in)* Z(:,in)*Z(:,in)';
    end
    alpha = alpha + d2ell\dell;
    loglike(kNR) = sum((alpha'*Z) .* (Yp==0))-...
                   sum(log(1+exp(alpha'*Z)));
    relativediff = abs(loglike(kNR)/ell_old-1);
    ell_old = loglike(kNR);
end
loglike = loglike(2:kNR);
Ztilde = Z' .* ((1 ./ (2*cosh(Z'*alpha/2)))*ones(1,p+1));
Covalpha = inv(Ztilde'*Ztilde);
```

Program testlogistic.m simulates a sample with a logistic distribution. It uses the function logisticNR.m to estimate α. Type:

```
%===== testlogistic.m
% Uses logisticNR
clear all
tol=1e-16; ITERMAX = 100;
alpha_true = [1;2;3;4];
p = length(alpha_true)-1;
N = 300; X = randn(N,p); Z = [ones(N,1) X];
Za = Z*alpha_true;
proba1 = 1 ./ (1+exp(Za));
Lruns = 200; alpha = zeros(Lruns,p+1);
for irun = 1:Lruns
    Y = rand(N,1) < proba1;
    [alphaaux,ell,covbeta]=logisticNR(Z,Y,tol,ITERMAX);
    alpha(irun,:) = alphaaux';
end
%=====
boxplot(alpha), set(gca,'ylim',[0,5])
```

Program ORing.m estimates the parameters of the data provided in table 2.5, along with the associated confidence intervals, given by expression (2.96). It calculates the p-value of the log-GLRT, expression (2.19), associated with the hypothesis $H_0 = \{\alpha_2 = 0\}$.

We see that the p-value of the log-GLRT associated with hypothesis H_0 is 0.01, leading us to reject H_0. Furthermore, the confidence interval at 95%, $0.01 \leq \alpha_2 \leq 0.34$, does not contain the value 0, which also leads us to reject H_0.

Type:

```
%===== ORing.m
% Uses logisticNR
clear all
K=[53;56;57;63;66;67;67;67;68;69;70;70; ...
   70;70;72;73;75;75;75;76;78;79;80;81];
S=[1;1;1;0;0;0;0;0;0;0;0;1; ...
   1;1;0;0;0;1;0;0;0;0;0;0];
Z = [ones(length(K),1) K];
tol=1e-16; ITERMAX = 100;
[alpha,ell,Covalpha]=logisticNR(Z,S,tol,ITERMAX);
%===== under H0=(alpha2=0)
[alpha0, ell0, Covalpha0] =...
          logisticNR(ones(length(K),1),S,tol,ITERMAX);
T = 2*(ell(end)-ell0(end));
pvalue = chi2pdf(T,1);
if pvalue<0.05, dec='H0 false';else dec='H0 true'; end
DC = diag(Covalpha);
Ib0(:,1) = alpha - 1.96*sqrt(DC);
Ib0(:,2) = alpha + 1.96*sqrt(DC);
fprintf('\t%5.2f \t< alpha1 =%5.2f < %5.2f\n',...
        Ib0(1,1),alpha(1),Ib0(1,2))
fprintf('\t%5.2f \t< alpha2 =%5.2f  < %5.2f\n',...
        Ib0(2,1),alpha(2),Ib0(2,2))
fprintf('\tp-value of H0={alpha2=0} = %4.2f ==> %s\n',pvalue,dec)
```

H2.17 (GLRT for the logistic model) (see page 65) Program `testGLRTlogistic.m` verifies the distribution of the GLRT for H_0. It uses the function `logisticNR.m` from solution H2.16. Type:

```
%===== testGLRTlogistic.m
clear
tol=1e-8; ITERMAX = 100;
N = 100; nb_zeros = 2;
alpha_true0 = [0.3;0.5;1;zeros(nb_zeros,1)];
p = length(alpha_true0)-1;
%===== design matrix randomly selected
X = randn(N,p); Z = [ones(N,1) X];
Z0 = [ones(N,1) X(:,1:p-nb_zeros)];
Za = Z*alpha_true0;
proba1 = 1 ./ (1+exp(Za));
Lruns = 500; GLRT = zeros(Lruns,1);
for irun = 1:Lruns
```

```
    Y = rand(N,1) < proba1;
    [alphaaux1, ell1]  = logisticNR(Z,Y, tol, ITERMAX);
    [alphaaux0, ell0]  = logisticNR(Z0,Y, tol, ITERMAX);
    GLRT(irun) = 2*(ell1(end)-ell0(end));
end
%=====
[hbn,xbin]=hist(GLRT,50);
hpdf = hbn/Lruns/(xbin(2)-xbin(1));
bar(xbin,hpdf), hold on
xpdf = linspace(0,12,50);
plot(xpdf,chi2pdf(xpdf,nb_zeros),'o-r');, hold off
```

H2.18 (GMM) (see page 72)

– Type:

```
function [x,states]=geneGMM(N,alphas,mus,sigma2s)
%!=============================================!
%! Generate a mixture of K gaussians            !
%! SYNOPSIS                                     !
%!    [x,states]=GENEGMM(N,alphas,mus,sigma2s)  !
%! N       = length of the sequence             !
%! alphas  = ponderation array (K x 1)          !
%! mus     = mean array (K x 1)                 !
%! sigma2s = variance array (K x 1)             !
%! x       = data array N x 1                   !
%! states  = state array N x 1                  !
%!=============================================!
sigma     = sqrt(sigma2s);
K         = length(mus);
cumalphas = zeros(K+1,1);
for ii=1:K
    cumalphas(ii+1) = cumalphas(ii)+alphas(ii);
end
U = rand(N,1); x = zeros(N,1); states = zeros(N,1);
for ii=1:N
    mm         = find(cumalphas>U(ii),1)-1;
    states(ii) = mm;
    x(ii)      = mus(mm)+ sigma(mm)*randn;
end
```

– Type:

```
function [alphas,mus,sigma2s,llike]=...
          estimGMM_EM(y,K,tol,ITERMAX)
%!============================================!
%! EM algorithm for GMM with S components and !
```

```
%! dimension D                                    !
%! SYNOPSIS                                       !
%!    [alphas,mus,sigma2s,llike]=...              !
%!     ESTIMGMM_EM(y,K,tol,ITERMAX)               !
%! Inputs                                         !
%!    y       = data N x 1                        !
%!    K       =                                   !
%!    tol     = tolerance                         !
%!    ITERMAX = maximal number of iterations      !
%! Outputs                                        !
%!    alphas  = final proportion                  !
%!    mus     = final means                       !
%!    sigma2s = final variances                   !
%!    llike  = log-likelihood                     !
%================================================!
N = size(y,1);
llike = zeros(ITERMAX,1);
likely_old = -inf; relativediff = +inf;
kEM = 1; logfact2pi = log(sqrt(2*pi));
% initialization
alphas = ones(K,1)/K;
ysort = sort(y);
NK = ceil(N/K)*K;
ysort(N+1:NK) = 0;
yK = reshape(ysort,NK/K,K);
mus = mean(yK); sigma2s = std(yK) .^2;
while relativediff>tol && kEM<ITERMAX
    logdet = log(sigma2s); bkn = zeros(K,N);
    for k=1:K
        difference = y-mus(k);
        aux1       = (difference .^2) / sigma2s(k);
        logbns     = -0.5*(aux1+logdet(k))-logfact2pi;
        bkn(k,:)   = exp(logbns)*alphas(k);
    end
    cnp            = sum(bkn,1);
    gamma          = bkn ./ (ones(K,1)*cnp);
    llike(kEM)     = sum(log(cnp));
    sum_gamma_nn   = sum(gamma,2);
    alphas         = sum_gamma_nn/N;
    %===== mean re-estimation
    mus            = (gamma*y) ./ sum_gamma_nn;
    %===== variance re-estimation
    aux2           = (y*ones(1,K)-(ones(N,1)*mus')) .^2;
    sigma2s        = sum(aux2 .* gamma',1) ./ sum_gamma_nn';
    relativediff   = abs((llike(kEM)/likely_old)-1);
    likely_old     = llike(kEM);
    kEM            = kEM+1;
end
llike = llike(2:kEM-1);
```

– Type:

```
%===== testEMonGMM.m
% Uses geneGMM and estimGMM_EM
clear all
N = 3000; K = 3;
mus     = [1;5;8];
sigma2s = [1;0.5;1.2];
alphas  = [0.5;0.3;0.2];
x    = geneGMM(N,alphas,mus,sigma2s);
tol = 1e-8; ITERMAX = 180; nbin = 50;
%=====
[hatalphas,hatmus,hatsigma2s,llike]=...
    estimGMM_EM(x,K,tol,ITERMAX);
%=====
[hbin,xbin] = hist(x,nbin);
hatpdf       = hbin/N/(xbin(2)-xbin(1));
logfact2pi  = log(sqrt(2*pi));
logdet       = log(hatsigma2s);
gamma        = zeros(K,nbin);
for k=1:K
    difference = xbin-hatmus(k);
    aux1       = (difference .^2) / hatsigma2s(k);
    logbns     = -0.5*(aux1+logdet(k))-logfact2pi;
    gamma(k,:) = exp(logbns)*hatalphas(k);
end
%=====
[mus, hatmus]
[sigma2s, hatsigma2s']
[alphas, hatalphas]
hatpdfGMM = sum(gamma,1);
figure(1), bar(xbin,hatpdf), hold on
plot(xbin,hatpdfGMM,'r','linew',2), hold off
figure(2), plot(llike), grid
```

H2.19 (Estimation of states of a GMM) (see page 72)

1. Equation (2.106) gives the probability of state S_n at instant n, conditional on the observation Y_n. Knowing parameter θ, we can therefore use the EM algorithm to calculate

$$\widehat{S} = \arg \max_{k \in \{1,\dots,K\}} \mathbb{P}_\theta \{S = k|Y\}.$$

2. Type the function:

```
function [S,PS]=estimState_GMM_EM(y,alphas,mus,sigma2s)
%!=====================================================!
```

```
%! EM algorithm for GMM                                  !
%! with S components and dimension D                     !
%! SYNOPSIS                                              !
%!    [S,PS]=estimState_GMM_EM(y,alphas,mus,sigma2s)     !
%! Inputs                                                !
%!     y       = data (Nx1)                              !
%!     alphas  = (Kx1) vector of proportion              !
%!     mus     = (Kx1) vector of means                   !
%!     sigma2s = (Kx1) vector of variances               !
%! Outputs                                               !
%!     S  = Nx1 state sequence, valued in (1,...,K)      !
%!     PS = probability of S                             !
%!=======================================================!
logfact2pi = log(sqrt(2*pi));
K = length(alphas); N = length(y);
logdet      = log(sigma2s);
bkn         = zeros(K,N);
for k=1:K
    difference = y-mus(k);
    aux1       = (difference .^2) / sigma2s(k);
    logbns     = -0.5*(aux1+logdet(k))-logfact2pi;
    bkn(k,:)   = exp(logbns)*alphas(k);
end
cnp     = sum(bkn,1);
gamma   = bkn ./ (ones(K,1)*cnp);
[PS, S] = max(gamma,[],1);
```

Type the program:

```
%===== testEstimState.m
% Uses geneGMM, estimGMM_EM and estimState_GMM_EM
mus     = [1;5;8];
sigma2s = [1;0.5;1.2];
alphas  = [0.5;0.3;0.2];
K       = length(alphas);
Nlearn  = 10000;
xlearn  = geneGMM(Nlearn,alphas,mus,sigma2s);
tol     = 1e-8;
ITERMAX = 180;
nbin    = 50;
%===== learning
[hatalphas,hatmus,hatsigma2s,llike]=...
    estimGMM_EM(xlearn,K,tol,ITERMAX);
%===== testing data
Ntest = 200;
[xtest,Strue]=geneGMM(Ntest,hatalphas,hatmus,hatsigma2s);
hatS=estimState_GMM_EM(xtest,hatalphas,hatmus, hatsigma2s);
errorrate = sum(not(Strue==hatS'))/Ntest;
fprintf('Error rate %4.2f\n',errorrate)
```

H2.20 (MLE on censored data) (see page 74)

1. We have $F_X(x;\theta) = 1 - e^{-\theta x}\mathbb{1}(x \geq 0)$. Using expression (2.110), we have:

$$Q(\theta, \theta') = N\log(\theta) - \theta\sum_{n=1}^{N} Y_n - \frac{\theta}{\theta'}\sum_{n=1}^{N}\mathbb{1}(c_n = 1)$$

2. Canceling the derivative with respect to θ, we obtain the recursion of the EM algorithm:

$$\frac{1}{\theta^{(p)}} = m_Y + \rho_1\frac{1}{\theta^{(p-1)}}$$

where $\rho_1 = N^{-1}\sum_{n=1}^{N}\mathbb{1}(c_n = 1)$ and $m_Y = N^{-1}\sum_{n=1}^{N} Y_n$. This equation converges, as $\rho_1 < 1$. The limit therefore verifies the equation, giving:

$$\theta_{\text{lim}} = \frac{1-\rho_1}{m_Y} \tag{7.10}$$

Using expression (2.108), we obtain a likelihood expressed as $\mathcal{L}_{\text{lim}} = N_0 \log N_0/S - N_0$

3. Program `censoredsimul.m` carries out a simulation with $N = 30$. 10 censored data points are obtained by random reduction of the "true" values. We see that the mean square error is better for estimator (7.10) than with the estimator which considers that none of the data points are censored, and than the estimator which only takes account of the uncensored data.

```
%===== censoredsimul.m
clear all
theta0=2; N=20; nc=10;
Lruns=300; tt=zeros(Lruns,3);
for ir=1:Lruns
    Y = -log(rand(N,1))/theta0;
    c = zeros(N,1); c(1:nc)=1;
    Y(1:nc)= 0.2*rand(nc,1) .* Y(1:nc);
    meany = mean(Y);rho0 = mean(c==0);
    tt(ir,1) = rho0/meany;
    tt(ir,2) = 1/mean(Y);
    tt(ir,3) = 1/mean(Y(nc+1:N));
end
std(tt-theta0)
```

4. Program `censoredHIV.m` estimates the constant θ for the proposed data.

```
%===== censoredHIV.m
clear all
Y = [1;30;7;4;8;5 ;10;2;9;36;3;9 ;...
      3;35;8;1;5;11;56;2;3;15;1;10];
c = [0;1;0;0;0;0;1;0;0;1;0;0;...
      0;1;0;0;0;1;0;0;0;1;0;1];
meany = mean(Y); rho0 = mean(c==0);
thetalim = rho0/meany
```

H2.21 (Estimation of a quantile) (see page 78)

1. An estimator is given by

$$\widehat{s}_N = X_{(\lfloor cN \rfloor)}$$

where $X_{(n)}$ is the n-th value of the series arranged in increasing order, known as the order statistic. This estimator may be refined by approximating the cumulative function locally around $X_{(\lfloor cN \rfloor)}$ by a polynomial.

2. Applying the δ-method to expression (2.113), we are able to deduce the asymptotic distribution of $\widehat{s}_N$:

$$\sqrt{N}(\widehat{s}_N - s) \to N(0, \eta) \text{ where } \eta = \frac{F(s)(1 - F(s))}{\left(\frac{dF}{ds}\right)^2}$$

3. From this, we deduce an approximate confidence interval at 95%:

$$I = \left(\widehat{s}_N - \frac{1.96\sqrt{\widehat{\gamma}}}{p(\widehat{s}_N)\sqrt{N}}, \widehat{s}_N + \frac{1.96\sqrt{\widehat{\gamma}}}{p(\widehat{s}_N)\sqrt{N}}\right)$$

4. Program:

```
%===== CumulInverseEstimate.m
clear all
Lruns = 1000; val = zeros(Lruns,1);
ICval = zeros(Lruns,1);
s_select = 0.05;
N        = 100000;
%===== a few percent of the values are taken into
% account when calculating the value of pdf
% around the value of interest
dxis = 0.08;
xinf = 1-dxis/2; xsup = 1+dxis/2;
calpha95 = (sqrt(2).*erfcinv(0.05) / sqrt(N));
cas = 'G';
```

```
switch cas
    case 'G'
        Xa = randn(N,Lruns);
        valtrue = sqrt(2).*erfcinv(2*s_select);
    case 'R'
        Xa = sqrt(randn(N,Lruns) .^2 + randn(N,Lruns) .^2);
        valtrue = sqrt(-2*log(s_select));
    case 'C'
        Xa = randn(N,Lruns) ./ randn(N,Lruns);
        valtrue = tinv(1-s_select,1);
end
degre           = 1;
VV              = exp(log(1-s_select)*(0:2*degre));
for ir = 1:Lruns,
    X           = Xa(:,ir); Xsort = sort(X);
    indp        = fix(N*(1-s_select));
    xxss        = (indp+(-degre:+degre)')/N;
    yyss        = Xsort(indp+(-degre:+degre)');
    MM          = exp(log(xxss)*(0:2*degre));
    alphass     = MM\yyss;
    %===== we interpolate a little bit
    val(ir)     = VV * alphass;
    pdf_select = sum(and(X> val(ir)*xinf,X<val(ir)*xsup))...
                 /N/(val(ir)*dxis);
    ICval(ir)   = calpha95 * sqrt(s_select .* (1-s_select))...
                 ./ pdf_select;
end
outofIC = sum(or(val-ICval>valtrue,val+ICval<valtrue))
%=====
plot(val), hold on, plot(val-ICval,':'), plot(val+ICval,':')
plot([1 Lruns],valtrue*ones(2,1),'--r')
plot([1 Lruns],mean(val)*ones(2,1),'--m')
hold off, set(gca,'xlim',[1 200])
title(sprintf('Nout = %i /%i about %4.2f',outofIC, Lruns,...
               outofIC/Lruns))
```

H2.22 (Image equalization) (see page 78) Type and run the program egalizeimage.m:

```
%===== egalizeimage.m
clear all
load lena
[NY,NX]=size(imglena);N=NX*NY;
tx = (0:255); hatp = zeros(256,1);
for it=1:256
    hatp(it)=sum(sum(imglena==it-1))/N;
end
cdfimg = cumsum(hatp);
imgequal = zeros(size(imglena));
for iX=1:NX
```

```
    for iY=1:NY
        itx = cdfimg(imglena(iY,iX)+1);
        imgequal(iY,iX) = round(255*itx);
    end
end
hatpequal = zeros(256,1);
for it=1:256
    hatpequal(it)=sum(sum(imgequal==it-1))/N;
end
cdfimgequal = cumsum(hatpequal);
%=====
subplot(221); imagesc(imglena/256), axis('image')
subplot(222); imagesc(imgequal/256), axis('image')
subplot(223); plot(tx,cdfimg), grid
subplot(224); plot(tx,cdfimgequal), grid
```

H2.23 (Bootstrap for a regression model) (see page 81) Type and run the program bootstraponregression.m:

```
%===== bootstraponregression.m
clear all
N   = 30;
mu  = [3;2];
p   = length(mu);
Z   = [ones(N,1) (1:N)'];
Zmu = Z*mu;
B = 200; mub = zeros(B,1);
Lruns = 1000; sigma2b = zeros(p,p,Lruns);
for irun = 1:Lruns
    X   = Zmu + randn(N,1);
    U   = fix(rand(N,B)*N)+1;
    mub = zeros(p,B);
    for ib=1:B
        mub(:,ib) = Z(U(:,ib),:) \ X(U(:,ib));
    end
    sigma2b(:,:,irun) = cov(mub');
end
theosigma2b = inv(Z'*Z);
msq2b11=squeeze(sigma2b(1,1,:)); m11 = mean(msq2b11);
msq2b12=squeeze(sigma2b(1,2,:)); m12 = mean(msq2b12);
msq2b22=squeeze(sigma2b(2,2,:)); m22 = mean(msq2b22);
%=====
subplot(131), boxplot(msq2b11), hold on
plot([0.5 1.5], ones(2,1)*theosigma2b(1,1),'--r')
hold off
subplot(132), boxplot(msq2b12), hold on
plot([0.5 1.5], ones(2,1)*theosigma2b(1,2),'--r')
hold off
subplot(133), boxplot(msq2b22), hold on
plot([0.5 1.5], ones(2,1)*theosigma2b(2,2),'--r')
```

```
hold off
fprintf('theor. cov(1,1) = %5.3f, boot-value = %5.3f\n',...
    theosigma2b(1,1),m11);
fprintf('theor. cov(1,2) = %5.3f, boot-value = %5.3f\n',...
    theosigma2b(1,2),m12);
fprintf('theor. cov(2,2) = %5.3f, boot-value = %5.3f\n',...
    theosigma2b(2,2),m22);
```

H2.24 (Model estimation by cross validation) (see page 83) Type and estimate the program `orderEstimCV.m`:

```
%===== orderEstimCV.m
clear all
N = 300; K = 10; L = N/K;
sigma = 2;
Ptrue = 10; Pmax  = 20;
X = randn(N,Pmax); Z = X(:,1:Ptrue);
beta = ones(Ptrue,1);
y = Z * beta + sigma*randn(N,1);
err = zeros(Pmax,K); err0 = zeros(Pmax,K);
for ik=1:K
    id1 = (ik-1)*L+1; id2 = id1+L-1;
    %===== testing DB
    Ty = y(id1:id2); TX = X(id1:id2,:);
    %===== learning DB
    Ly = y([1:id1-1,id2:N]);
    LX = X([1:id1-1,id2:N],:);
    for ip=1:Pmax
        LH = LX(:,1:ip); TH = TX(:,1:ip);
        hbeta = LH\Ly;
        err(ip,ik)  = norm(Ty-TH*hbeta) .^2 /(L-ip);
        err0(ip,ik) = norm(Ly-LH*hbeta) .^2 /((K-1)*L-ip);
    end
end
err_p  = sum(err,2)/K; err0_p = sum(err0,2)/K;
%=====
figure(1), plot([err0_p,err_p],'.-'), grid on
```

Figure H2.2 shows the results of a simulation. The curve marked using ‘x-’ represents the prediction error calculated over the learning base. The mean decreases as the number of regressors increases. The curve marked using ‘o-’, which represents the prediction error calculated using the test base, passes through a minimum for the true value $p = 10$. Thus, when the number of regressors is increased above the true value, we “learn” noise. This is known as overtraining. Furthermore, note that the error for the test base is greater than that for the learning base, as predicted by equation (2.59).

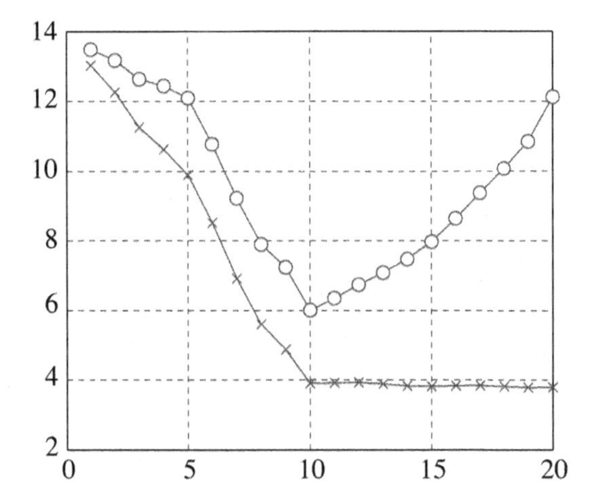

Figure H2.2 – *Prediction errors as a function of the supposed order of the model: 'x-' for the learning base, 'o-' for the test base. The observation model is of the form* $y = Z\beta + \sigma\epsilon$*, where* Z *includes* $p = 10$ *column vectors. As the number of regressors increases above and beyond the true value, we "learn" noise; this is known as overtraining*

H3 Monte-Carlo simulation

H3.1 (Multinomial law) (see page 90)

Type:

```
%===== multinomial.m
clear all
N=10000;
mu=[0.1;0.2;0.3;0.2;0.2];
F=cumsum(mu);
u=rand(N,1); x=zeros(N,1);
for ii=1:N, x(ii)=find(F>=u(ii),1); end
figure(1), [hx,dx]=hist(x,(1:5));
bar(dx,hx/N)
```

H3.2 (Homogeneous Markov chain) (see page 91)

Type:

```
%===== MC.m
clear all
pi1 = [0.5 0.2 0.3]; F1 = cumsum(pi1);
A   = [0.3, 0  , 0.7;
       0.1, 0.4, 0.5;
       0.4, 0.2, 0.4];
```

```
N   = 10000; X = zeros(N,1); X(1) = find(F1>=rand,1);
F = zeros(size(A));
for ia = 1:size(A,1)
    F(ia,:) = cumsum(A(ia,:));
end
for n = 2:N,
    c = F(X(n-1),:); U = rand; X(n)=find(c>=U,1);
end
for ia=1:3
    vcurr = find(X(1:N-1)==ia);
    succ  = X(vcurr+1); hy = hist(succ,(1:3));
    subplot(3,1,ia), bar((1:3),hy/length(succ))
    mstr=sprintf('Ligne %d',ia); title(mstr)
end
```

H3.3 (2D Gaussian) (see page 91)

Type:

```
%===== gauss2d.m
clear
N = 1000000;
R = [2, 0.95;0.95, 0.5];
S = sqrtm(R); W = randn(N,2); X = W*S;
plot(X(:,1),X(:,2),'.')
set(gca,'xlim',[-8 8],'ylim',[-8 8]), axis('square')
[U,D] = eig(R);
hold on, plot(U(2,2)*[-8 8],-U(1,2)*[-8 8],'r','linew',2)
plot(U(2,1)*[-8 8],-U(1,1)*[-8 8],'y','linew',2), hold off
```

H3.4 (Box-Muller method) (see page 92) The generation algorithm is written:

1. draw two independent samples (U, V) with a uniform distribution over $(0, 1)$,
2. calculate the variable pair:

$$\begin{cases} X = \sigma\sqrt{-2\log(U)}\cos(2\pi V) \\ Y = \sigma\sqrt{-2\log(U)}\sin(2\pi V) \end{cases}$$

Type:

```
%===== boxmuller.m
clear
N = 100000;
sigma = 1;
U = rand(N,1); V = rand(N,1);
X = sigma*sqrt(-2*log(U)) .* cos(2*pi*V);
Y = sigma*sqrt(-2*log(U)) .* sin(2*pi*V);
plot(X,Y,'.'), axis('square')
```

H3.5 (The Cauchy distribution) (see page 92)

1. The cumulative function of Z is written:

$$\mathbb{P}\{Z \leq z\} = \frac{1}{2} + \frac{1}{\pi}\text{atan}\frac{z - z_0}{a}$$

The density is therefore expressed:

$$p_Z(z) = \frac{a}{\pi(a^2 + (z - z_0)^2)}$$

2. The cumulative function of Z is written:

$$\mathbb{P}\{Z \leq z\} = \int_{\{(x,y):ay/x<z-z_0\}} \frac{1}{2\pi} e^{-(x^2+y^2)/2} dxdy$$

$$\begin{aligned}
\mathbb{P}\{Z \leq z\} &= \frac{1}{2\pi}\int_{-\infty}^{0} e^{-x^2/2} \int_{x(z-z_0)/a}^{+\infty} e^{-y^2/2} dydx \\
&\quad + \frac{1}{2\pi}\int_{0}^{+\infty} e^{-x^2/2} \int_{-\infty}^{x(z-z_0)/a} e^{-y^2/2} dydx \\
&= \frac{1}{2\pi}\int_{0}^{+\infty} e^{-x^2/2} \int_{-x(z-z_0)/a}^{+x(z-z_0)/a} e^{-y^2/2} dydx
\end{aligned}$$

From the derivative with respect to z we obtain the density:

$$\begin{aligned}
p_Z(z) &= \frac{1}{a\pi}\int_{0}^{+\infty} x e^{-x^2(1+(z-z_0)^2/a^2)/2} dx \\
&= \frac{1}{a\pi}\int_{0}^{+\infty} e^{-u(1+(z-z_0)^2/a^2)} du = \frac{1}{\pi}\frac{a}{a^2 + (z - z_0)^2}
\end{aligned}$$

3. Type:

```
%===== cauchylaw.m
clear all
z0=10; a=0.8; N=10000;
z=(-10:0.1:30);
pztheo = a ./(a*a+(z-z0) .^2)/pi;
%===== atan(U)
z1       = z0+a*tan(pi*(rand(N,1)-0.5));
[h1,b1]  = hist(z1,z);
p1       = h1 / N / (b1(2)-b1(1));
subplot(211), bar(b1,p1,'c'), hold on
```

```
plot(z,pztheo), hold off
set(gca,'xlim',[0 20],'ylim',[0 0.5]), grid on
%===== Y/X
X       = randn(N,2);
z2      = z0+a*(X(:,2) ./ X(:,1));
[h2,b2] = hist(z2,z);
p2      = h2 / N / (b2(2)-b2(1));
subplot(212), bar(b2,p2,'c')
hold on, plot(z,pztheo), hold off
set(gca,'xlim',[0 20],'ylim',[0 0.5]), grid on
```

H3.6 (Metropolis-Hastings algorithm) (see page 98)

Type:

```
%===== applicationMetropolis.m
clear all
N      = 10000;
sigma2 = 5;
U0     = 10*sqrt(sigma2);
x = zeros(N,1); pprevious = 1;
for n=2:N
    xproposal = U0*(rand-1/2);
    pproposal = exp(-(xproposal .^2)/2/sigma2);
    rho       = pproposal / pprevious;
    if rho > 1
        x(n) = xproposal; pprevious = pproposal;
    else
        b = rand<rho; x(n) = xproposal*b+x(n-1)*(1-b);
        pprevious = pproposal*b+pprevious*(1-b);
    end
end
%===== integral approx. value
I = mean(x .^2,1)'
%=====
[hx,dx] = hist(x,30); gx = hx/N/(dx(2)-dx(1));
bar(dx,gx,'y');
gtheo = exp(- (dx .^2)/2/sigma2)/sqrt(2*pi*sigma2);
hold on, plot(dx,gtheo,'linew',1.5), hold off
```

H3.7 (Gibbs sampler) (see page 98)

1. The conditional distribution is given by expression (1.47):

$$p_{X_1|X_2}(x_1, x_2) = \mathcal{N}\left(\mu_1 + \rho\frac{\sigma_1}{\sigma_2}(X_2 - \mu_2), \sigma_1^2(1-\rho^2)\right)$$

2. Type:

```
%===== applicationGibbs.m
clear all
mus = [2;1]; sigmas = [2;3]; rho = 0.4;
C12=rho*sigmas(1)*sigmas(2);
C=[sigmas(1)^2 , C12;
   C12, sigmas(2)^2];
N = 10000; X = zeros(N,2);
for n=2:N
    mucond = mus(1)+...
        rho*sigmas(1)*(X(n-1,2)-mus(2))/sigmas(2);
    sigmacond = sigmas(1)*sqrt(1-rho*rho);
    X(n,1) = mucond+sigmacond*randn;
    %=====
    mucond = mus(2)+...
        rho*sigmas(2)*(X(n,1)-mus(1))/sigmas(1);
    sigmacond = sigmas(2)*sqrt(1-rho*rho);
    X(n,2) = mucond+sigmacond*randn;
end
Nburn = 200; Nval = N - Nburn;
X = X(Nburn+1:N,:);
plot(X(:,1),X(:,2),'.')
Xc = X - ones(Nval,1)*mean(X);
[C , Xc'*Xc/(Nval-1)]
```

H3.8 (Direct draws and importance sampling) (see page 103)
Type:

```
%===== IS_GaussfromCauchy.m
% Importance sampling for estimation of
% int_{alpha}^{+infty} p(x)dx with p(x)=N(0,1)
%     (1) directly from N(0,1)
%     (2) from Cauchy distribution and weights
%     (3) from Cauchy without knowing p(x) up to
%         a multiplicative constant
%=====
clear all
N                     = 1000;
alpha                 = 3;
palpha_th             = 0.5 * erfc(alpha/sqrt(2));
palpha_th2            = palpha_th*palpha_th;
Lruns                 = 500;
P1_direct             = zeros(Lruns,1);
P1_IS_withCteNorm1    = zeros(Lruns,1);
P1_IS_withoutCteNorm1 = zeros(Lruns,1);
dpitdemi              = sqrt(2/pi);
for ii=1:Lruns
```

```
    x               = randn(N,1);
    u               = rand(N,1);
    x_IS            = tan(pi*(u-0.5));
    P1_direct(ii) = sum(x>alpha)/N;
    weight_IS_withoutCteNorm1  = exp(-x_IS.*x_IS/2)...
        .* (1+x_IS.*x_IS);
    P1_IS_withoutCteNorm1(ii) =...
        (x_IS>alpha)'*weight_IS_withoutCteNorm1/...
            sum(weight_IS_withoutCteNorm1);
    poids_IS_withCteNorm1     =...
        weight_IS_withoutCteNorm1/dpitdemi;
    P1_IS_withCteNorm1(ii)    =...
        (x_IS>alpha)'*poids_IS_withCteNorm1/N;
end
%===== display
figure(1), subplot(1,3,1)
h_direct = boxplot(P1_direct);
axa      = get(gca,'ylim');
set(gca,'xticklabel','(a)')
hold on, plot([0 2],palpha_th*ones(1,2),'g:'), hold off
%=====
subplot(1,3,2)
h_IS_with=boxplot(P1_IS_withCteNorm1);
set(gca,'xticklabel','(b)')
hold on, plot([0 2],palpha_th*ones(1,2),'g:')
set(gca,'ylim',axa);, hold off
%=====
subplot(1,3,3)
h_IS_without=boxplot(P1_IS_withoutCteNorm1);
set(gca,'xticklabel','(c)')
hold on, plot([0 2],palpha_th*ones(1,2),'g:')
set(gca,'ylim',axa); hold off
%=====
ect_direct_estime = std(P1_direct-palpha_th);
ect_direct_theo   = sqrt((palpha_th-palpha_th2)/N);
ect_IS_estime     = std(P1_IS_withCteNorm1-palpha_th);
ect_IS_theo       = sqrt((sqrt(pi)*3*0.5 * erfc(alpha)/4+...
    +0.25*alpha*exp(-alpha*alpha)-palpha_th2)/N);
disp('*********************')
disp('Draw from Gaussian distribution')
resul_direct = ...
    sprintf('\tTheorical std = %5.2e, Estimated std = %5.2e',...
    ect_direct_theo,ect_direct_estime);
disp(resul_direct)
disp('Draw from Cauchy distribution and IS (not normalized)')
resul_IS = ...
    sprintf('\tTheorical std = %5.2e, Estimated std = %5.2e',...
    ect_IS_theo,ect_IS_estime);
```

```
disp(resul_IS)
```

Instead of boxplot, readers may wish to use myboxplot (see section A1) or any other more sophisticated realization freely available.

H3.9 (Stratification) (see page 105)

1. Following (1.32) for $\mu = 0$ and $\sigma^2 = 1$:

$$\mathbb{E}\left\{e^{juX}\right\} = e^{-u^2/2} = \mathbb{E}\left\{\cos(ux) + j\sin(ux)\right\}$$

   Thus $\mathbb{E}\left\{\cos(uX)\right\} = e^{-u^2/2}$.

2. Type the program:

```
function stratification(u)
%!================================================!
%! I(u) = int_(-infty)^(+infty)cos(ux)N(0,1)dx !
%!      = exp(-u^2/2)                              !
%! SYNOPSIS: STRATIFICATION(u)                    !
%!================================================!
I_exact = exp(-u^2/2);
%===== using S N-sample strata
N             = 1000;
S             = 200;
ni            = fix(N/S);
vect_n        = ones(S,1)*ni;
N             = sum(vect_n);
nb_runs       = 300;
hatDirect_I   = zeros(nb_runs,1);
hatStratif_I = zeros(nb_runs,1);
for i_run=1:nb_runs,
    Udirect  = rand(N,1);
    Xdirect  = norminv(Udirect,0,1); % normal inverse cdf
    hatDirect_I(i_run)   = sum(cos(u*Xdirect))/N;
    Ustrate              = stratif_uniform(vect_n);
    Xstrate              = norminv(Ustrate,0,1);
    hatStratif_I(i_run) = sum(cos(u*Xstrate))/N;
end
figure(1)
boxplot([hatDirect_I hatStratif_I])
td=sprintf('bias DIRECT = %5.2e, std DIRECT = %5.2e',...
    abs(I_exact-mean(hatDirect_I)),std((hatDirect_I)));
ts=sprintf('bias STRATIF = %5.2e, std STRATIF = %5.2e',...
    abs(I_exact-mean(hatStratif_I)),std((hatStratif_I)));
disp('***************************')
disp(td),disp(ts)
%===== stratif_uniform function
function U = stratif_uniform(vect_n)
```

```
%!======================================================!
%! SYNOPSIS: U = STRATIF_UNIFORM(vect_n)                !
%! vect_n = k-length sequence to be drawn uniformly     !
%!          in the intervals of length 1/k of [0,1]     !
%!======================================================!
vect_n = vect_n(:); k = length(vect_n); ampl = 1/k;
N = sum(vect_n); U = zeros(N,1);
Sjm1 = 1; Sj = 0;
for j=1:k
    Sj         = Sj+vect_n(j);
    U(Sjm1:Sj) = (rand(vect_n(j),1)+(j-1))*ampl;
    Sjm1       = Sj+1;
end
```

H3.10 (Antithetic variates approach) (see page 106)

1. Using the estimators given by (3.12) and (3.30) respectively, we have

$$\begin{aligned}
\text{cov}\left(f(X), f(X)\right) &= \int_0^1 \frac{1}{(1+x)^2}dx - I^2 = 0.5 - \log^2(2) \\
\text{cov}\left(f(X), f(X)\right) &= \int_0^1 \frac{1}{(2-x)(2-x)}dx - I^2 = 0.5 - \log^2(2)
\end{aligned}$$

and

$$\text{cov}\left(f(X), f(\widetilde{X})\right) = \int_0^1 \frac{1}{(1+x)(2-x)}dx - I^2 = \frac{2}{3}\log(2) - \log^2(2)$$

For $N = 100$, we therefore obtain:

$$\text{var}\left(\widehat{I}_1\right) \approx 1.95\,10^{-4} \text{ and } \text{var}\left(\widehat{I}_a\right) \approx 1.19\,10^{-5}$$

2. Type the program:

```
%===== antithetic.m
clear all
Lruns = 2000; N = 100;
X = rand(N,Lruns); Xtilde = 1-X;
fX = 1 ./ (1+X); fXtilde = 1 ./ (1+Xtilde);
I1 = mean(fX,1);
I2 = (mean(fX(1:N/2,:),1)+mean(fXtilde(1:N/2,:),1))/2;
Nvar1th = 0.5-log(2)*log(2);
Nvar2th = 2*log(2)/3-log(2)*log(2);
fprintf('***\t\t\t\t\tdirect\t\tantithetic\n')
fprintf('***\ttheoretical variances : \t%4.2e| \t%4.2e\n',...
    Nvar1th/N, (Nvar1th+Nvar2th)/N);
fprintf('***\tempirical variances : \t\t%4.2e| \t%4.2e\n',...
    std(I1)^ 2, std(I2)^ 2);
```

H4 Second order stationary process

H4.1 (Levinson algorithm) (page 120) Type the function:

```
function [a,eps2]=levinsonA(X,N)
%!=============================================!
%! SYNOPSIS: [a,eps2]=LEVINSONA(X,N)          !
%!      X    = signal                          !
%!      N    = prediction order                !
%!      a    = prediction coeffs ((N+1)*(N+1)) !
%!      eps2 = prediction errors ((N+1)*1)     !
%!=============================================!
R=zeros(N+1,1); T=length(X);
for in = 1:N+1
    R(in) = X(1:T-in+1)'*X(in:T)/T;
end
ki=zeros(N,1); a=zeros(N,N+1); eps2=zeros(N+1,1);
ki(1)=0; eps2(1)=R(1);
for p=1:N
    if p==1,
        ki(p+1) = R(p+1)/eps2(p);
    else
        ki(p+1) = (R(p+1)-a(p-1:-1:1,p)'*R(2:p))/eps2(p);
    end
    a(p,p+1) = ki(p+1);
    a(1:p-1,p+1) = a(1:p-1,p)-ki(p+1)*a(p-1:-1:1,p);
    eps2(p+1) = eps2(p)*(1-ki(p+1)^2);
end
a = a(:,2:N+1);
```

Run the following program:

```
%===== testLevinson.m
clear all
T = 1000; N = 8;
atrue = [1;-1.7;0.8];
X = filter(1,atrue,randn(T,1));
[alevinson, eps2levinson] = levinsonA(X,N);
[aM,eM] = lpc(X,N);
max(abs(aM(2:end)' + alevinson(1:end,end)))
```

This program compares the calculated values with those supplied by the `lpc.m` function in MATLAB®. Note that this function produces coefficient values with the opposite sign, preceded by the value 1. This corresponds to the coefficients of the AR model (see theorem 4.1).

H4.2 (Lattice filtering) (page 123)

1. Type the function:

```
function [epsF, epsB]=lattice_analysis(xn, ki)
%!=================================================!
%! SYNOPSIS: [epsF, epsB]=LATTICE_ANALYSIS(xn,ki) !
%!    xn   = signal                                !
%!    ki   = reflection coefficients (k1 ... kP)   !
%!    epsF = forward error                         !
%!    epsB = backward error                        !
%!=================================================!
N=length(xn); epsF=zeros(N,1); epsB=zeros(N,1);
P=length(ki); eB=zeros(P,1); eBm1=zeros(P,1);
for nn=1:N
    eF=xn(nn);
    for pp=1:P
        eFp=eF+ki(pp)*eBm1(pp);
        eB(pp)=ki(pp)*eF+eBm1(pp);
        eF=eFp;
    end
    eBm1=[xn(nn); eB(1:P-1)];
    epsF(nn)=eFp; epsB(nn)=eB(P);
end
```

2. Type the function:

```
function [xn, epsB]=lattice_synthesis(epsF, ki)
%!==================================================!
%! SYNOPSIS: [epsF, epsB]=LATTICE_SYNTHESIS(xn,ki) !
%!    epsF = Forward Error                          !
%!    ki   = Reflection coefficients (k1 ... kP)    !
%!    xn   = Reconstructed signal                   !
%!    epsB = Backward Error                         !
%!==================================================!
N=length(epsF); xn=zeros(N,1); epsB=zeros(N,1);
P=length(ki); eB=zeros(P,1); eBm1=zeros(P,1);
for nn=1:N
    eF=epsF(nn);
    for pp=P:-1:1
        eF=eF-ki(pp)*eBm1(pp);
        eB(pp)=eBm1(pp)+ki(pp)*eF;
    end
    xn(nn)=eF; eBm1=[eF; eB(1:P-1)];
    epsB(nn)=eB(P);
end
```

3. Run the following program:

```
%===== testLattice.m
clear all
```

```
ai=[1 -1.8 0.9]; N=1000;
wn=randn(N,1); xn=filter(1,ai,wn);
ki=atok(ai);
[eF,eB]=lattice_analysis(xn,ki);
[xn_s, eB_s]=lattice_synthesis(eF,ki);
[max(abs([wn-eF])) max(abs([xn-xn_s]))]
%=====
figure(1); plot(xn); hold on; plot(xn_s,'--r'); hold off
figure(2); plot(eF); hold on; plot(wn,'--r'); hold off,
figure(3); plot(eB); hold on; plot(eB_s,'--r'); hold off
```

H4.3 (Smoothed periodogram (bias/variance compromise)) (see page 132) Type the program:

```
%===== smoothperio1dMSE.m
% MSE of the smoothed periodogram
clear all
Lruns  = 20;
N = 4096; K = N/2-1;
aAR = [1;-1.7;0.9]; sAR = 1;
Stheo  = (sAR^2) ./ abs(fft(aAR,N)) .^2;
listM  = 2:80;
LM     = length(listM);
bias2E = zeros(LM,Lruns);
varE   = zeros(LM,Lruns);
MSE    = zeros(LM,Lruns);
known  = false;
for ir=1:Lruns
    for im=1:LM
        wn = sAR*randn(N,1);
        xn = filter(1,aAR,wn);
        Xk = fft(xn)/sqrt(N);
        Pk = abs(Xk .^2);
        M  = listM(im);
        twoMplus1    = 2*M+1;
        smoothWindow = hamming(twoMplus1);
        smoothWindow = smoothWindow / sum(smoothWindow);
        suitek2      = (-M:M).^2;
        sumW2        = smoothWindow' * smoothWindow;
        sumk2W       = suitek2*smoothWindow;
        smoothperio  = conv(Pk,smoothWindow,'same');
        if known
            smperioseconde = diff(diff(Stheo))*N*N;
        else
            smperioseconde = diff(diff(smoothperio))*N*N;
        end
        varEk          = sumW2 .* (smoothperio) .^2;
        bias2k         = (smperioseconde.* sumk2W/2/N/N ) .^2;
        varE(im,ir)    = mean(varEk);
        bias2E(im,ir) = mean(bias2k);
```

```
            MSE(im,ir)      = varE(im,ir) + bias2E(im,ir);
        end
end
%=====
figure(2), plot([Stheo, smoothperio]), grid
figure(4), plot(listM,mean(bias2E,2), 'o-'), grid
hold on, plot(listM,mean(varE,2), '--')
plot(listM,mean(MSE,2), '-'), hold off
figure(3), plot(listM,MSE, '.','color',0.9*ones(3,1))
hold on, plot(listM,mean(MSE,2), 'k'), grid, hold off
```

We have chosen to draw a new trajectory for each value of M. It is also possible to use the same trajectory for each value of M; if N is sufficiently large, the results are equivalent. The Boolean variable `known` allows us to compare results depending on whether the spectrum is presumed to be known or unknown.

Figure H4.1 shows the variations of the mean square integrated bias, the mean integrated variance and the mean integrated MSE for 20 trajectories as a function of M. We see that the bias increases, the variance decreases and the MSE passes through a minimum around 15.

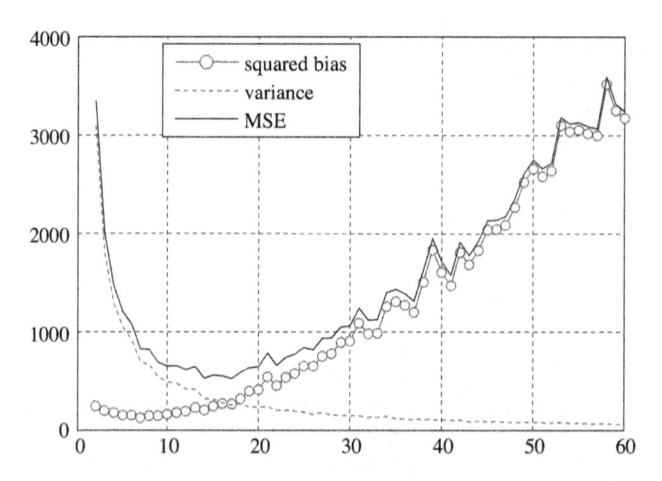

Figure H4.1 – *Variations of the mean square integrated bias, the mean integrated variance and the mean integrated MSE for 20 trajectories as a function of M. The signal is made up of* 4096 *samples, and smoothing is carried out using the Hamming window*

We recommend modifying the value of N, the form of the window or the spectrum under consideration in order to visualize the effects of these parameters on the estimation.

H4.4 (Smoothed periodogram - confidence interval) (see page 135) Type the program:

```
%===== smoothperioCI.m
% Confidence interval of the Smoothed Periodogram
% Uses pwelch from the signal toolbox
clear all
beta = 0.5; alpha = 0.95;
N    = 4096; freq = (0:N-1)/N; K = N/2-1;
aAR = [1;-1.7;0.9]; sAR    = 1;
Stheo  = (sAR^2) ./ abs(fft(aAR,N)) .^2;
Stheo_dB = 10*log10(Stheo);
wn = sAR*randn(N,1); xn = filter(1,aAR,wn);
Xk = fft(xn)/sqrt(N); Pk = abs(Xk .^2);
M  = fix(N^(beta/2));
%=====
twoMplus1       = 2*M+1;
smoothWindow    = rectwin(twoMplus1);
LW              = sum(smoothWindow .^2);
dof             = 2*LW;
smoothWindow    = smoothWindow / sum(smoothWindow);
smoothperio     = conv(Pk,smoothWindow,'same');
smoothperio_dB = 10*log10(smoothperio);
ICmin_dB        = 10*log10(2*LW)-...
    10*log10(chi2inv(0.5+alpha/2, dof));
ICmax_dB        = 10*log10(2*LW)-...
    10*log10(chi2inv(0.5-alpha/2, dof));
%=====
figure(1), subplot(121)
plot(freq,[Stheo_dB, smoothperio_dB]), hold on
plot(freq,[smoothperio_dB+ICmin_dB,...
    smoothperio_dB+ICmax_dB],':')
hold off
set(gca,'xlim',[0,1/2],'ylim',[-20 30]), grid on
subplot(122)
[Pxx,Fxx,Pxxc] = pwelch(xn,fix(N/M),[],[],1,...
    'twosided','ConfidenceLevel',alpha);
plot(Fxx,10*log10(Pxx))
hold on, plot(Fxx,10*log10(Pxxc),':'), hold off
set(gca,'xlim',[0,1/2],'ylim',[-20 30]), grid on
%===== compare confidence interval
figure(2), plot(Fxx,10*log10(Pxxc),'b'), hold on
plot(freq,[smoothperio_dB+ICmin_dB...
    smoothperio_dB+ICmax_dB],'r')
hold off
set(gca,'xlim',[0,1/2],'ylim',[-20 30])
```

The function `pwelch` is in the `signal` toolbox of the most recent versions of MATLAB®.

H4.5 (Magnitude square coherence) (see page 137) Let $G_1(f)$ and $G_2(f)$ be the frequency responses of the two filters.

- We obtain:

$$\begin{aligned}\Gamma(f) &= \gamma(f)\begin{bmatrix}G_1(f)\\ G_2(f)\end{bmatrix}\begin{bmatrix}G_1^*(f) & G_2^*(f)\end{bmatrix}+\begin{bmatrix}\sigma_1^2 & 0\\ 0 & \sigma_2^2\end{bmatrix}\\ &= \begin{bmatrix}|G_1(f)|^2\gamma(f)+\sigma_1^2 & s(f)G_1(f)G_2^*(f)\\ \gamma(f)G_1^*(f)G_2(f) & |G_2(f)|^2\gamma(f)+\sigma_2^2\end{bmatrix}\end{aligned}$$

The MSC is expressed

$$\begin{aligned}\text{MSC}(f) &= \frac{\gamma^2(f)|G_1(f)G_2(f)|^2}{(|G_1(f)|^2\gamma(f)+\sigma_1^2)(|G_2(f)|^2\gamma(f)+\sigma_2^2)} &(7.11)\\ &= \frac{1}{(1+\text{iSNR}_1)(1+\text{iSNR}_2)} &(7.12)\end{aligned}$$

where

$$\text{iSNR}_j = \frac{\sigma_j^2}{\gamma(f)|G_j(f)|^2} \tag{7.13}$$

is interpreted as the inverse of a signal to noise ratio. In the case of an AR-1, $\gamma(f) = s_a^2/|1+ae^{-2j\pi f}|^2$.

- Type the program:

```
%===== MSCexo.m
clear all
alphaCIpercent = 90; % confidence interval
calpha = norminv((1+alphaCIpercent/100)/2);
beta = 0.4; N = 64*1024; M = fix(N^(beta)/2);
twoMplus1 = 2*M+1;
smWindow  = rectwin(twoMplus1);
smWindow  = smWindow / sum(smWindow);
LW        = 1/sum(smWindow.*smWindow);
sigma1 = 2; sigma2 = 1; a = -0.9; sa = 2;
g1 = [1 ; -1.2; 0.9]; g2 = [1 ; -0.4 ; 0.3];
sftheo  = (sa*sa) ./ (abs(fft([1;a],N)) .^2);
G1f2 = abs(fft(g1,N)) .^2; G2f2 = abs(fft(g2,N)) .^2;
iSNR1   = sigma1*sigma1 ./ (sftheo .* G1f2);
iSNR2   = sigma2*sigma2 ./ (sftheo .* G2f2);
MSCtheo = 1 ./ ((1+iSNR1).*(1+iSNR2));
cpICoutside=0; Lruns = 100;
```

```
for ir=1:Lruns
    en       = sa*filter(1,[1 a],randn(N,1));
    x(:,1)   = filter(g1,1,en) + sigma1*randn(N,1);
    x(:,2)   = filter(g2,1,en) + sigma2*randn(N,1);
    X        = fft(x)/sqrt(N);
    perio12 = X(:,1) .* conj(X(:,2));
    perio11 = X(:,1) .* conj(X(:,1));
    perio22 = X(:,2) .* conj(X(:,2));
    smperio11 = conv(perio11,smWindow,'same');
    smperio22 = conv(perio22,smWindow,'same');
    smperio12 = conv(perio12,smWindow,'same');
    smMSC = (abs(smperio12) .^2) ./ (smperio11 .* smperio22);
    deltaIC = calpha*((1-smMSC))/sqrt(2*LW);
    cpICoutside = cpICoutside+...
        sum(or(sqrt(MSCtheo)>sqrt(smMSC)+deltaIC,...
               sqrt(MSCtheo)<sqrt(smMSC)-deltaIC));
end
CIoutsidepercent = cpICoutside/Lruns/N
%=====
% [MSC_ML,W] =...
%      mscohere(x(:,1),x(:,2),fix(N/M),[],N,'twosided');
figure(1), plot((0:N-1)/N,[sqrt(smMSC)]), hold on
plot((0:N-1)/N, ...
    [sqrt(smMSC)-deltaIC sqrt(smMSC)+deltaIC],':')
% plot(W/pi/2,MSC_ML,'--')
hold off, set(gca,'xlim',[0 1/2])
```

H5 Inferences on HMM

H5.1 (Kalman recursion in the scalar case) (see page 148)

In the Gaussian case, relationship $\mathbb{E}\{X_n|Y_{1:n}\} = (Y_{n+1}|Y_{1:n})$ takes the following form:

1. Using the linearity of the expectation in the evolution equation, we have:

$$X_{n+1|n} = a_n X_{n|n} + \mathbb{E}\{B_n|Y_{1:n}\}$$

   Noting that B_n is independent of $Y_{1:n}$ and is centered, we deduce:

$$X_{n+1|n} = a_n X_{n|n}$$

2. If we replace Y_{n+1} with $(c_{n+1}X_{n+1}+U_{n+1})$ and use the hypothesis stating that U_{n+1} and B_{n+1} are orthogonal to $Y_1, \ldots, Y_n$, we get:

$$\begin{aligned}(Y_{n+1}|Y_{1:n}) &= c_{n+1}(X_{n+1}|Y_{1:n}) \\ &= c_{n+1}X_{n+1|n} \qquad (7.14)\end{aligned}$$

3. Using the property (1.31) we write:

$$\begin{aligned} X_{n+1|n+1} &= (X_{n+1}|Y_{1:n}, Y_{n+1}) \\ &= (X_{n+1}|Y_{1:n}) + G_{n+1} i_{n+1} \end{aligned}$$

where $G_{n+1} = (X_{n+1}, i_n)/(i_n, i_n)$ and $i_{n+1} = Y_{n+1} - (Y_{n+1}|Y_{1:n})$. Using (7.14), we also get $i_n = Y_{n+1} - c_{n+1} X_{n+1|n}$. Therefore:

$$X_{n+1|n+1} = X_{n+1|n} + G_{n+1}(Y_{n+1} - c_{n+1} X_{n+1|n}) \qquad (7.15)$$

4. We have:

$$\begin{aligned} i_{n+1} &= \underbrace{c_{n+1} X_{n+1} + U_{n+1}}_{= Y_{n+1}} - c_{n+1} X_{n+1|n} \\ &= c_{n+1}(X_{n+1} - X_{n+1|n}) + U_{n+1} \end{aligned}$$

Let $P_{n+1|n} = (X_{n+1} - X_{n+1|n}, X_{n+1} - X_{n+1|n})$. Because $(X_{n+1} - X_{n+1|n})$ and U_{n+1} are orthogonal, we can write:

$$\|i_{n+1}\|^2 = \text{var}(i_{n+1}) = c_{n+1}^2 P_{n+1|n} + \sigma_U^2(n+1)$$

5. The process

$$i_n = Y_n - (Y_n|Y_{1:n-1})$$

is known as the innovation process. We verify that $\mathbb{E}\{i_n\} = 0$. By definition, i_n belongs to the linear space spanned by the $Y_{1:n}$. In accordance with the projection theorem, i_{n+1} is orthogonal to the linear space generated by the $Y_{1:n}$. Consequently, $i_n \perp i_{n+1}$ and, being Gaussian, they are independent.

Moreover, the linear space spanned by the $Y_{1:n}$ corresponds with the linear space generated by $i_{1:n}$. The joint distribution of the $Y_{1:n}$ is therefore equal to the joint distribution of the $i_{1:n}$. Hence:

$$-2 \log p_{Y_{1:n}}(y_{1:n}) = n \log(2\pi) + \sum_{k=1}^{n} \frac{i_k^2}{\text{var}(i_k)}$$

6. Furthermore:

$$\begin{aligned}(X_{n+1}, i_{n+1}) &= (X_{n+1}, X_{n+1} - X_{n+1|n})c_{n+1} + \underbrace{(X_{n+1}, U_{n+1})}_{=0} \\ &= (X_{n+1} - X_{n+1|n}, X_{n+1} - X_{n+1|n})c_{n+1} \\ &= P_{n+1|n}c_{n+1}\end{aligned}$$

And hence:

$$G_{n+1} = \frac{P_{n+1|n}c_{n+1}}{\sigma_U^2(n+1) + c_{n+1}^2 P_{n+1|n}} \tag{7.16}$$

7. Let us now determine the expression of $P_{n+1|n}$. We have:

$$X_{n+1} - X_{n+1|n} = (a_n X_n + B_n) - a_n X_{n|n}$$

This leads us to:

$$P_{n+1|n} = a_n^2 P_{n|n} + \sigma_B^2(n) \tag{7.17}$$

involving that $X_{n|n}$ and B_n are orthogonal. Using (7.15) we obtain:

$$\begin{aligned}X_{n+1} - X_{n+1|n+1} &= (X_{n+1} - X_{n+1|n}) - G_{n+1}(Y_{n+1} - c_{n+1}X_{n+1|n}) \\ &= (X_{n+1} - X_{n+1|n}) - G_{n+1}c_{n+1}(X_{n+1} - X_{n+1|n}) + G_{n+1}U_{n+1}\end{aligned}$$

Using the fact that $X_{n+1} - X_{n+1|n}$ and U_{n+1} are orthogonal, the expression (7.16) leads to:

$$P_{n+1|n+1} = (1 - G_{n+1}c_{n+1})P_{n+1|n} \tag{7.18}$$

If we group expressions (7.15), (7.16), (7.17) and (7.18) together, we get the following algorithm in accordance with the Kalman algorithm (3):

$$\left\{\begin{aligned} X_{n+1|n} &= a_n X_{n|n} \\ P_{n+1|n} &= a_n^2 P_{n|n} + \sigma_B^2(n) \\ G_{n+1} &= \frac{P_{n+1|n}c_{n+1}}{\sigma_U^2(n+1) + c_{n+1}^2 P_{n+1|n}} \\ X_{n+1|n+1} &= X_{n+1|n} + G_{n+1}\left(Y_{n+1} - c_{n+1}X_{n+1|n}\right) \\ P_{n+1|n+1} &= (1 - G_{n+1}c_{n+1})P_{n+1|n} \end{aligned}\right. \tag{7.19}$$

with the initial conditions $x_{1|1} = 0$ and $P_{1|1} = \mathbb{E}\left\{X_1^2\right\}$.

H5.2 (Denoising an AR-1 using Kalman) (see page 149)

1. The state equation shows that the stationary solution is an AR-1 process. Therefore, its power has the expression:

$$\mathbb{E}\left\{X_n^2\right\} = \frac{\sigma_b^2}{1-a^2}$$

2. According to (5.13) and (5.14), we have:

$$G_n = \frac{P_{n|n-1}}{P_{n|n-1}+\sigma_u^2} \tag{7.20}$$

which leads us to $P_{n|n-1}(1-G_n) = \sigma_u^2 G_n$. Using (5.12) and (5.17):

$$\begin{aligned} P_{n|n-1} &= a^2(1-G_{n-1})P_{n-1|n-2} + \sigma_b^2 \\ &= a^2\sigma_u^2 G_{n-1} + \sigma_b^2 \end{aligned}$$

Carrying this result in the expression (7.20) that gives the recursive formula:

$$G_n = \frac{\rho + a^2 G_{n-1}}{1+\rho+a^2 G_{n-1}} \tag{7.21}$$

In this case, the initial conditions lead to $K_1 = \sigma_b^2/(1-a^2)$. Therefore $G_1 = \rho/(1+\rho-a^2)$. From the calculation point of view, everything happens as if we started out with formula (7.21) and the initial values $\widehat{X}_{0|0} = 0$ and $G_0 = \rho/(1-a^2)$.

We can easily verify that the series G_n is increasing monotone and bounded by 1. It therefore converges and the limit verifies the recursive equation, giving $G_{\lim} = \frac{-(1+\rho-a^2)+\sqrt{(1+\rho-a^2)^2+4\rho a^2}}{2a^2}$.

The Kalman algorithm can be summed up as follows:

- Initial conditions: $\widehat{X}_{0|0} = 0$ et $G_0 = \rho/(1-a^2)$.
- For n from 1 to N:

$$\left\{\begin{aligned} G_n &= \frac{\rho + a^2 G_{n-1}}{1+\rho+a^2 G_{n-1}} \\ \widehat{X}_{n|n} &= a\widehat{X}_{n-1|n-1} + G_n\left(Y_n - a\widehat{X}_{n-1|n-1}\right) \end{aligned}\right.$$

3. The following program is designed to test the algorithm:

```
%===== kalmannoisyAR1.m
clear all
N=100; a=0.9; sigmab=2;
x=filter(1,[1 -a],sigmab*randn(N,1));
sigmau=3; y=x+sigmau*randn(N,1);
%===== tracking
xch=zeros(N,1); G = zeros(N,1);
a2=a*a; rho=(sigmab/sigmau)^2; G(1)=rho/(1-a2);
for nn=2:N
    G(nn)= (rho+a2*G(nn-1))/(1+rho+a2*G(nn-1));
    xch(nn)=a*xch(nn-1)+G(nn-1)*(y(nn)-a*xch(nn-1));
end
mtime=(0:N-1);
plot(mtime,x,'.-g',mtime,y,':k',mtime,xch,'.-b');
Glim = (-(1+rho-a2)+sqrt((1+rho-a2)^2+4*rho*a2))/(2*a2);
[G(N), Glim]
```

The results are shown in Figure H5.1.

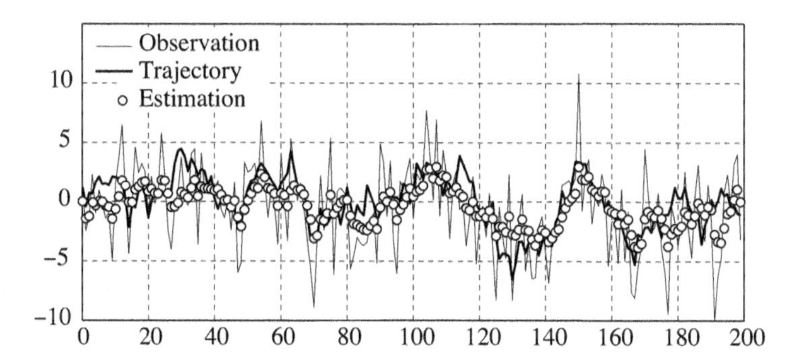

Figure H5.1 – *Results for the study of the filtering*

When we presented the Kalman filter, and implemented it in the previous program, we assumed that the model as well as the characteristic features of the noise were known. However, this is usually not the case. For example, if in our case the signal X_n is an AR process, the choice of a and of σ_b^2 requires that we compromise between the ability of X_n to track the trajectory and the elimination of the noise. Choosing a too close to 1 means that the model does not take into account the rapid variations of the signal X_n. Therefore, the filter has difficulties "keeping up" with such variations. Likewise, if we choose σ_b^2 too high, we assume that we expect significant variations of the signal X_n with respect to the equation $X_n \approx aX_{n-1}$. You can check by using the previous algorithm and changing the parameters.

H5.3 (2D tracking) (see page 149)

1. Using a second-order Taylor expansion, for each component i we have:

$$x_i(nT+T) \approx x_i(nT) + T\dot{x}_i(nT) + \frac{T^2}{2}\ddot{x}_i(nT) \tag{7.22}$$

$$\dot{x}_i(nT+T) \approx \dot{x}_i(nT) + T\ddot{x}_i(nT) \tag{7.23}$$

Hence:

$$X_{n+1} \approx \begin{bmatrix} 1 & 0 & T & 0 \\ 0 & 1 & 0 & T \\ 0 & 0 & 1 & 0 \\ 0 & 0 & 0 & 1 \end{bmatrix} X_n + \begin{bmatrix} T^2/2 & 0 \\ 0 & T^2/2 \\ T & 0 \\ 0 & T \end{bmatrix} \begin{bmatrix} \ddot{x}_1(nT) \\ \ddot{x}_2(nT) \end{bmatrix}$$

Considering that the pairs $(\ddot{x}_1(nT), \ddot{x}_2(nT))$, representing the acceleration components, form a series of independent, centered, Gaussian random variables with a covariance matrix $\sigma^2 I_2$, we have:

$$X_{n+1} = AX_n + B_n$$

with

$$R^B = \sigma^2 \begin{bmatrix} T^4/4 & 0 & T^3/2 & 0 \\ 0 & T^4/4 & 0 & T^3/2 \\ T^3/2 & 0 & T^2 & 0 \\ 0 & T^3/2 & 0 & T^2 \end{bmatrix}$$

σ is expressed in m/s^2.

2. Let the speed v_0 be 30 m/s, and consider that it may vary by a quantity proportional to v_0 of the form λv_0. We can therefore take $\sigma \approx \lambda v_0/T$ as the dispersion of the acceleration.

3. Type the following functions:

```
function [Xtt,Ptt,loglikeli, loglikeli_k]= ...
             KalmanFilter(Y,A,RB,C,RU,mu0,R0)
%!=====================================================!
%! Kalman filter                                       !
%! SYNOPSIS                                            !
%!   [Xtt,Ptt,loglikeli]=KALMANFILTER(Y,A,RB,C,RU,mu0,R0) !
%! Inputs                                              !
%!      Y    = observations (dY x T)                   !
%!      A    = state matrix (dX x dX)                  !
%!      RB   = state covariance (dX x dX)              !
%!      RU   = observation covariance (dY x dY)        !
```

```
%!     mu0 = initial state mean (dX x 1)                    !
%!     R0  = initial state covariance (dX x dX)             !
%! Outputs                                                  !
%!     Xtt = filtered state (dX x T)                        !
%!     Ptt = covariance of estimate (dX x dX x T)           !
%!     loglikeli = log p(Y_1:T)                             !
%!==========================================================!
dimX           = size(A,1);
T              = size(Y,2);
Xtt            = zeros(dimX,T);
Xtt(:,1)       = mu0;
Ptt            = zeros(dimX,dimX,T);
Ptt(:,:,1)     = 0;
Pttm1          = zeros(dimX,dimX,T);
inov_1         = Y(:,1);
cov_inov       = C * R0 * C'+RU;
loglikeli_k    = zeros(T,1);
loglikeli_k(1) = -(log(det(cov_inov)) ...
    + inov_1'*(cov_inov\inov_1) )/2;
for k=2:T
    Xttm1    = A * Xtt(:,k-1);
    Rttm1    = (A * Ptt(:,:,k-1) * A') + RB;
    cov_inov = (C * Rttm1 * C') + RU;
    inov_k   = (Y(:,k) - C*Xttm1);
    Kn       = (Rttm1*C') / cov_inov;
    Xtt(:,k) = Xttm1 + Kn * inov_k;
    Pttm1(:,:,k)    = Rttm1;
    Ptt(:,:,k)      = Rttm1 - Kn*C*Rttm1;
    loglikeli_k(k) = -(log(det(cov_inov)) ...
        + inov_k'*(cov_inov\inov_k) )/2;
end
loglikeli = sum(loglikeli_k);
```

and

```
function confidenceellipse(X0,C,alpha)
%!=====================================================!
%! SYNOPSIS: CONFIDENCEELLIPSE(X0, E, c)               !
%! Ellipse equation:                                   !
%!    (X-X0)'*(C^-1)*(X-X0)= c(alpha)                  !
%!    X0 = coordinates of the ellipse's center (2x1)   !
%!    C  = positive (2x2) matrix                       !
%!    alpha = confidence level in(0,1)                 !
%!=====================================================!
N=100; theta = 2*pi*(0:N)/N;
c = -2*log(1-alpha);
Y = sqrt(c)*[cos(theta);sin(theta)];
X = sqrtm(C)*Y;
plot(X(1,:)+X0(1),X(2,:)+X0(2),'g','linew',2);
```

Type and run the following program:

```
%===== kalmantraj2D.m
clear all
T  = 1/100; % in second
v0 = 30;    % m/s
%===== trajectory generation
M = 100; M0 = 2; N = M*M0; L0 = N*v0*T;
Xtrue = resample(L0*randn(M0,2),M,1);
Xtrue = Xtrue';
sigmaUdB  = -10;
sigma2obs = 10^(-sigmaUdB/10);
A = [1 0 T 0; 0 1 0 T; 0 0 1 0; 0 0 0 1];
C = [1 0 0 0; 0 1 0 0];
D = [T^2/2 0; 0 T^2/2; T 0; 0 T];
%===== observation generation
RUobs = ((v0*T)^2)*sigma2obs*eye(2);
Y = Xtrue + sqrtm(RUobs)*randn(2,N);
%===== a priori knowledge
lambda = 0.03; sigma  = lambda*v0/T; % m/s^2
RBstate = sigma*sigma*(D*D');
%===== initial conditions
X0 = [Y(1,1);Y(2,1);0;0];
R0 = sigma*sigma*eye(4);
[Xfilt,Pnn] = KalmanFilter(Y, A, RBstate, C, RUobs, X0, R0);
%=====
figure(2);
plot(Y(1,:),Y(2,:),'x--b');
hold on
plot(Xtrue(1,:),Xtrue(2,:),'--r')
plot(Xtrue(1,1),Xtrue(2,1),'or','markerfa','r')
plot(Xfilt(1,:),Xfilt(2,:),'.-k');
% we plot only 1/10 of the confidence regions
for t=1:N
   if mod(t,10)==0
      confidenceellipse([Xfilt(1,t);Xfilt(2,t)], ...
          Pnn(1:2,1:2,t), 0.90);
   end
end
hold off
```

COMMENTS: Modifying the value of σ using λ, we modify our initial ideas concerning the acceleration, and thus concerning the fact that the trajectory more or less follows a straight line. Hence, if σ is small, the mobile element is easy to follow when the trajectory is close to a straight line, but harder to follow if the trajectory curves. If σ is large, the observation noise is difficult to remove. The Kalman filter establishes a good compromise between prior knowledge, i.e. σ, and observations, i.e. Y_n.

H5.4 (Calibration of an AR-1) (see page 150)
Type and run the following program:

```
%====== calibratenoisyAR1.m
clear all
a=0.9;sigmaB=0.3;sigmaU=0.2;N=100;
x=filter(sigmaB,[1 -a],randn(1,N));
y=x+sigmaU*randn(1,N);
nba=16;lista=a+linspace(-0.1,0.1,nba);
nbB=15;listsB=sigmaB+linspace(-0.1,0.1,nbB);
nbU=15;listsU=sigmaU+linspace(-0.1,0.1,nbU);
LL=zeros(nba,nbB,nbU);
for ia=1:nba
    ai=lista(ia);
    for iB=1:nbB
        sBi=listsB(iB);
        for iU=1:nbU
            sUi=listsU(iU);
            [Xtt,Ptt,loglikeli,loglikeli_k] = ...
                KalmanFilter(y,ai,sBi*sBi,1,...
                sUi*sUi,y(1),sBi*sBi/(1-ai*ai));
            LL(ia,iB,iU)=loglikeli;
        end
    end
end
idt=find(LL==max(max(max(LL))))-1;
id3=fix(idt/(nba*nbB));idtp=idt-id3*nba*nbB;id2=fix(idtp/nba);
id1=idtp-id2*nba;
aestim=lista(id1+1); sBestim=listsB(id2+1); sUestim=listsU(id3+1);
[aestim sBestim sUestim;a sigmaB sigmaU]
```

H5.5 (Calculating the likelihood of an ARMA) (see page 151)

1. The first equation of expression (5.20) is written:

$$\begin{matrix} 1 & \times \\ \vdots & \vdots \\ z^{-(r-2)} & \times \\ z^{-(r-1)} & \times \end{matrix} \qquad \begin{bmatrix} X_{1,n} \\ \vdots \\ X_{r-1,n} \\ X_{r,n} \end{bmatrix} = \begin{bmatrix} -a_1 \\ \vdots \\ -a_{r-1} \\ -a_r \end{bmatrix} X_{1,n-1} + \begin{bmatrix} X_{n-1,2} \\ \vdots \\ X_{n-1,r} \\ 0 \end{bmatrix} + \begin{bmatrix} 1 \\ \vdots \\ b_{r-2} \\ b_{r-1} \end{bmatrix} Z_n \tag{7.24}$$

Applying a series of delays such as those given in the first column of (7.24), we obtain:

$$X_{1,n} = -\sum_{k=1}^{r} a_k X_{1,n-k} + Z_n + \sum_{m=1}^{r-1} b_m Z_{n-m}$$

Noting that, following (5.20), $X_{1,n} = Y_n$, we deduce that Y_n is an ARMA-p, q.

2. For t ranging from 2 to n, the Kalman algorithm giving the log-likelihood is written:

$$\begin{aligned}
X_{t|t-1} &= AX_{t-1|t-1} & (7.25)\\
P_{t|t-1} &= AP_{t-1|t-1}A^T + \sigma^2 RR^T & (7.26)\\
\gamma_t &= C^T P_{t|t-1} C & (7.27)\\
K_t &= \frac{1}{\gamma_t} P_{t|t-1} C & (7.28)\\
I_t &= Y_t - C^T X_{t|t-1} & (7.29)\\
X_{t|t} &= X_{t|t-1} + K_t I_t & (7.30)\\
P_{t|t} &= P_{t|t-1} - K_t C^T P_{t|t-1} & (7.31)\\
\ell_t &= \ell_{t-1} + \log(\gamma_t) + \frac{1}{\gamma_t} I_t^2 & (7.32)
\end{aligned}$$

with initial values

$$\begin{aligned}
P_{1|1} & \quad \text{square matrix of size } r \text{ (see remark)} & (7.33)\\
X_{1|1} &= 0 \quad \text{null vector of size } r & (7.34)\\
\ell_1 &= 0 & (7.35)
\end{aligned}$$

3. A tilde is used to indicate quantities calculated using the algorithm with $\sigma = 1$. We consider that, on initialization, $P_{1|1} = \sigma^2 \widetilde{P}_{1|1}$. We deduce that $P_{t|t-1} = \sigma^2 \widetilde{P}_{T|t-1}$, that$\gamma_t = \sigma^2 \widetilde{\gamma}_t$, $K_t = \widetilde{K}_t$ and that $X_{t|t} = \widetilde{X}_{t|t}$. The log-likelihood is therefore written:

$$\ell_n = \sum_{t=1}^{n} \left(\log(\sigma^2 \widetilde{\gamma}_t) + \frac{I_t^2}{\sigma^2 \widetilde{\gamma}_t} \right) \tag{7.36}$$

We must therefore simply apply the algorithm for $\sigma = 1$ then use formula (7.36) to calculate the likelihood. This relationship is highly useful for maximum likelihood-based estimation problems, as maximization in relation to σ^2 possesses an analytical solution.

REMARK: The choice of $P_{1|1}$ requires further explanation. According to question 3, we can consider that $\sigma^2 = 1$. One choice for $P_{1|1}$ is the matrix which corresponds to the stationary form of X_t. Noting $P_1 = \text{cov}(X_t)$ and using the evolution equation and the stationarity we have:

$$P_1 = AP_1A^T + Q$$

taking $Q = RR^T$. This is known as the Lyapunov equation. Using the identity

$$\text{vec}(ABC) = (C^T \otimes A)\text{vec}(B)$$

where the operator noted "vec" denotes the column vectorization operation applied to a matrix. This last equation may be rewritten

$$\text{vec}(P_1) = (A \otimes A)\text{vec}(P_1) + \text{vec}(Q)$$

where $\otimes$ denotes the Kronecker product. Hence:

$$\text{vec}(P_1) = (I_{r^2} - A \otimes A)^{-1}\text{vec}(Q)$$

4. Type:

```
function L=LLarmaKalman(Y,a,b,sigma2)
%!=====================================================!
%! Likelihood of ARMA by Kalman filtering              !
%! SYNOPSIS: L=LLARMAKALMAN(Y,a,b,sigma2)              !
%! Inputs                                              !
%!      Y_n + \sum_k=1^p a_k Y_n-k =                   !
%!                 Z_n  + \sum_k=1^q b_k Z_n-k         !
%!      a = [1,a_1,...,a_p]                            !
%!      b = [1,b_1,...,b_q]                            !
%!      Z_n white gaussian noise with variance sigma2 !
%! Outputs                                             !
%!      L = likelihood maximized on the scale factor  !
%!=====================================================!
T = length(Y); p = length(a)-1;
q = length(b)-1; r = max([p,q+1]);
A = zeros(r,r); C = eye(r,1);
A(1:p,1) = -a(2:p+1);
A(1:r-1,2:r) = eye(r-1);
R = b; Q = sigma2*(R*R');
%=====
Lcurr = 0; Xhat_tt = zeros(r,1);
% steady state
P_tt = reshape( (eye(r^2)-kron(A,A) ) \ Q(:), r, r );
%=====
for t=1:T
    Xhat_t1t = A*Xhat_tt;
    P_t1t    = A*P_tt*A' + Q;
    gamma_t  = C'*P_t1t*C;
    Yhat_t1t = C'*Xhat_t1t;
    K_t1     = P_t1t*C/gamma_t;
    I_t      = Y(t) - Yhat_t1t;
```

```
    Xhat_tt  = Xhat_t1t + K_t1*I_t;
    P_tt     = P_t1t - K_t1*C'*P_t1t;
    s_t      = I_t^2/gamma_t;
    %===== current likelihood
    Lcurr    = Lcurr + log(gamma_t)+s_t;
end
L = Lcurr/2;
```

5. Type:

```
function gamma=arma2ACF(a,b,sigma2,n)
%!==================================================!
%! SYNOPSIS: gamma=ARMA2ACF(a,b,sigma2,n)           !
%!    Compute ARMA autocovariance sequence          !
%! Input:                                           !
%!   a      = AR param sans 1 [p x 1]               !
%!   b      = MA param sans 1 [q x 1]               !
%!   sigma2 = Innovation variance                   !
%!   n      = Number of autocovariances             !
%! Output:                                          !
%!   gamma  = Autocovariances 0 to (n-1) [n x 1] !
%!==================================================!
p     = length(a)-1;
q     = length(b)-1;
r     = max(p,q+1);
J     = flipud(eye(p+1,1));
a1    = -a(2:p+1);
%===== first p elements of AR autocovariance
A1    = toeplitz(eye(p+1,1), a);
A2    = hankel(J*a(p+1),a(p+1:-1:1));
R     = J'/(A1+A2)';
R     = fliplr(R);
R(1) = 2*R(1);
for ii = p+2:n+r+1
    R(ii) = R(ii-1:-1:ii-p)*a1;
end
gamma = zeros(n,1);
for kk = 1:n
    gamma(kk) = 0;
    for ii = 0:q
        for jj = 0:q
            gamma(kk) = gamma(kk) + ...
                b(ii+1)*b(jj+1) * ...
                R(abs((kk-1) - ii + jj) + 1);
        end
    end
end
gamma = gamma * sigma2;
```

Type:

```
%===== testKarmaOnARMApq.m
clear all
N = 50;
%===== ARMA causal and inversible
a = [1;-0.9;0.8]; b = [1;0.6;0.1;0.3];
sigma = 2; sigma2 = sigma^2;
w = sigma*randn(N,1); x = filter(b,a,w);
%===== with Kalman
L1 = LLarmaKalman(x,a,b,sigma2);
%===== direct ACFs
V = arma2ACF(a,b,sigma2,N);
R = toeplitz(V);
kappa = x'*(R\x)/N;
L2 = (log(det(R))+x'*(R\x))/2;
L1 ./ L2
```

H5.6 (Discrete HMM generation) (see page 152)

1. Let us show that:

$$X_{n+1} = \sum_{j=1}^{S} j \times \mathbb{1}(U_n \in [F_{X_n}(j-1), F_{X_n}(j)])$$

where U_n is a series of independent r.v.s with values in $(0,1)$. To do this, let us determine $\mathbb{P}\{X_{n+1} = s|X_n = k\}$. As, according to our hypothesis, U_n is independent of X_n, we can write:

$$\mathbb{P}\{X_{n+1} = s|X_n = k\} \quad = \quad \mathbb{P}\{U_n \in [F_k(s-1), F_k(s)[\} = p_{k|s}$$

using the fact that U_n is uniform. Note that this expression may be written:

$$X_{n+1} = f(X_n, U_n)$$

where U_n is an i.i.d. series with uniform distribution over $(0,1)$. It is similar to the expression of the state equation evolution expression(5.8), except that it is neither linear, nor Gaussian.

2. Type and run the program:

```
%===== testHMMGgenerate.m
clear all
d = 2; S = 4; N = 300;
Sigma2s = zeros(d,d,S);
```

```
mus = zeros(d,S);
for is = 1:S
    Sigma2s(:,:,is) = rand * diag(ones(1,d));
    mus(:,is) = randn * ones(d,1);
end
P = [0.4 0.1 0.3 0.2; ...
     0.1 0.4 0.3 0.2; ...
     0.3 0.1 0.4 0.2; ...
     0.1 0.3 0.1 0.5];
omega  = [1/2;1/4;1/8;1/8];
[X,Y] = HMMGaussiangenerate(N,omega,P,mus,Sigma2s);
```

We use the function:

```
function [X,Y]=HMMGaussiangenerate(N,omega,P,mus,sigma2s)
%!==============================================!
%! Generate an HMM of S gaussians               !
%! SYNOPSIS                                     !
%!    [x,states]=HMMGAUSSIANGENERATE(...        !
%             N,omega,P,mus,sigma2s)            !
%!     N       = length of the sequence         !
%!     omega   = initial distribution S x 1     !
%!     P       = transition distribution S x S !
%!     mus     = mean array K x S               !
%!     sigma2s = variance array K x K x S       !
%!     X       = state sequence 1 x N           !
%!     Y       = observation array K x N        !
%!==============================================!
K = size(sigma2s,1);
S = size(P,1);
C1on2 = zeros(K,K,S);
F = zeros(S);
for is=1:S
    F(is,:)=cumsum(P(is,:));
    C1on2(:,:,is) = sqrtm(sigma2s(:,:,is));
end
Fomega = cumsum(omega);
Y = zeros(K,N);
X = zeros(1,N);
X(1) = find(Fomega>=rand,1);
Y(:,1) = mus(:,X(1))+C1on2(:,:,X(1))*randn(K,1);
for n=2:N
    c=F(X(n-1),:);U=rand;X(n)=find(c>=U,1);
    Y(:,n)=mus(:,X(n))+C1on2(:,:,X(n))*randn(K,1);
end
```

H5.7 (EM algorithm for HMM) (see page 158)

1. Type the function:

```
function [alpha,beta,gamma,ell,ct]= ...
         ForwBackwGaussian(Y,omega,P,C,mu)
%!=================================================!
%! SYNOPSIS                                        !
%!    [alpha,beta,gamma, ell]= ...                 !
%!         FORWBACKWGAUSSIAN(Y,omega,P,C,mu)       !
%! Inputs                                          !
%!     Y = array  d x N                            !
%!     omega = intitial distribution array S x 1   !
%!     P     = transition distribution S x S       !
%!         where P(j,i) = Prob(X_n = i|X_n-1 = j)  !
%!         independent of n                        !
%!     C     = observation covariance d x d x S    !
%!     mu    = observation mean d x S              !
%! Outputs                                         !
%!     alpha = array N x 1 - forward propagation   !
%!     beta = array N x 1 - backward propagation   !
%!     gamma = array N x 1                         !
%!     ell = log likelihood                        !
%!     ct = normalisation constant N x 1           !
%!=================================================!
[d,N] = size(Y);
unsur2pid = (2*pi)^(-d/2);
S = length(omega);
ct = zeros(N,1); alpha = zeros(S,N);
tildealpha = zeros(S,1); like = zeros(S,N);
for is=1:S
    Cis = squeeze(C(:,:,is));
    detCis = sqrt(det(Cis));
    muis = mu(:,is);
    for in=1:N
        yin  = Y(:,in)-muis;
        aux1 = yin'*(Cis\yin);
        like(is,in) = unsur2pid * exp(-aux1/2) / detCis;
    end
end
for is=1:S
    tildealpha(is) = like(is,1)*omega(is);
end
ct(1) = sum(tildealpha);
alpha(:,1) = tildealpha / ct(1);
ell = log(ct(1));
for in=2:N
    gin = like(:,in);
    for is=1:S
        tildealpha(is) = gin(is)*alpha(:,in-1)'*P(:,is);
    end
    ct(in) = sum(tildealpha);
    alpha(:,in) = tildealpha / ct(in);
```

```
    ell = ell+log(ct(in));
end
beta = ones(S,N);
for in=N-1:-1:1
    ginp1 = like(:,in+1);
    for is=1:S
        beta(is,in) = ...
            sum((beta(:,in+1) .* P(is,:)') .* ginp1) ...
            / ct(in+1);
    end
end
gamma = beta .* alpha;
```

2. Type the function:

```
function [omega_new,P_new,C_new,mu_new]= ...
    EMforHMM(Y,P_old,C_old,mu_old,alpha,beta,gamma,ct)
%!=========================================================!
%! SYNOPSIS:                                                !
%!   [omega_new,P_new,C_new,mu_new] = ...                   !
%!    EMFORHMM(Y, P_old,C_old,mu_old,alpha,beta,gamma,ct)   !
%! Inputs                                                   !
%!      Y = array  d x N                                    !
%!      P_old = transition distribution S x S               !
%!          where P(j,i) = Prob(X_n = i|X_n-1 = j)          !
%!          independent of n                                !
%!      C_old  = observation covariance d x d x S           !
%!      mu_old = observation mean d x S                     !
%!      alpha  = array N x 1                                !
%!      beta   = array N x 1                                !
%!      gamma  = array N x 1                                !
%!      (see ForwBackwGaussian.m)                           !
%! Outputs                                                  !
%!   new parameters                                         !
%!=========================================================!
log2pi = log(2*pi);
S = size(P_old,1);
[d,N] = size(Y);
omega_new = gamma(:,1);
xi = zeros(S,S,N-1);
g  = zeros(N,S);
for si=1:S
    for in=1:N
        aux_g = d*log2pi+log(det(C_old(:,:,si)))+...
            (Y(:,in)-mu_old(:,si))'*...
            (C_old(:,:,si)\(Y(:,in)-mu_old(:,si)));
        g(in,si) = exp(-aux_g/2);
    end
end
for in=1:N-1
```

```
    for sf=1:S
        for si=1:S
            xi(sf,si,in) = alpha(si,in)*...
                beta(sf,in+1)*P_old(si,sf)*g(in+1,sf);
        end
    end
    xi(:,:,in) = xi(:,:,in) / ct(in+1);
end
%=====
sumgamma = sum(gamma,2);
Paux = zeros(S); P_new = zeros(S);
for si=1:S
    for sf=1:S
        Paux(si,sf)=sum(xi(sf,si,:));
    end
    P_new(si,:)=Paux(si,:)/sum(Paux(si,:));
end
%=====
mu_new = zeros(d,S); C_new = zeros(d,d,S);
for si=1:S
    mu_new(:,si)=Y*gamma(si,:)'/sumgamma(si);
    C_newaux=0;
    for in=1:N
        C_newaux = C_newaux + ...
            (Y(:,in)-mu_new(:,si))*...
            (Y(:,in)-mu_new(:,si))'*gamma(si,in);
    end
    C_new(:,:,si)=C_newaux/sumgamma(si);
end
```

3. Type and run the program:

```
%===== testEMforHMMG.m
clear all
%===== generation HMM gaussian with S hidden states
%    observation of size d
d = 2; S = 4; N = 1000;
Sigma2s = zeros(d,d,S);
for s = 1:S
    Sigma2s(:,:,s) = rand * eye(d);
end
mus = randn(d,S);
P = [0.4 0.1 0.3 0.2; ...
     0.1 0.4 0.3 0.2; ...
     0.3 0.1 0.4 0.2; ...
     0.1 0.3 0.1 0.5];
omega  = [1/2;1/4;1/8;1/8];
[X,Y] = HMMGaussiangenerate(N,omega,P,mus,Sigma2s);
%===== EM initialisation
omega_old =ones(S,1)/S;
C_old = zeros(d,d,S);
```

```
for si=1:S
    C_old(:,:,si)=eye(d);
end
P_old = ones(S)/S; mu_old = randn(d,S);
%=====
Liter=25; LL=zeros(Liter,1);
update_omega = 0;
for ip=1:Liter
    [alpha, beta, gamma, ell, ct] = ...
        ForwBackwGaussian(Y,omega_old,P_old,C_old,mu_old);
    LL(ip)=ell;
    %===
    [omega_new, P_new, C_new, mu_new] = ...
        EMforHMM(Y,P_old,C_old,mu_old,alpha,beta,gamma,ct);
    if update_omega
        omega_old=omega_new;
    end
    P_old=P_new; mu_old = mu_new; C_old = C_new;
end
plot(LL,'.-')
if not(sum(diff(LL)<0)==0), disp('pb'); end
```

H6 Selected Topics

H6.1 (MUSIC 2D) (see page 185)

1. The 2D-MUSIC function is given by

$$S_{\mathrm{MUSIC}}(\zeta,\phi) = \frac{1}{\boldsymbol{a}^H(\zeta,\phi)\boldsymbol{G}\boldsymbol{G}^H\boldsymbol{a}(\zeta,\phi)}$$

where $\boldsymbol{G}\boldsymbol{G}^H$ refers to the orthogonal projector onto the noise subspace.

2. Type:

```
%===== music2D.m
clear all
N=100; K=3; M=25; sigma=0.3;
%===== dzeta from 0 to 180, phi from -180 to 180
dzeta=[30 40 70]; phi=[60 50 20];
%===== array sensor location
loc_array=zeros(3,M);
for xx=1:5
    for yy=1:5
        loc_array(1:2,(xx-1)*5+yy)=[xx-1,yy-1]'/2-1;
    end
end
```

```
Aalpha=array2D(dzeta,phi,loc_array); sn=randn(K,N);
bn=sigma*(randn(M,N)+j*randn(M,N)); xn=Aalpha*sn+bn;
%===== covariance estimation
Rest=zeros(M);
for nn=1:N, Rest=Rest + xn(:,nn)*xn(:,nn)'; end
Rest=Rest/N; [Uest,Dest,Vest]=svd(Rest);
%===== noise subspace projector
Vnoise=Vest(:,(K+1):M); PIest=Vnoise*Vnoise';
plage_theta=(20:0.5:90);ltheta=length(plage_theta);
plage_phi=(10:0.5:80);lphi=length(plage_phi);
imageM=zeros(ltheta,lphi);
for iphi=1:lphi
    ph=plage_phi(iphi);
    for itheta=1:ltheta
        th=plage_theta(itheta);
        aij = array2D(th,ph,loc_array);
        imageM(itheta,iphi)=-log10(abs(aij'*PIest*aij));
    end
end
imagesc(plage_phi,plage_theta,imageM); grid
xlabel(['\phi = [' sprintf('%3.2g',phi) ']']);
ylabel(['\dzeta = [' sprintf('%3.2g',dzeta) ']']);
```

```
function A=array2D(dzeta,phi,loc_array)
%!==========================================!
%! SYNOPSIS: A=ARRAY2D(dzeta,phi,loc_array) !
%!==========================================!
DtoRAD=pi/180; M=size(loc_array,2); K=length(dzeta);
A=zeros(M,K);
for kk=1:K
    tk=dzeta(kk)*DtoRAD; pk=phi(kk)*DtoRAD;
    beta_kk=[cos(pk)*sin(tk);sin(pk)*sin(tk);cos(tk)];
    A(:,kk)=exp(2*pi*j*(loc_array.')*beta_kk);
end
```

H6.2 (Phase modulator) (see page 190)

1. Because the binary rate is equal to $1,500$ bps, each bit lasts $1/1,500$ s. Since the bits are arranged in groups of three, the transmission of each elementary signal lasts a duration of $T = 3 \times 1/1,500 = 2$ ms.

2. Figure H6.1 shows a Gray code solution.

3. Type (see Figure H6.2):

```
%===== CPSK.M
M = 3;          % bits per symbol
```

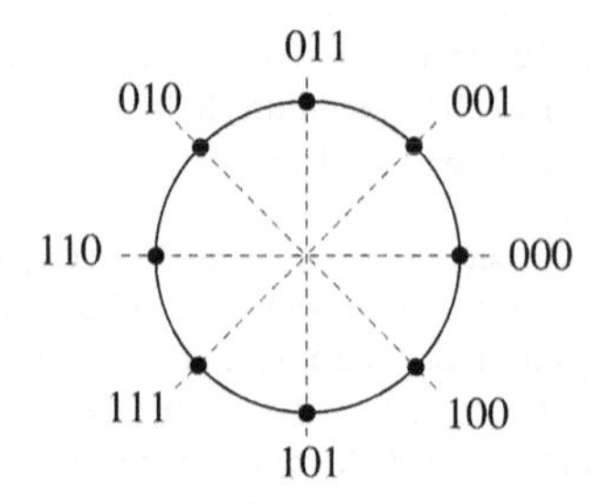

Figure H6.1 – *Gray code for an 8 state phase modulation*

```
dpM = 2^M;      % number of symbols
%===== Gray code
gray = [ 0 1 3 2 7 6 4 5 ];
% exemple : 101=5 -> gray(5+1)=6
ak = exp(2*j*pi*gray/dpM);
%=====
Fe  = 20000;  % frequency for displaying the results
F0  = 2000;   % carrier frequency
F0r = F0/Fe;
Db  = 1500;   % binary rate
T   = M/Db;   % symbol duration
NT  = Fe*T;   % number of display points
%===== 3K bits
seqbit = [0 0 0 1 0 1 0 0 1 1 0 0 0 1 1];
nbsymb=length(seqbit)/M;
seqgr = reshape(seqbit,M,nbsymb);
%===== indices for the  symbols sequence
matcod = [4 2 1]; incod = matcod * seqgr + 1;
seqsymb = ak(incod);
%===== complex envelope
ec = ones(NT,1)*seqsymb; ec = reshape(ec,1,nbsymb*NT);
tsa = (0:nbsymb*NT-1)/Fe;
sig = real(ec .* exp(2*j*pi*F0*tsa));
%=====
subplot(311); plot(tsa,real(ec)); grid
subplot(312); plot(tsa,imag(ec)); grid
subplot(313); plot(tsa,sig); grid
```

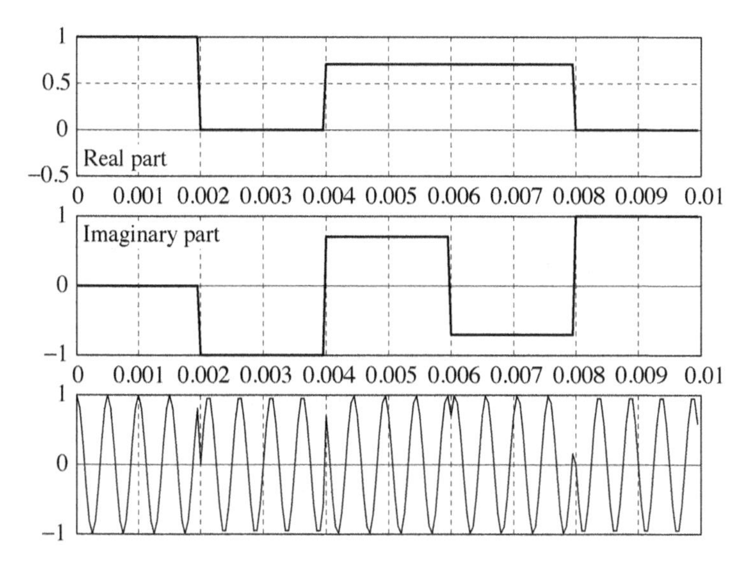

Figure H6.2 – *The real and imaginary parts of the complex envelope and the signal*

H6.3 (AMI code) (see page 195)

1. Because d_k is a uniform, i.i.d. sequence with possible values in $\{0, 1\}$:

$$\mathbb{E}\{d_k\} = 1/2,\ \mathbb{E}\{d_k^2\} = 1/2 \text{ and } \mathbb{E}\{d_k d_j\} = \mathbb{E}\{d_k\}\mathbb{E}\{d_j\} = 1/4$$

for $k \neq j$. According to the coding rule:

$$\begin{cases} s_{k+1} = (1 - 2d_k)s_k \\ a_k = d_k s_k \end{cases}$$

therefore $a_k = d_k(1 - 2d_{k-1})(1 - 2d_{k-2})...$

Because the d_k are independent, $\mathbb{E}\{a_k\} = 0$. By expanding, we get $\mathbb{E}\{(1 - 2d_k)^2\} = 1$ and $\mathbb{E}\{(1 - 2d_{k-1})d_{k-1}\} = -1/2$.

Using the fact that the d_k are independent, we get $\mathbb{E}\{a_k^2\} = 1/2$ and:

$$\mathbb{E}\{a_k a_{k-1}\} = \mathbb{E}\{d_k\}\mathbb{E}\{(1 - 2d_{k-1})d_{k-1}\}\mathbb{E}\{(1 - 2d_{k-2})^2\}, ...$$

Therefore, $R_a(\pm 1) = -1/4$. We have to check that $\mathbb{E}\{a_k a_{k-j}\} = 0$ for $|j| \geq 2$. We infer that:

$$S_a(f) = -\frac{1}{4}e^{2j\pi fT} + \frac{1}{2} - \frac{1}{4}e^{-2j\pi fT} = \sin^2(\pi fT)$$

2. Type:

```
%===== AMI.M
N=10000; bits=(rand(1,N)>0.5); N=length(bits);
symb=zeros(1,N);
%===== Coder
vp=1;
for ii=1:N
    if (bits(ii)==1),
       symb(ii)=vp; vp=-vp;
       else symb(ii)=0;
    end;
end
srate=1000; Lfft=256; fq=srate*(0:Lfft-1)/Lfft;
blksz=50; ps=welch(symb,blksz,'rec',Lfft,0.95);
pt=sin(pi*fq/srate) .^2;
plot(fq,[ps pt']); grid; axis([0 srate/2 0 1.1]);
```

H6.4 (HDB3 code) (see page 196)

1. Starting with the initial values $p_v = +1$ and $p_1 = -1$, we have:

d_k	$\emptyset$	0	1	1	1	0	0	0	0	1	0	0	0	0	0	1	0
p_v	+1	⋯	⋯	⋯	⋯	⋯	⋯	⋯	−1	⋯	⋯	⋯	⋯	+1	⋯	⋯	⋯
a_k	$\emptyset$	0	+1	−1	+1	−1	0	0	−1	+1	0	0	0	+1	0	−1	0
p_1	−1	⋯	+1	−1	+1	⋯	⋯	⋯	−1	+1	⋯	⋯	⋯	+1	⋯	−1	⋯

2. Type:

```
function an=hdb3(dn,pv,p1)
%%=====================================================%
%% SYNOPSIS : an=HDB3(dn,pv,p1)                        %
%%     dn = Binary sequence to be coded                %
%%     pv = Initial value of the bipolar violation bit %
%%     p1 = Initial value of the bipolar bit           %
%%     an = Coded sequence                             %
%%=====================================================%
N=length(dn); nz=0;
for ii=1:N
    if (dn(ii)==1)
        nz=0; an(ii)=-p1;p1=-p1;
    elseif (nz<3)
        nz=nz+1; an(ii)=0;
    else
        nz=0; an(ii)=-pv;
        if (pv==p1) an(ii-3)=-pv; end;
        pv=-pv; p1=pv;
```

```
    end
end
return
```

By typing:

```
hdb3([0 1 1 1 0 0 0 0 1 0 0 0 0 0 1 0],+1,-1)
```

check the result of the previous question.

3. Because the symbol sequence is centered in AMI coding, formula (6.35), which gives us the digital signal's spectrum, amounts to only the first term. The `welch` function then allows us to estimate the periodic part between 0 and $1/T$, corresponding to the correlations of the symbols a_n, that is to say:

$$S_a(f) = \sum_{\ell} R_a(\ell) e^{-2j\pi f \ell T}$$

at the frequency points $f = m/TL$ where $m = 0, \ldots, L-1$ and where L refers to the number of frequency points between 0 and $1/T$. All we have to do after that is multiply by the square modulus of the pulse spectrum, given, except for a multiplication factor (related to the amplitude), by $G(f) = \sin(\pi f T)/\pi f T$. If we restrict ourselves to the $(0, 1/T)$ frequency band, we have:

$$S_x(m/TL) = \frac{\sin(\pi m/L)}{\pi m/L} S_a(m/TL)$$

where the $S_a(m/TL)$ are estimated by applying the `welch` function to the sequence a_n. Type:

```
%===== TESTHDB3.M
clear; N=10000; bn=(rand(1,N)>0.5);
an=hdb3(bn,1,-1);
%===== estimation of the spectrum of the sequence an
Lfft=256; fq=(0:Lfft-1)/Lfft; tblocs=200;
Saf=welch(an,tblocs,'rec',Lfft,0.95);
%===== rectangular pulse on (0,T)
fq1=pi*(1:Lfft-1)'/Lfft;
Gf=[1;sin(fq1) ./ fq1]; Gf2=abs(Gf) .^2;
Sx= Saf .* Gf; plot(fq,Sx); grid
```

H6.5 (Linear equalization of a communications channel) (see page 198)

1. The constellation contains 16 symbols that can be coded using four bits. Remember that if F_0 refers to the carrier frequency, the transmitted signal can be written $s(t) = \text{Re}(\alpha(t)e^{2j\pi F_0 t})$.

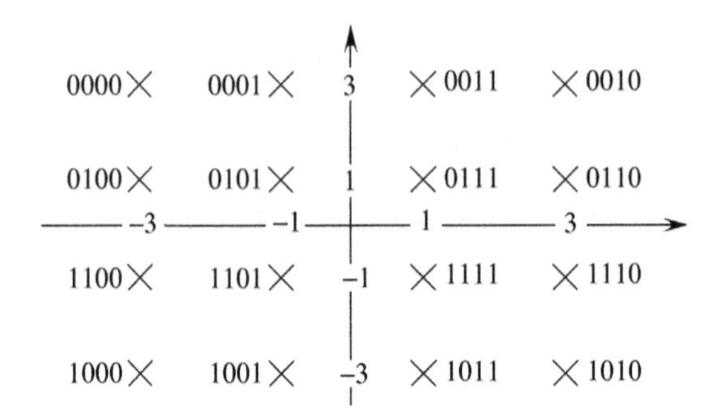

Figure H6.3 – *Constellation and Gray code*

2. The filter $G(z) = 1/(h_0+h_1z^{-1})$ is stable if the convergence area contains the unit circle. Therefore, if the square root of the denominator has a modulus greater than 1, the stable filter is anti-causal.

   For $h_0 = 1$ and $h_1 = -1.6$:

$$G(z) = \frac{1}{1-1.6z^{-1}} = -\frac{z}{1.6} - \frac{z^2}{1.6^2} - \frac{z^3}{1.6^3} - \cdots$$

   A length 21 causal approximation leads to a delay of 21 (see Figure H6.6).

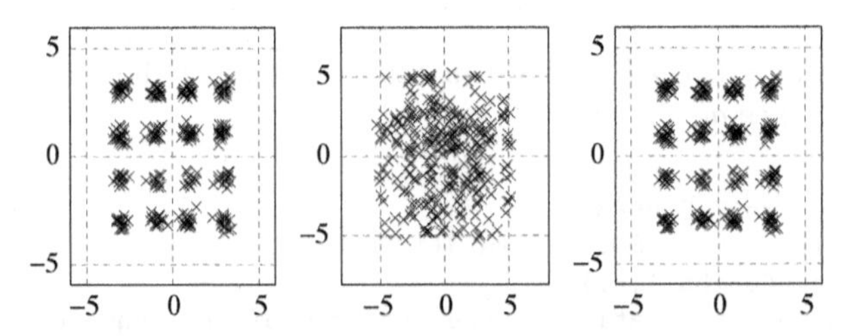

Figure H6.4 – *Noised signal, the signal-to-noise ratio is equal to 20 dB; signal containing the ISI caused by the FIR filter, $h_0 = 1$ and $h_1 = -1.6$; signal after equalization*

3. Type:

```
%===== CMAQ.M
N=300; M=4;
```

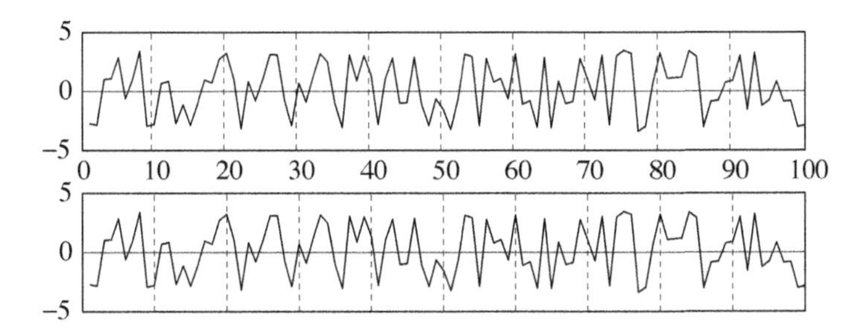

Figure H6.5 – *Comparison for a minimum phase channel*

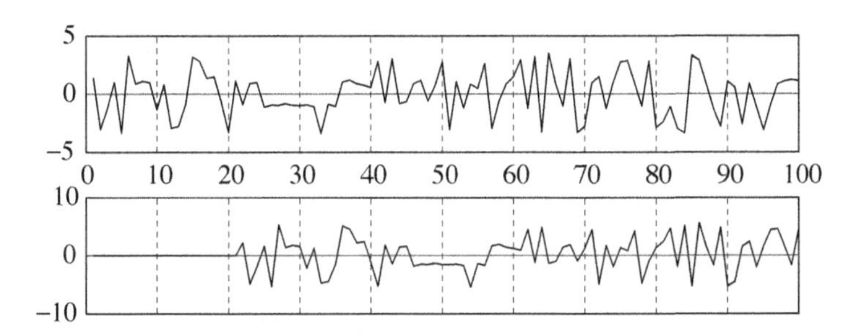

Figure H6.6 – *Comparison for a non-minimum phase channel*

```
%===== alphabet
ar=2*(0:M-1)-M+1; un=ones(M,1); pa=2*ar*ar'/M;
ac=(un*ar+j*ar'*un'); M2=M^2; ind=ceil(M2*rand(N,1));
symb=zeros(N,1); symb(:)=ac(ind);
SNR=20; rho=10^(SNR/20); sb=sqrt(pa)/(rho*sqrt(2));
br=sb*(randn(N,1)+j*randn(N,1)); xt=symb+br;
subplot(231); axis('square'); plot(xt,'x'); grid;
axis(1.2*[-1 1 -1 1]*max(abs(xt)));
%===== non minimum phase  channel
% hc=[1 -1.6]; he=-(1/1.6) .^(20:-1:0);
%===== minimum phase channel
hc=[1 -0.6]; he=0.6 .^(0:20);
yt=filter(hc,1,xt); subplot(232); axis('square')
plot(yt,'x'); grid; axis(1.2*[-1 1 -1 1]*max(abs(yt)));
zt=filter(he,1,yt); subplot(233); axis('square')
plot(zt,'x'); grid; axis(1.2*[-1 1 -1 1]*max(abs(zt)));
subplot(413);plot(real(xt(1:100))); grid
subplot(414);plot(real(zt(1:100))); grid
```

H6.6 (2-PAM modulation) (see page 204)

1. Generating the coded signal:

```
%===== CMIA1.M
Fe=20000;        % display frequency
Daff=500;        % nb of display points
Db=1000;         % binary rate = symbol rate
T=1/Db;          % interval between symbols
NT=fix(Fe*T);
nbsymb=300; seqbits=round(rand(1,nbsymb));
seqsymb=2*seqbits-1; he=ones(NT,1); xe=he*seqsymb;
lx=NT*nbsymb; xe=reshape(xe,1,lx);    % NT pts/bit
tx=(0:lx-1)/Fe;
%===== displaying only Daff points
subplot(221); plot(tx(1:Daff),xe(1:Daff)); grid
axis([tx(1) tx(Daff) -1.2 1.2]);
save miadata
```

2. Received signal:

```
%===== CMIA2.M
%       Output signal
clear; load miadata; hold off
lhsT=3.5;     % hc length
bc=0.06;      % channel band
hc=rif(lhsT*NT-1,bc);
xr=filter(hc,1,xe);        % filtering by the channel
subplot(222); plot(tx(1:Daff),xr(1:Daff))
axis([tx(1) tx(Daff) -1.3 1.3]); grid
save miadata
```

3. Type:

```
%===== CMIA3.M
% Eye pattern
clear; load miadata; hold off
%===== matched filter output
h=conv(he,hc); lh=length(h); h=h/sum(h);
xa=filter(h(lh:-1:1),1,xr);
xo=zeros(2*NT,nbsymb/2);
%===== length 2*NT  window (2 symbols)
for ii=1:nbsymb/2
    tdeb=(ii-1)*2*NT+1; tfin=tdeb+2*NT-1;
    xo(:,ii)=xa(tdeb:tfin)';
end
%===== displaying the trajectories
subplot(223); plot((0:2*NT-1)/Fe,xo(:,5:nbsymb/2));
```

```
axis('square'); grid
%===== choice of the decision time
disp('Choose the sampling time')
disp('corresponding to the widest aperture')
[aa bb]=ginput(1);
co=round(aa*Fe);              % eye center
co=co-fix(co/NT)*NT           % between 0 and NT-1
save miadata
```

4. Type:

```
%===== CMIA4.M
% Displaying the result of sampling
clear; load miadata
hold off; subplot(224); axis('normal')
plot(tx(1:Daff),xa(1:Daff))
hold on; plot(tx(co+1:NT:lx),xa(co+1:NT:lx),'o')
hold off; axis([tx(1) tx(Daff) -1.3 1.3]); grid
```

5. Type:

```
%===== CMIA5.M
% Estimation of the error probability Pe
clear; load miadata
retard=ceil(lhsT)-round(co/NT)+1;
Px=xa*xa'/lx; SNR=1;
Pb=(NT/2) * Px/(10 ^(SNR/10));
zc=xr + sqrt(Pb)*randn(1,lx);
%===== matched filtering
za=filter(h(lh:-1:1),1,zc(1:lx));
%===== threshold detection
sbest=(sign(za(co+1:NT:lx))+1)/2;
sbest=sbest(retard:length(sbest));
seqbits=seqbits(1:length(sbest));
merr=find(sbest~=seqbits); nbe=length(merr);
pe=nbe/length(sbest);
%=====
disp(sprintf('SNR: %g dB',SNR))
disp(sprintf('Pe: %1.2f',pe))
```

H6.7 ("Zero Forcing" linear equalization) (see page 213)

1. If $g(n)$ refers to the channel's impulse response, the suggested equalizer has the impulse response $w(n)$, such that $w(n) \star g(n) = \delta(n)$. The equalizer's output signal therefore has the expression:

$$\begin{aligned} y(n) &= w(n) \star x(n) = w(n) \star g(n) \star a(n) + w(n) \star b(n) \\ &= a(n) + w(n) \star b(n) \end{aligned}$$

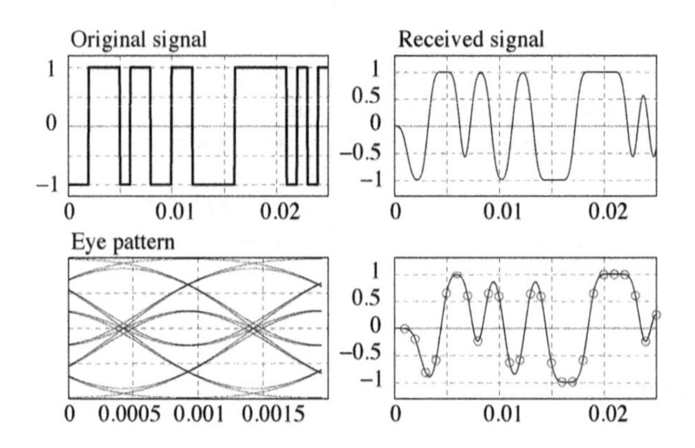

Figure H6.7 – *Signals at different points of the communication line*

Notice that $y(n)$ only contains the contribution of the symbol $a(n)$. Interference due to other symbols is completely eliminated. This is why this equalizer is said to be "Zero Forcing", in the sense that it forces the ISI to be equal to zero.

2. The output noise $u(n) = w(n) \star b(n)$ of the filter $w(n)$ is Gaussian and centered. Using formula (6.49) for $P = 2$, we get the variance:

$$\sigma_u^2 = \mathbb{E}\left\{u^2(n)\right\} = \sigma^2 \frac{g_0 + g_2}{(g_0 - g_2)(g_0^2 + g_2^2 + 2g_0g_2 - g_1^2)}$$

3. We have $y(n) = a(n) + u(n)$. Therefore, under the hypothesis that $a_n = -1$, $y(n)$ is a Gaussian variable with the mean -1 and the variance σ_u^2. Under the hypothesis that $a_n = +1$, $y(n)$ is a Gaussian variable with the mean $+1$ and the variance σ_u^2.

4. The error probability is:

$$\begin{aligned}
P_e &= \Pr(\text{Decide } -1 \text{ knowing } a_n = 1) \times \Pr(a_n = 1) \\
&\quad + \Pr(\text{Decide } 1 \text{ knowing } a_n = -1) \times \Pr(a_n = -1) \\
&= \Pr(y < 0 | a_n = +1)\frac{1}{2} + \Pr(y > 0 | a_n = -1)\frac{1}{2} \\
&= \frac{1}{2}\int_{-\infty}^{0} \frac{1}{\sigma_u\sqrt{2\pi}} \exp\left(-\frac{(y-1)^2}{2\sigma_u^2}\right) dy \\
&\quad + \frac{1}{2}\int_{0}^{+\infty} \frac{1}{\sigma_u\sqrt{2\pi}} \exp\left(-\frac{(y+1)^2}{2\sigma_u^2}\right) dy \\
&= \int_{\rho}^{+\infty} \frac{1}{\sqrt{2\pi}} \exp\left(-\frac{v^2}{2}\right) dv = Q(\rho)
\end{aligned}$$

with $\rho = 1/\sigma_u$. In MATLAB®, $Q(\rho)$ is obtained by typing `Q=(1-erf(rho/sqrt(2)))/2`.

5. Type:

```
%===== EGALLIN.M
clear
N=5000; g=[1 -1.4 0.8];
%===== calculating sigma
Ch2=(g(1)+g(3))...
      /(g(1)^2+g(3)^2+2*g(1)*g(3)-g(2)^2)/(g(1)-g(3));
Ch=sqrt(Ch2);
%===== several values for the SNR
SNRdB=(5:17); longSNR=length(SNRdB);
Pelin=zeros(longSNR,1); PeTheo=zeros(longSNR,1);
ak=sign(randn(1,N)); sk=filter(g,1,ak);
vs=sqrt(sk*sk'/N);
%=====
for jj=1:longSNR
   %=====
   RSB=10^(SNRdB(jj)/20); sigma_b=vs/RSB;
   bk=sigma_b*randn(1,N);
   %=====
   xk=sk+bk; ykegal=filter(1,g,xk);
   aklin=sign(ykegal(1:N));
   Pelin(jj)=sum(abs(ak-aklin)/2);
   %=====
   sigma_u=sigma_b*Ch;
   rho=1/sigma_u;
   %=====
   PeTheo(jj)=(1-erf(rho/sqrt(2)))/2;
end
Pelin=Pelin/N;
semilogy(SNRdB, Pelin, 'x', SNRdB, PeTheo, '-'); grid
```

Figure H6.8 shows the results. They are in perfect agreement with the theoretical values.

H6.8 (Wiener equalization) (see page 214)

1. Let us assume that $\boldsymbol{w} = (w(0), \ldots, w(N-1))^T$ and $\boldsymbol{x}(n) = (x(n), ldots, x(n-N+1))^T$. The expression we have to minimize with respect to $\boldsymbol{w}$, is written $\mathbb{E}\left\{|a(n-d) - \boldsymbol{w}^T\boldsymbol{x}(n)|^2\right\}$. Using the projection principle, we have $a(n-d) - \boldsymbol{w}^T\boldsymbol{x}(n) \perp x(p)$ for $p \in \{n, \ldots, n-N+1\}$, which can be written $\mathbb{E}\left\{(a(n-d) - \boldsymbol{w}^T\boldsymbol{x}(n))x(n-k)^*\right\} = 0$ where $k \in \{0, \ldots, N-1\}$, or also, in matrix form:

$$\mathbb{E}\left\{a(n-d)\boldsymbol{x}^*(n)\right\} = \mathbb{E}\left\{\boldsymbol{x}(n)\boldsymbol{x}^H(n)\right\}\boldsymbol{w} \tag{7.37}$$

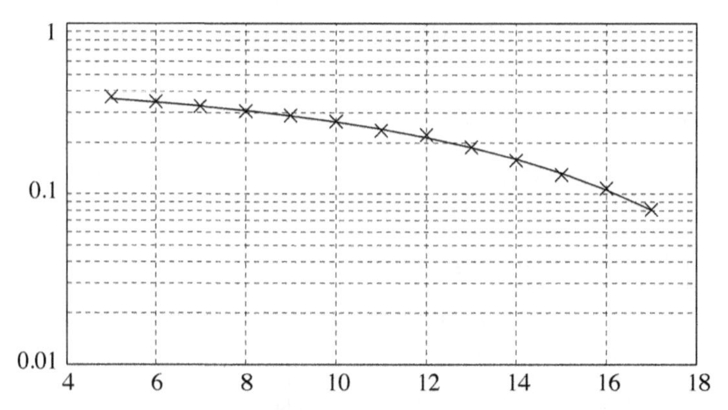

Figure H6.8 – *Symbol-by-symbol detection of a Zero Forcing equalizer's output for a binary transmission. The equivalent channel has the coefficients* $g_0 = 1$, $g_1 = -1.4$, $g_2 = 0.8$. *The* **x***'s indicate the probabilities obtained through* 5,000 *simulations for different values of the signal-to-noise ratio in dB*

2. If we assume that $a(n)$ is an identically and independently distributed sequence with possible values in $\{-1, +1\}$, we have $\mathbb{E}\{a_n\} = 0$ and:

$$R_{aa}(k) = \sigma_a^2 \delta(k) \quad \text{avec} \quad \sigma_a^2 = 1$$

Let $s(n)$ be the channel output. According to the filtering formulas, we have:

$$R_{ss}(k) = g(k) \star g^*(-k) \star R_{aa}(k) = g(k) \star g^*(-k)$$

This sequence only has $2L - 1$ non-zero terms since $g(k)$ is of length L.

By assuming that $b(n)$ is independent of $a(n)$, $b(n)$ is independent of $s(n)$ and we have:

$$\begin{aligned} R_{xx}(k) &= \mathbb{E}\{(s(n+k) + b(n+k))(s^*(n) + b^*(n))\} \\ &= \mathbb{E}\{s(n+k)^* s(n)\} + \mathbb{E}\{b(n+k)b^*(n)\} \\ &= R_{ss}(k) + \sigma_b^2 \delta(k) \\ &= g(k) \star g^*(-k) + \sigma_b^2 \delta(k) \end{aligned}$$

Likewise, if we use the input/output filtering formula, we have:

$$\begin{aligned} R_{ax}(k) &= \mathbb{E}\{a(n+k)x^*(n)\} = \mathbb{E}\{a(n+k)(s^*(n) + b^*(n))\} \\ &= \mathbb{E}\{a(n+k)s^*(n)\} = g^*(-k) \end{aligned}$$

3. If we refer to expression (7.37), $\mathbb{E}\left\{\boldsymbol{x}(n)\boldsymbol{x}^H(n)\right\} = \boldsymbol{R}_{ss} + \sigma_b^2\boldsymbol{I}$ where $\boldsymbol{R}_{ss}$ is a Toeplitz matrix constructed from the sequence $R_{ss}(k)$. The vector $\mathbb{E}\left\{a(n-d)\boldsymbol{x}^*(n)\right\}$ is comprised of the coefficients $g(k)$:

$$\mathbf{r}_{ax} = [\underbrace{0\ldots0}_{d}\,\underbrace{g^*(L-1)\ldots g^*(0)}_{L}\,\underbrace{0\ldots0}_{N-L-d}]^T$$

Therefore, the filter we are trying to determine is $\boldsymbol{w} = (\boldsymbol{R}_{ss}+\sigma_b^2\boldsymbol{I}_N)^{-1}\mathbf{r}_{ax}$.

4. The `mmse.m` program allows us to examine the histograms of the received values as well as the histogram after equalization. As you can see, the intersymbol interference leads to a more spread-out histogram:

```
%===== MMSE.M
clear; N=5000; gc=[1 -1.4 0.8]; lg=length(gc);
ak=sign(randn(1,N));     % sequence of symbols
sk=filter(gc,1,ak);      % emitted signal
vsth=sqrt(gc*gc'); RSBdB=20;
%=====
sigma_b=vsth*10^(-RSBdB/20); bk=sigma_b*randn(1,N);
xk=sk+bk;                 % received signal
%===== MMSE (order-50 FIR)
LW=50; d=23;
rss=conv(gc,gc(lg:-1:1));
rsspos=[rss(lg:2*lg-1) zeros(1,LW-lg)];
Rxx=toeplitz(rsspos)+sigma_b*sigma_b*eye(LW);
ras=[zeros(1,d) gc(lg:-1:1) zeros(1,LW-lg-d)];
w=inv(Rxx)*ras';
ykwiener=filter(w,1,xk);
%===== SNR after equalization
roMMSE=max(conv(w,gc))/std(ykwiener);
%===== displaying the results
points=50; subplot(211); hist(xk,points); grid
subplot(212); hist(ykwiener,points); grid
```

5. We can now perform the equalization with the Wiener filter, then, by simple detection with the threshold 0, estimate the binary sequence and measure the error probability. We can then compare the results for a Zero Forcing equalization. If the filter $g(n)$ is minimum phase, we can perform the Zero Forcing equalization simply by typing `filter(1,gc,xk)` and comparing the error probabilities. On the other hand, if the filter $g(n)$ is not minimum phase, the Zero Forcing equalization can only be achieved using the command `filter`, which implements the causal solution that is not stable. We then have to determine a causal and stable approximation of the inverse of $g(n)$.

The following program implements the performance comparison in terms of error probability. After having run the `mmse.m` program, run the following program (Figure H6.9):

```
%===== EGALLINCMP.M
% Program to to be run after MMSE.M
SNRdB=(5:17); SNRlgth=length(SNRdB); retardW=d+lg-1;
%=====
for jj=1:SNRlgth
    RSB=10^(SNRdB(jj)/20); sigma_b=vsth/RSB;
    bk=sigma_b*randn(1,N); xk=sk+bk;
    ykegalZF=filter(1,gc,xk); aklinZF=sign(ykegalZF(1:N));
    PelinZF(jj)=sum(abs(ak-aklinZF)/2);
    %=====
    Rxx=toeplitz(rsspos)+sigma_b*sigma_b*eye(LW);
    w=inv(Rxx)*ras'; ykegalW=filter(w,1,xk);
    aklinW=sign(ykegalW(1:N));
PelinW(jj)=sum(abs(ak(1:N-retardW)-aklinW(retardW+1:N))/2);
end
PelinZF=PelinZF/N; PelinW=PelinW/N;
semilogy(SNRdB, PelinZF,'x'); hold on;
semilogy(SNRdB,PelinW,'o'); hold off; grid
```

Notice that the results are significantly better with the Wiener filter.

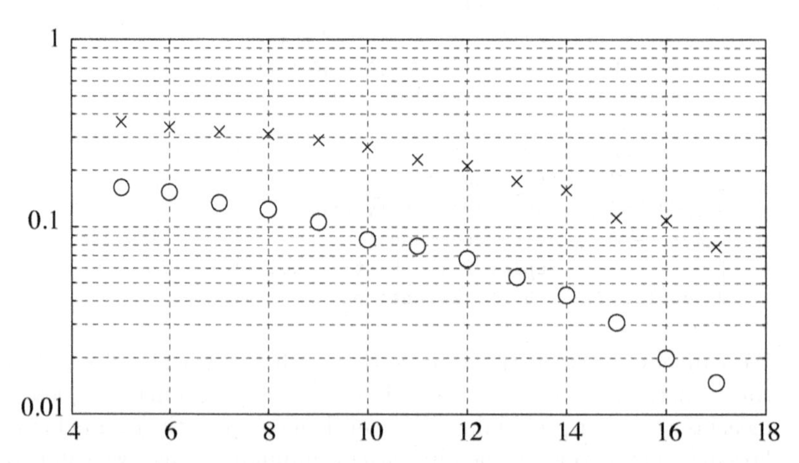

Figure H6.9 – *Error probability plotted against the signal-to-noise ratio in dB after equalizing with the Wiener filter ('o') and with the Zero Forcing filter ('×'). The results are obtained through simulation using* 5,000 *symbols. The channel filter has the finite impulse response* $(1 \;\; -1.4 \;\; 0.8)$

H6.9 (Performances with two sub-codebooks) (see page 232)

1. Type:

```
%===== LBG64.M
N=5000; w=randn(2*N,1);
xAR=filter(1,[1 0.9],w);
s=zeros(2,N); s(:)=xAR;
[CfD,CiD,ED]=lbg(s,64);
```

2. Type:

```
%===== LBGCOMP.M
% Comparison of the dictionaries
%===== first dictionary
[CfC1,CiC1,EC1]=lbg(s,4);
%===== calculating the difference
diff=zeros(2,4*N);
for ii=1:4
   diff(:,(ii-1)*N+1:ii*N)=s-CfC1(:,ii)*ones(1,N);
end
%===== second dictionary
[CfC2,CiC2,EC2]=lbg(diff,16);
```

In the first case, we obtained a mean square error `ED=0.1189`. In the second one, we obtained for the first (very crude) codebook `EC1=1.4684` and for the second one, which is "more subtle", `EC2=0.5890`. Hence the second one gives the order of magnitude of the distortion caused by this second type of partition. We therefore have to compare, in terms of square deviation, the values 0.1189 and 0.5890. The loss is significant, but the total size of the two sub-codebooks is 20 instead of 64 for the maximum size dictionary.

Chapter 8

Appendices

A1 Miscellaneous functions

```
function h=myboxplot(x,w)
%!==========================================!
%! SYNOPSIS: h=myboxplot(x,w)              !
%!      x = data                            !
%!      w = defines the limits for outliers !
%!          wH=q3+w*IQR; wL=q1-w*IQR;       !
%!          (default 1.5)                   !
%!      h = handle of the gca               !
%!==========================================!
if nargin<2, w=1.5; end
llm=-5; llp=5;
idx=find(~isnan(x)); x=x(idx); x=sort(x);
Lx=length(x); iQ1=round(Lx/4); iQ3=round(3*Lx/4);
q2=median(x);
q1=x(iQ1); q3=x(iQ3); IQR=q3-q1;
% IQR=interquantile range = q3-q1
q1=x(iQ1); q3=x(iQ3); IQR=q3-q1;
wH=q3+w*IQR; wL=q1-w*IQR;
plot([llm llp],[q2 q2],':r',[-1 +1],[q2 q2],'r'), hold on
plot([llm llp],[q1 q1],':r',[-1 +1],[q1 q1],'b')
plot([llm llp],[q3 q3],':r',[-1 +1],[q3 q3],'b')
plot([-1 -1],[q1 q3],'b',[1 1],[q1 q3],'b')
plot([-.5 +.5],[wH wH],'b',[-.5 .5],[wL wL],'b')
plot([0,0],[q3,wH],'b',[0,0],[q1,wL],'b')
plot([llm llp],[wH wH],':r',[llm llp],[wL wL],':r')
idxH=find(x>wH); idxL=find(x<wL);
if ~isempty(idxH), plot(zeros(length(idxH),1),x(idxH),'xr'), end
if ~isempty(idxL), plot(zeros(length(idxL),1),x(idxL),'xr'), end
set(gca,'XTickLabel',{' '}); h=gca; hold off
```

A2 Statistical functions

A.2.1 Notable functions

The following functions are used in defining the most widespread distributions in the field of statistics:

- gamma function (`gamma`):

$$\Gamma(x) = \int_0^{+\infty} e^{-t} t^{x-1} dt \tag{8.1}$$

- beta function (`beta`):

$$B(z, w) = \int_0^1 t^{z-1}(1-t)^{w-1} dt = \frac{\Gamma(z)\Gamma(w)}{\Gamma(z+w)} \tag{8.2}$$

A.2.2 Beta distribution

The beta distribution $f_x(z, w)$ is defined as:

$$f_x(z, w) = \frac{1}{B(z, w)} x^{z-1}(1-x)^{w-1} \tag{8.3}$$

The cumulative of $f_x(z, w)$ is called the *regularized incomplete beta function* or RIBF, $I_x(z, w)$.

$$I_x(z, w) = \frac{\int_0^x t^{z-1}(1-t)^{w-1} dt}{B(z, w)} \tag{8.4}$$

$$= \frac{B_x(z, w)}{B(z, w)} = 1 - I_{1-x}(w, z) \tag{8.5}$$

where $B_x(z, w)$ is called the *incomplete beta function.*

In MATLAB® the function `betainc` implements the RIBF $I_x(z, w)$.

A.2.3 Student's distribution

The Student's t-distribution has the following probability density function:

$$f(t) = \frac{1}{\sqrt{k\pi}} \frac{\Gamma((k+1)/2)}{\Gamma(k/2)} (1 + t^2/k)^{-(k+1)/2} \tag{8.6}$$

$$= \frac{1}{B(1/2, k/2)} \frac{1}{\sqrt{k}} (1 + t^2/k)^{-(k+1)/2} \tag{8.7}$$

With MATLAB®, we obtain:

```
cst=1/beta(.5,k/2)/sqrt(k);
dTS=cst*(1+(t.^2)/k).^(-(k+1)/2);
```

The Student's t cumulative distribution function is given by:

$$\begin{aligned} F(t) &= \frac{1}{2} - \frac{1}{2} I_v(1/2, k/2) \quad \text{if} \quad -\infty < t < 0 \\ &= \frac{1}{2} + \frac{1}{2} I_v(1/2, k/2) \quad \text{if} \quad 0 \le t < +\infty \\ &\quad \text{with } v = \frac{t^2}{k+t^2} \end{aligned}$$

With MATLAB®, we obtain:

```
vt=k./(k+t.^2);
rbf=betainc(1,k/2,1/2)-betainc(vt,k/2,1/2);
cfTS=.5+.5*rbf.*sign(t);
```

Student's t-distribution implementation

MATLAB® provides five functions for the Student's law. They can be replaced, up to a point, by the functions listed below:

– `tcdf`: computes Student's t cumulative distribution:

```
function cTS=mytcdf(t,k)
%!================================================!
%! Student's t cumulative distribution            !
%! synopsis: cTS=MYTCDF(t,k)                      !
%!    k   = degrees of freedom (vector or scalar) !
%!    t   = variable (vector)                     !
%!    cTS = cdf (length(t) x length(k))           !
%!================================================!
k=round(k);
tt=zeros(length(t),1); tt(:)=t;
cTS=zeros(length(t),length(k)); lp=0;
for n=k
    vt=n./(n+t.^2);
    rbf=betainc(1,n/2,1/2)-betainc(vt,n/2,1/2);
    c=.5+.5*rbf.*sign(t);
    lp=lp+1; cTS(:,lp)=c;
end
```

– `tpdf`: provides the Student's t probability density function:

```
function dTS=mytpdf(t,k)
%!================================================!
%! Student's t distribution function              !
%! synopsis: dTS=MYTPDF(t,k)                      !
```

```
%!    k   = degrees of freedom (vector or scalar) !
%!    t   = variable                              !
%!    dTS = density (length(t) x length(k)        !
%!============================================!
k=round(k);
tt=zeros(length(t),1); tt(:)=t;
dTS=zeros(length(t),length(k)); lp=0;
for n=k
    cst=1/beta(.5,n/2)/sqrt(n);
    d=cst*(1+(tt.^2)/n).^(-(n+1)/2);
    lp=lp+1; dTS(:,lp)=d;
end
```

Application (Figure A2.1):

```
%===== testmystudent.m
clear
t=[-5:.1:5].'; k=[1,3,5,10];
dst=mytpdf(t,k); cft=mytcdf(t,k);
%=====
subplot(121), plot(t,dst), grid on
%===== gaussian limit (k=infty)
dG=1/sqrt(2*pi)*exp(-t.^2/2);
hold on, plot(t,dG,'k','linew',1), hold off
%=====
subplot(122), plot(t,cft), grid on
%===== normal cdf
g=(1/2)*(1+erf(t/sqrt(2))); % normal cdf (0,1)
% norminv: t = sqrt(2)*erfinv(2*g - 1)
hold on, plot(t,g,'k','linew',1), hold off
```

– `tinv`: function `betaincinv` is used to calculate the Student's t inverse cumulative distribution function:

```
function tn=mytinv(T,k)
%!=======================================!
%! SYNOPSIS: T=MYTINV(T,k)                !
%! Student's inverse cdf                  !
%!    k  = number of degrees of freedom !
%!    T  = input value (vector)           !
%!    tn = output value (vector)          !
%!=======================================!
k=round(k);
TT=zeros(length(T),1); TT(:)=T;
idxm=find(TT<1/2); idxM=find(TT>=1/2);
TTm=TT(idxm); TTM=TT(idxM);
%=====
vtm=betaincinv(1-2*TTm,1/2,k/2);
```

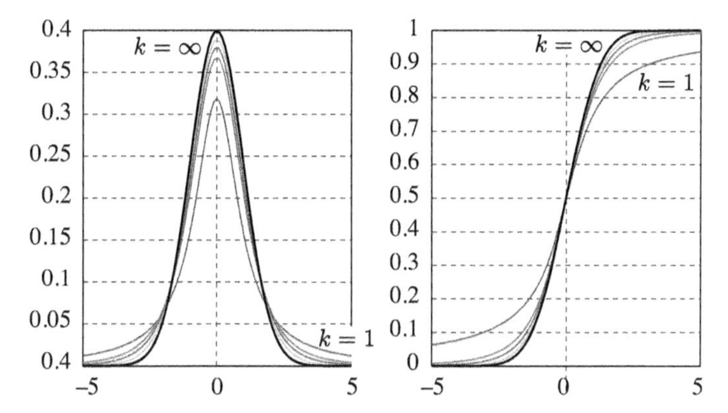

Figure A2.1 – *Distribution (left) and cumulative distribution (right) for values of k 1, 3, 5 and 10. The limiting Gaussian is shown alongside with the distributions*

```
t2=k*vtm ./ (1-vtm); tnm=-sqrt(t2);
vtM=betaincinv(2*TTM-1,1/2,k/2);
t2=k*vtM ./ (1-vtM); tnM=sqrt(t2);
%=====
tn=[tnm;tnM];
```

```
%===== testmytinv.m
k=7;              % nb of d.o.f.
T=[(.01:.02:.1),(.2:.1:.9)]; % cdf values
tp=[-5:.1:5].'; cft=mytcdf(tp,k);
plot(tp,cft), grid on
g=(1/2)*(1+erf(t/sqrt(2))); % normal cdf (0,1)
hold on, plot(t,g,'k','linew',1), hold off
%=====
tic, tn=mytinv(T,k); toc
hold on, plot(tn,T,'or'), hold off
```

The `mytinvd` function directly applies Newton's method to calculate the inverse:

```
function tn=mytinvd(T,k)
%!=======================================!
%! SYNOPSIS: tn=MYTINVD(T,k)             !
%! Student's inverse cdf                 !
%!    k  = number of degrees of freedom !
%!    T  = input value (vector)          !
%!    tn = output value (vector)         !
%!=======================================!
```

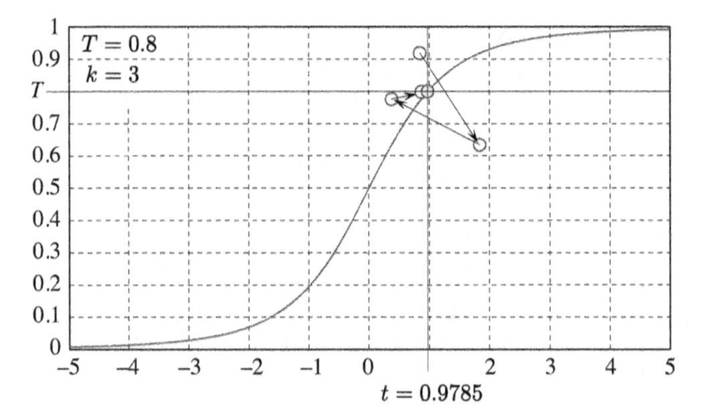

Figure A2.2 – *Application of Newton's method to find t from $T = 0.8$ and $k = 3$*

```
k=round(k);
myeps=1.0e-6;
LT=length(T); tn=zeros(LT,1);
%===== initial values
tn = sqrt(2)*erfinv(2*T-1); % norminv
for n=1:LT
    tnp1 = tn(n)+1;
    %===== Newton(s method
    while abs(tnp1-tn(n))>myeps
        tn(n)=tnp1;
        yn=mytcdf(tn(n),k)-T(n); ydotn=mytpdf(tn(n),k);
        tnp1=tn(n)-yn/ydotn;
    end
end
```

– `trnd`: generates random numbers from Student's t-distribution. See generation `testmychi2rnd.m` for the method [2].

– and `tstat`: Student's t mean and variance.

```
function [mu,sigma2]=mytstat(k)
%!============================================!
%! SYNOPSIS: [mu,sigma2]=MYTSTAT(k)           !
%!    k      = number of degrees of freedom !
%!    mu     = mean                           !
%!    sigma2 = variance                       !
%!============================================!
k=round(k);
if k<1
    error('nb deg. of freedom must be > 0');
end
```

```
switch k
    case 1
        mu=NaN; sigma2=NaN;
    case 2
        mu=0; sigma2=NaN;
    otherwise
        mu=0; sigma2=k/(k-2);
end
```

A.2.4 Chi-squared distribution

The *chi-squared distribution* or χ^2_k-distribution is the distribution of a sum of the squares of k ("degrees of freedom") independent $\mathcal{N}(0,1)$ r.v.:

– chi2pdf: the probability density function is given by:

$$t \in [0, +\infty),\ f(t) = \frac{1}{2^{k/2}\Gamma(k/2)} t^{k/2-1} e^{-t/2} \tag{8.8}$$

which may be writen:

```
function dCH=mychi2pdf(t,k)
%!==================================================!
%! chi2 distribution function                       !
%! synopsis: dTS=MYCHI2PDF(t,k)                     !
%!    k   = degrees of freedom (vector or scalar) !
%!    t   = variable                                !
%!    dCH = density (size(t))                       !
%!==================================================!
k=round(k);
tt=zeros(length(t),1); tt(:)=t;
dCH=zeros(length(t),length(k)); lp=0;
for n=k
    cst=1 ./ 2.^(n/2) ./ gamma(n/2);
    d=cst * t.^(n/2-1).*exp(-t/2);
    lp=lp+1; dCH(:,lp)=d;
end
```

– chi2cdf: the cumulative distribution function is given by:

$$t \in [0, +\infty),\ F(t) = \frac{1}{\Gamma(k/2)} \gamma(k/2, t/2) \tag{8.9}$$

which may be writen:

```
function cCH=mychi2cdf(t,k)
%!==================================================!
%! chi2 cumulative distribution function            !
%! synopsis: dTS=MYCHI2CDF(t,k)                     !
```

```
%!    k   = degrees of freedom (vector or scalar) !
%!    t   = variable                               !
%!    cCH = cdf (size(t))                          !
%!=============================================!
k=round(k);
tt=zeros(length(t),1); tt(:)=t;
cCH=zeros(length(t),length(k)); lp=0;
for n=k
    c=gammainc(t/2,n/2);
    lp=lp+1; cCH(:,lp)=c;
end
```

The testmychi2 program displays the densities and the χ^2_k distributions for $k = 1$, 3, 5, 7 and 10 (Figure A2.3).

```
%===== testmychi2.m
clear
t=[0:.1:15].'; k=[1,3,5,7,10]; Lk=length(k);
A=char(ones(Lk,1)*[107 61]);
C=num2str(k'); D=[A C];
dst=mychi2pdf(t,k);
subplot(121), plot(t,dst), grid on
legend(D)
set(gca,'ylim',[0 1])
cft=mychi2cdf(t,k);
%=====
subplot(122), plot(t,cft), grid on
%legend('k=1','k=3','k=5','k=7','k=10')
legend(D)
```

– chi2inv: inverse cdf:

```
function t=mychi2inv(T,k)
%!=====================================!
%! SYNOPSIS: t=MYCHI2STAT(T,k)          !
%!    k = number of degrees of freedom  !
%!    T = cdf value                     !
%!    t = output (vector (size(T))      !
%!=====================================!
k=round(k);
if k<1
    error('deg. of freedom must be > 0');
end
t=2*gammaincinv(T,k/2);
```

– chi2rnd: r.v.s are generated using the mychi2inv function (Figure A2.4). This method is illustrated in program testmychi2rnd.m.

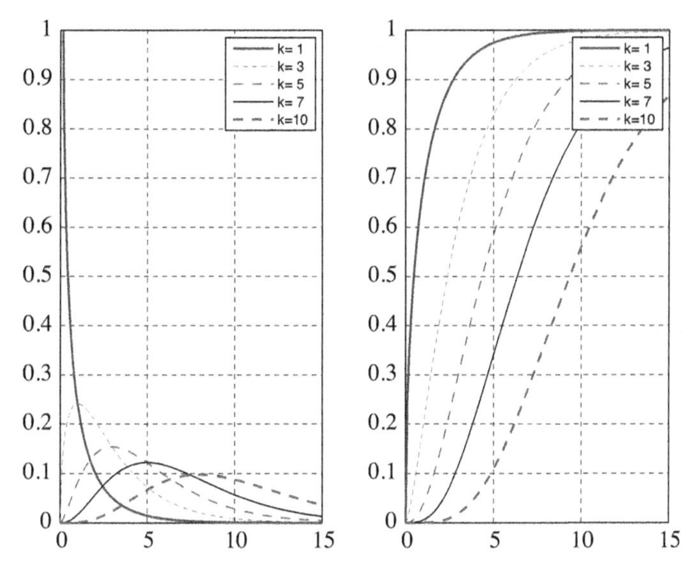

Figure A2.3 – *A few probability densities and cumulative distributions of the* χ^2 *distribution*

```
%===== testmychi2rnd.m
clear
L  = 20000;          % sample size
lk = .5;             % bin size
t  = [0.25:lk:30];   % bin centers
k  = 5;              % d.o.f.
%=====
U=rand(1,L); V=mychi2inv(U,k);
N=hist(V,t);
plot(t,N/L/lk,'o'), grid
Vt=mychi2pdf(t,k);
hold on, plot(t,Vt,'r'), hold off
```

– chi2stat

```
function [mu,sigma2]=mychi2stat(k)
%!==========================================!
%! SYNOPSIS: [mu,sigma2]=MYCHI2STAT(k)      !
%!    k      = number of degrees of freedom !
%!    mu     = mean                          !
%!    sigma2 = variance                      !
%!==========================================!
k=round(k);
if k<1
```

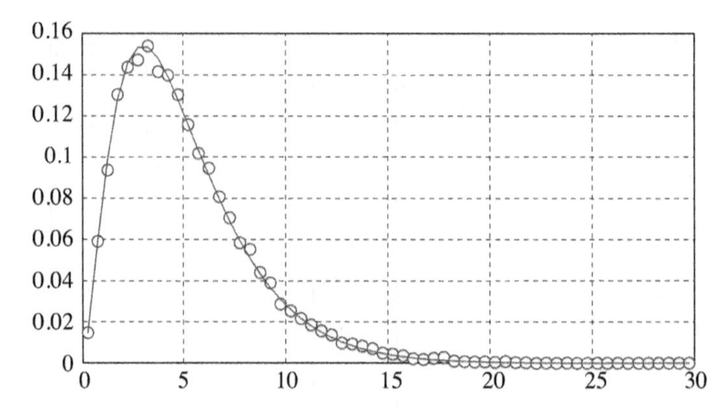

Figure A2.4 – *Use of the* `mychi2inv` *function to generate a random sequence, histogram and theoretical pdf*

```
    error('nb deg. of freedom must be > 0');
end
mu=k; sigma2=2*k;
```

A.2.5 Fisher's distribution

The Fisher's centered distribution has the following probability density function:

$$f(t) \quad = \quad \frac{1}{\beta(d1/2, d2/2)} \sqrt{\frac{(k_1 t)^{k_1} k_2^{k_2}}{(k_1 t + k_2)^{(k_1+k_2)}}} \frac{1}{t} \tag{8.10}$$

where k_1 and k_2 are degrees of freedom (integers > 0). They correspond to an r.v. generated by the relationship $\frac{X_1/k_1}{X_2/k_2}$, where X_1 and X_2 are r.v.s with a χ^2 type distribution.

The Fisher's cumulative distribution function is given by the RIBF:

$$F(t) = I_{\frac{k_1 t}{(k_1 t + k_2)}}(k_1/2, k_2/2) = \frac{B_{\frac{k_1 t}{(k_1 t + k_2)}}(k_1/2, k_2/2)}{B(k_1/2, k_2/2)}$$

Fisher's distribution implementation

– `fpdf`: Fisher probability density function:

```
function dFS=myfpdf(t,k1,k2)
%!===================================================!
%! Fishert's distribution function                   !
```

```
%! synopsis: dTS=MYTPDF(t,k1,k2)                       !
%!    k1,k2 = degrees of freedom (vector or scalar) !
%!    t     = variable                                 !
%!    dFS   = density (size(t))                        !
%!=====================================================!
k1=round(k1); k2=round(k2);
if (k1<1) | (k2<1), error('Incorrect d.o.g.'); end
cst=k1.^(k1/2) * k2^(k2/2) / beta(k1/2,k2/2);
vt=t^(k1/2-1) ./ (k1*t+k2).^((k1+k2)/2);
dFS=vt * cst;
```

– fcdf: Fisher cumulative distribution function:

```
function cFS=myfcdf(t,k1,k2)
%!=========================================!
%! Fisher's cumulative distribution        !
%! synopsis: cTS=MYFCDF(t,k1,k2)           !
%!    k1,k2 = degrees of freedom (scalars) !
%!    t   = variable (vector)              !
%!    cFS = cdf (length(t) x 1)            !
%!=========================================!
k1=round(k1); k2=round(k2);
vt=k1*t ./ (k1*t+k2);
cFS=betainc(vt,k1/2,k2/2);
```

```
%===== testmyfisher.m
clear
t=[0:.05:8].'; k1=5; k2=7;
dFS=myfpdf(t,k1,k2); cFS=myfcdf(t,k1,k2);
%=====
subplot(121), plot(t,dFS), grid on
%=====
subplot(122), plot(t,cFS), grid on
```

– finv:

```
function t=myfinv(T,k1,k2)
%!=========================================!
%! Fisher's cumulative distribution        !
%! synopsis: t=MYFINV(T,k1,k2)             !
%!    T     = cdf values                   !
%!    k1,k2 = degrees of freedom (scalars) !
%!    t     = vector (size(T))             !
%!=========================================!
k1=round(k1); k2=round(k2);
vt = betaincinv(T,k1/2,k2/2);
t = (k2/k1) * vt ./ (1-vt);
```

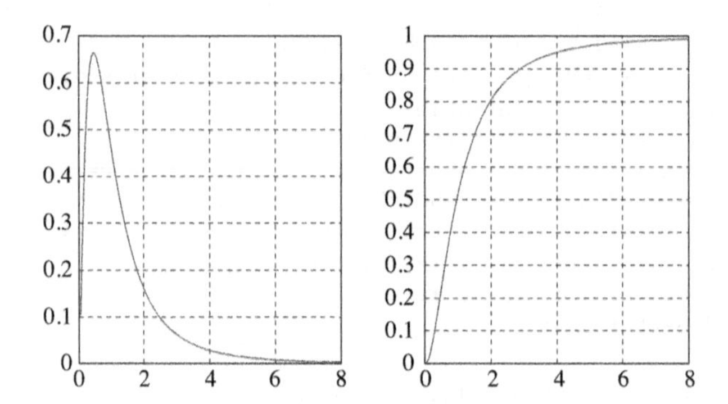

Figure A2.5 – *pdf and cdf of the Fisher distribution for* $k_1 = 5$ *and* $k_2 = 7$

- `frnd`: generates random numbers from Fisher distribution. See generation `testmychi2rnd.m` for the method.
- and `fstat`: the Fisher mean and variance are given by:

$$\frac{k_2}{k_2 - 2} \text{ if } k_2 > 2$$

and

$$\frac{2k_2^2(k_1 + k_2 - 2)}{k_1(k_2 - 2)^2(k_2 - 4)} \text{ if } k_2 > 4$$

```
function [mu,sigma2]=myfstat(k1,k2)
%!=======================================!
%! SYNOPSIS: [mu,sigma2]=MYTSTAT(k1,k2) !
%!     k1,k2  = degrees of freedom       !
%!     mu     = mean                     !
%!     sigma2 = variance                 !
%!=======================================!
k1=round(k1); k2=round(k2);
if (k2<3) | (k1<1)
    error('parameter 2 must be > 2, pram. 1 must be > 0');
end
switch k2
    case 1
        mu=NaN; sigma2=NaN;
    case {3,4}
        mu=k2/(k2-2); sigma2=NaN;
    otherwise
        mu=k2/(k2-2);
        sigma2=(2*k2^2*(k1+k2-2))/(k1*(k2-2)^2*(k2-4));
end
```

Bibliography

[1] P. Billingsley. *Probability and Measure*. Probability and Statistics. John Wiley and Sons, 2012.

[2] G. Blanchet and M. Charbit. *Digital Signal and Image Processing, Volume 1: Fundamentals*, volume 1. ISTE Ltd, London and John Wiley & Sons, New York, 2nd edition, 2014.

[3] G. Blanchet and M. Charbit. *Digital Signal and Image Processing, Volume 2 - Advances and Applications: The Deterministic Case*, volume 2. ISTE Ltd, London and John Wiley & Sons, New York, 2nd edition, 2015.

[4] P. Brockwell and R. Davies. *Time Series: Theory and Methods*. Springer Verlag, 1990.

[5] J. Capon. "High-resolution frequency-wavenumber spectrum analysis". *Proc. IEEE*, 57, no. 8:1408–1418, August 1969.

[6] O. Cappé, E. Moulines, and T. Ryden. *Inference in Hidden Markov Models*. Springer Series in Statistics. Springer-Verlag New York, 2005.

[7] B. Efron. "Bootstrap methods: Another look at the jackknife". *Annals of Statistics*, 7(1):1–26, 1979.

[8] N. J. Gordon, D. J. Salmond, and A.F.M. Smith. "Novel approach to nonlinear/non-gaussian bayesian state estimation". *IEE Proceedings-F on Radar and Signal Processing*, 140(2):107–113, 1993.

[9] M. Joindot and A. Glavieux. *Introduction aux Communications Numériques*. Collection Pédagogique de Télécommunication. Ellipses, 1995.

[10] R. E. Kalman. "A new approach to linear filtering and prediction problems". *Transactions of the ASME–Journal of Basic Engineering*, 82(Series D):35–45, 1960.

[11] S. M. Kay. *Fundamentals of Statistical Signal Processing: Estimation Theory*. Prentice Hall, 1993.

[12] Y. Linde, A. Buzzo, and R. Gray. "An algorithm for vector quantizer design". *IEEE Trans. on Communications*, COM-28:84–95, January 1980.

[13] S.P. Lloyd. "Least squares quantization in PCM". *IEEE Trans. on Information Theory*, pages 129–137, March 1982.

[14] M. Matsumoto and T. Nishimura. "Mersenne twister: A 623-dimensionally equidistributed uniform pseudorandom number generator". *ACM Trans., Modeling and Computer Simulations*, 1998.

[15] J. Max. "Quantizing for minimum distorsion". *IRE Trans. on Information Theory*, pages 7–12, March 1960.

[16] F.J. Mc Clure. "Ingestion of fluoride and dental caries: Quantitative relations based on food and water requirements of children one to twelve years old". *American Journal of Diseases of Children*, 66(4):362–369, 1943.

[17] R. G. Miller. "The jackknife – a review". *Biometrika*, 61(1):1–15, 1974.

[18] D. C. Montgomery and G. C. Runger. *Applied Statistics and Probability for Engineers.* John Wiley and Sons, Inc., March 2010.

[19] N. Moreau. *Techniques de Compression des Signaux.* Masson, 1995.

[20] J. Neyman and E. S. Pearson. "On the problem of the most efficient tests of statistical hypotheses". *Philosophical Transactions of the Royal Society A: Mathematical, Physical and Engineering Sciences*, 231:694–706, 1933.

[21] J. Pearl. *Probabilistic Reasoning in Intelligent Systems: Networks of Plausible Inference.* Number 1558604790. Morgan Kaufmann Publishers Inc., 1988.

[22] J.G. Proakis and Masoud Saheli. *Digital Communications.* Electrical and Computer Engineering. McGraw-Hill, 5 edition, 2007.

[23] M. H. Quenouille. "Notes on bias in estimation". *Biometrika*, 43, 1956.

[24] R. Roy and T. Kailath. "ESPRIT-estimation of signal parameters via rotational invariance techniques". *IEEE Trans. on Acoust. Speech, Signal Processing*, ASSP-37:984ñ–995, August 1989.

[25] W. Rudin. *Real and Complex Analysis.* 3rd edition, International Series in Pure and Applied Mathematics. McGraw-Hill Science/Engineering/Math, 3rd edition, 1986.

[26] R. H. Shumway and D. S. Stoffer. *Time Series Analysis and Its Applications: With R Examples (Springer Texts in Statistics).* 2nd edition, Springer, May 2006.

[27] S. Stigler. "Studies in the history of probability and statistics. xxxii: Laplace, fisher and the discovery of the concept of sufficiency". *Biometrika*, pages 439–445, December 1973.

[28] J. W. Tuckey. "Bias and confidence in not-quite large samples". *Ann. Math. Statist.*, 29, 1958.

[29] F. J. Harris, "On the Use of Windows for Harmonic Analysis with the Discrete Fourier Transform". *Proc. IEEE*, 66, pages 51-83, January, 1978.

Index

Other titles from

in

Digital Signal and Image Processing

2015

CLARYSSE Patrick, FRIBOULET Denis
Dynamic Cardiac Imaging

BLANCHET Gérard, CHARBIT Maurice
Digital Signal and Image Processing using MATLAB®
Volume 2 – Advances and Applications:The Deterministic Case – 2nd edition

GIOVANNELLI Jean-François, IDIER Jérôme
Regularization and Bayesian Methods for Inverse Problems in Signal and Image Processing

2014

AUGER François
Signal Processing with Free Software: Practical Experiments

BLANCHET Gérard, CHARBIT Maurice
Digital Signal and Image Processing using MATLAB®
Volume 1 – Fundamentals – 2nd edition

DUBUISSON Séverine
Tracking with Particle Filter for High-dimensional observation and State Spaces

ELL Todd A., LE BIHAN Nicolas, SANGWINE Stephen J.
Quaternion Fourier Transforms for Signal and Image Processing

FANET Hervé
Medical Imaging Based on Magnetic Fields and Ultrasounds

MOUKADEM Ali, OULD Abdeslam Djaffar, DIETERLEN Alain
Time-Frequency Domain for Segmentation and Classification of Non-stationary Signals: The Stockwell Transform Applied on Bio-signals and Electric Signals

NDAGIJIMANA Fabien
Signal Integrity: From High Speed to Radiofrequency Applications

PINOLI Jean-Charles
Mathematical Foundations of Image Processing and Analysis Volumes 1 and 2

TUPIN Florence, INGLADA Jordi, NICOLAS Jean-Marie
Remote Sensing Imagery

VLADEANU Calin, EL ASSAD Safwan
Nonlinear Digital Encoders for Data Communications

2013

GOVAERT Gérard, NADIF Mohamed
Co-Clustering

DAROLLES Serge, DUVAUT Patrick, JAY Emmanuelle
Multi-factor Models and Signal Processing Techniques: Application to Quantitative Finance

LUCAS Laurent, LOSCOS Céline, REMION Yannick
3D Video: From Capture to Diffusion

MOREAU Eric, ADALI Tulay
Blind Identification and Separation of Complex-valued Signals

PERRIN Vincent
MRI Techniques

WAGNER Kevin, DOROSLOVACKI Milos
Proportionate-type Normalized Least Mean Square Algorithms

FERNANDEZ Christine, MACAIRE Ludovic, ROBERT-INACIO Frédérique
Digital Color Imaging

FERNANDEZ Christine, MACAIRE Ludovic, ROBERT-INACIO Frédérique
Digital Color: Acquisition, Perception, Coding and Rendering

NAIT-ALI Amine, FOURNIER Régis
Signal and Image Processing for Biometrics

OUAHABI Abdeljalil
Signal and Image Multiresolution Analysis

2011

CASTANIÉ Francis
Digital Spectral Analysis: Parametric, Non-parametric and Advanced Methods

DESCOMBES Xavier
Stochastic Geometry for Image Analysis

FANET Hervé
Photon-based Medical Imagery

MOREAU Nicolas
Tools for Signal Compression

2010

NAJMAN Laurent, TALBOT Hugues
Mathematical Morphology

2009

BERTEIN Jean-Claude, CESCHI Roger
Discrete Stochastic Processes and Optimal Filtering / 2nd edition

CHANUSSOT Jocelyn *et al.*
Multivariate Image Processing

DHOME Michel
Visual Perception through Video Imagery

GOVAERT Gérard
Data Analysis

GRANGEAT Pierre
Tomography

MOHAMAD-DJAFARI Ali
Inverse Problems in Vision and 3D Tomography

SIARRY Patrick
Optimization in Signal and Image Processing

2008

ABRY Patrice *et al.*
Scaling, Fractals and Wavelets

GARELLO René
Two-dimensional Signal Analysis

HLAWATSCH Franz *et al.*
Time-Frequency Analysis

IDIER Jérôme
Bayesian Approach to Inverse Problems

MAÎTRE Henri
Processing of Synthetic Aperture Radar (SAR) Images

MAÎTRE Henri
Image Processing

NAIT-ALI Amine, CAVARO-MENARD Christine
Compression of Biomedical Images and Signals

NAJIM Mohamed
Modeling, Estimation and Optimal Filtration in Signal Processing

QUINQUIS André
Digital Signal Processing Using Matlab

2007

BERTEIN Jean-Claude, CESCHI Roger
Discrete Stochastic Processes and Optimal Filtering

BLOCH Isabelle
Information Fusion in Signal and Image Processing

GLAVIEUX Alain
Channel Coding in Communication Networks

OPPENHEIM Georges *et al.*
Wavelets and their Applications

2006

CASTANIÉ Francis
Spectral Analysis

NAJIM Mohamed
Digital Filters Design for Signal and Image Processing

Lightning Source UK Ltd.
Milton Keynes UK
UKOW06n2354080617
302973UK00008B/39/P